D0049394

# AMERICAN SPARTAN

# AMERICAN SPARTAN

## THE PROMISE, THE MISSION, AND THE BETRAYAL
## OF SPECIAL FORCES MAJOR JIM GANT

# ANN SCOTT TYSON

*wm*

WILLIAM MORROW
*An Imprint of* HarperCollins*Publishers*

For my beloved mother and father, Joy and Haney,
and my parents-in-law, James and Judy,
wise teachers all, who have touched many lives with their "little,
nameless, unremembered acts of kindness and of love."

and

In memory of Malik Noor Afzhal, a great man.

The names and identifying characteristics of three individuals in this book have been changed to protect their privacy or maintain their current position in the military.

Grateful acknowledgment is made for permission to reprint from the following:

"Book 23: The Great Rooted Bed," from *The Odyssey*, by Homer, translated by Robert Fagles, translation copyright © 1996 by Robert Fagles. Used by permission of Viking Penguin, a division of Penguin Group (USA) LLC.

HarperCollins books may be purchased for educational, business, or sales promotional use. For information please e-mail the Special Markets Department at SPsales@harpercollins.com.

FIRST EDITION

*Designed by Lisa Stokes*

*Frontispiece: Panjshir valley, Afghanistan © Lebedev/Shutterstock, Inc.*

*Map by Nick Springer, copyright © 2014 Springer Cartographics LLC.*

Library of Congress Cataloging-in-Publication Data has been applied for.

ISBN 978-0-06-211498-3

14 15 16 17 18   OV/RRD   10 9 8 7 6 5 4 3 2 1

It is their war, and you are to help them, not to win it for them.

—T. E. LAWRENCE

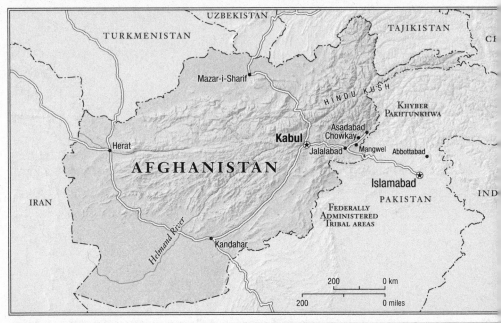

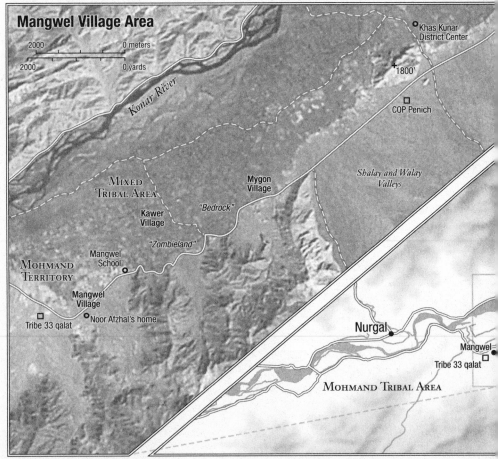

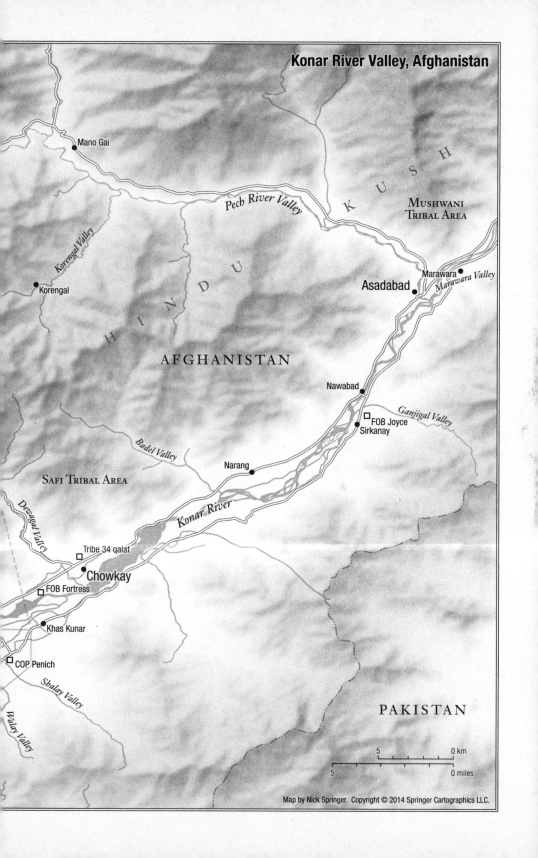

# Konar River Valley, Afghanistan

Mano Gai

*Pech River Valley*

*Korengal Valley*

H I N D U    K U S H

**MUSHWANI
TRIBAL AREA**

Korengal

Marawara

**Asadabad**

*Marawara Valley*

**AFGHANISTAN**

Nawabad

*Ganjigal Valley*

FOB Joyce
Sirkanay

*Badel Valley*

Narang

**SAFI TRIBAL AREA**

*Dewagal Valley*

*Konar River*

Tribe 34 qalat

**Chowkay**

FOB Fortress

Khas Kunar

COP Penich

*Shalay Valley*

*Walay Valley*

**PAKISTAN**

5        0 km

5        0 miles

Map by Nick Springer. Copyright © 2014 Springer Cartographics LLC.

# PROLOGUE

NIGHT WAS FALLING ON the valley.

Beneath the Hindu Kush Mountains, the terraced fields, and the rock-strewn grazing lands, an isolated band of American soldiers and Afghans returned from patrol, navigating the rugged terrain. They wore night vision goggles to see in the darkness, and passed without a whisper through mud-brick farming villages clustered along Afghanistan's Konar River. They'd been summoned by their commander, a man who never let them forget that Taliban insurgents could be watching from the high ground. The sky was pitch black by the time the men reached their walled outpost, a typical Afghan compound, or *qalat*, and made their way to the makeshift operations center inside.

Dressed in traditional tunics and baggy pants, most of the men were Pashtuns, members of the powerful ethnic group whose tribes had dominated the border with western Pakistan for thousands of years. But the Taliban was Pashtun, too. The remote outpost was situated on the edge of the village of Mangwel in Afghanistan's Konar Province. Konar was considered by the U.S. military to contain a witches' brew of hard-core foreign and local insurgents, and its valleys had proven some of the deadliest battlefields in Afghanistan. In a 2005 ambush, insurgents in Konar left only a lone survivor, Marcus Luttrell, when they killed three members of a four-man Navy SEAL team and then shot down a quick reaction force helicopter with a rocket-propelled grenade (RPG), sending sixteen more U.S. troopers to their deaths. It was one of the biggest American losses of the war. Intense fighting had led U.S. forces to abandon some Konar outposts in 2009 and 2010, including one named Restrepo.

A few miles to the east lay Pakistan's tribal territory, and another hundred miles beyond that was the teeming Pakistani city of Abbottabad, the hideout of Al Qaeda terrorist group leader Osama bin Laden. Members of Al Qaeda as well as the Afghan and Pakistani factions of the Taliban and other radical Islamic groups had sanctuaries on the Pakistani side of the border. Insurgent fighters moved unfettered along mountain trails into Konar Province and other parts of eastern Afghanistan to stage attacks on U.S. forces. Al Qaeda and the Taliban alike sought to strengthen their grip on Konar. The Pashtun tribes that held sway in these villages and valleys were the only force that in the long run could stand in their way. Osama bin Laden understood the power of tribes at a strategic and visceral level. He took refuge in tribal areas of Afghanistan and Pakistan and often appeared in propaganda videos wearing the traditional dagger of his Yemeni tribal ancestors.

Two tribesmen, Hakim Jan and Umara Khan, shouldered their AK-47 rifles at the end of their guard shift and climbed down wooden ladders to join the other men in the central courtyard of the small camp. They were members of the force of twenty Afghan tribal police that lived in the Mangwel qalat together with a dozen U.S. soldiers. The American mission was simple: to empower the tribe to push the insurgents out. The lives of everyone in the camp depended on it.

Conditions in the camp were austere. The commander made sure they stayed that way. It had crude outhouses and no running water. All of the men slept on cots in canvas tents. They ate the same food, too, mainly beans, rice, and flatbread. The commander knew that the hardship could bring these strangers together. Soon the men were conversing in broken Pashto and English as they went about their work—cleaning weapons, loading bullets into magazines, repairing vehicles, standing guard.

Staff Sgt. Robert Chase, a thirty-two-year-old U.S. infantry squad leader, and Pfc. Jeremiah "Miah" Hicks appeared before the assembled soldiers and Afghan tribesmen. They carefully unfolded an eight-foot-long American flag and, each holding one upper corner, unfurled it before the group.

Then their commander, Special Forces Maj. Jim Gant, wearing a black Afghan tunic and pants with a fitted maroon *kandari* cap, stepped in front of the flag. The bearded forty-three-year-old Green Beret addressed his men.

"Today, we had our revenge, our *badal*," he said, using the Pashto word for "retribution." "I am proud to fight alongside you," he went on. "Tonight, in honor of that, I will bleed."

Gant drew an eight-inch Spartan Harsey knife, a gift from the father of a fallen Special Forces teammate. As a captain, Gant had embraced the Spartan warrior ethos of sacrifice and courage and used it to inspire every unit he'd commanded. Gant believed in the depth of his being that men had to be willing to die for one another without hesitation if they were to be victorious in battle. He also believed that the ancient code of honor that Spartans lived by was, at its core, no different from the one that underpinned Pashtun tribal law.

He had carefully planned this meeting to inspire both his American and Afghan men, and he had already asked that a goat be slaughtered and prepared in the Pashtun tradition.

Gripping the Spartan knife in his right hand, Gant slowly slit long, deep gashes between the thumb and index finger of his left hand—one cut each for seven of his friends killed in Afghanistan. The blood ran down his hand and dripped onto the broken ground.

He stared into the shocked faces of his men.

"Today was a long time in coming for us," Gant said, his face drawn and voice tense. Some eighteen hours earlier, on May 2, 2011, U.S. Navy SEALs had stormed a fortified three-story house in Abbottabad. They'd gunned down Osama bin Laden and packed up a treasure trove of Al Qaeda's strategic and tactical plans as well as its leader's personal effects.

"Osama bin Laden was a coward and a murderer," Gant continued. "He killed innocent men, women, and children in our country. Thousands of them. His actions have cost your great country thousands of lives and hurt relations between Muslims and Christians all over the world."

He watched the shock become hushed recognition. Gant was shedding his blood in honor of the dead, and it was vital that his men understood this.

"Now if we could just have a moment of silence for all of our friends, for all of the people who have been killed . . ."

The Afghans and Americans, indistinguishable one from the other in the darkness, stood shoulder to shoulder and bowed their heads. They remembered their fallen comrades and the homes they were fighting for near and far, and they smelled the rich scent of simmering goat that awaited them.

# CHAPTER 1

SEVEN THOUSAND MILES AWAY from that otherworldly corner of Afghanistan, the lights flickered off and on in my Bethesda, Maryland, rental house. Rain pelted against the cellar window and dripped onto the sill through a crack in the glass.

I watched a small pool of rain swell until it overflowed in a thin stream down the wall, the trickle triggering in me a disproportionate flood of doom.

It was September 2012. I was living temporarily in a one-bedroom basement apartment with my fiancé, my fourteen-year-old daughter Kathryn, a stray cat, and God knew how many camelback crickets.

Kathryn's makeshift "room" had no walls, only crimson bedsheets attached by clothespins to a rope I strung along the ceiling. The setup cost $13.91 from the hardware store—not bad. There was no kitchen. We boiled macaroni in a microwave and ate on paper plates. Putting the best spin I could on the latest Dickensian twist in our lives, I told Kathryn hardship was good for her—she would thank me one day. She shot me a look that said maybe, maybe not.

As a war correspondent turned author, I'd been surviving on savings for most of the past year. I was near the end of a torturous four-year separation and divorce and was trying to help my four children survive the fallout from a failed relationship that I could never make up for. I was having to assert my right of media privilege to keep my book materials out of divorce court.

All of which explained why we were holed up in the basement. I had sublet the upstairs to four foreign students—three Chinese and one Dutch—who were covering my rent. Everything had been going all right until one of the Chinese had a chauvinistic fit and posted a sign on his bedroom door warning: "No dogs or Europeans allowed."

Under other circumstances, I would have kicked him out, but I could not—I needed his money. So instead, I kept trying to make peace, which was one of my bigger mistakes in life.

A sharp rap on the door at the top of the musty stairwell alerted me that one of the students wanted something. I ran up the stairs, opened the door, and came face-to-face with the rich, overweight Chinese bully holding an overflowing hamper of dirty laundry.

"I'll wash that," I said, determined to keep his toxicity out of our space. "Do you want everything in the dryer?"

He nodded.

"I'll put it back at the top of the stairs when it's done," I said.

I closed the door.

*Peace out, bitches.*

I dumped his laundry into the washer, braced myself against the machine with both hands, and took a deep breath.

Then, when I thought nothing else could go wrong, the storm hit.

I rushed upstairs to shut the windows, just in time to see a huge gust of wind blow a sheet of rain across the backyard and send a sixty-foot-tall oak tree crashing through the roof of the house, inches from where I stood.

It wasn't supposed to be this way.

I was supposed to be sitting in a lounge chair on a beach, sipping a glass of wine, enjoying my family and professional success.

How had I gotten here?

A Kansas native along with my parents, I drew inspiration from the pioneer roots and open prairies of my childhood. Growing up in Seattle, Ireland, and Greece, I was given every advantage a middle-class family of six could offer. My dad, an economics professor, had high expectations for me and never handicapped me because I was a girl. He taught me to sail, ski, do math, and change a tire. He had the wisdom to make me work at a local Texaco gas station at the age of fourteen—my first task was to clean the men's bathroom. My mother, who had a master's degree in education but chose to be a stay-at-home mom, taught me courage and advised me to always have the ability to be independent. And my parents let me run. I traveled alone in the Greek islands at the age of fifteen, and at seventeen worked as a paramedic in Nicaragua.

As a Harvard undergraduate, I earned an honors degree in government and East Asian studies, spent a year studying social sciences in French at the selective Institut d'Etudes Politiques in Paris, and did stints researching China at the State Department and Central Intelligence Agency. From there I went to Hong Kong to study Mandarin Chinese on a graduate fellowship from the Rotary Club. Bridling at the constraints of government work, I decided I was better cut out to serve the public through journalism. I landed a job at United Press International in Hong Kong. My mentor there, a chain-smoking veteran Vietnam War correspondent named Sylvana Foa, gave me the break I sought and assigned me to the UPI Beijing bureau.

In China, I learned how to be a journalist the hard way, cutting my teeth against one of the most secretive and oppressive governments on earth. I gained fluency in Mandarin and learned the social customs and mannerisms so well that in some places native Chinese didn't believe I was an American, mistaking me for a member of one of western China's minority groups. Despite being detained twice in mainland China and kicked out of Tibet, I was able to make my way into ordinary people's homes, underground churches, ancient clan temples, political dissident networks, migrant tenements, and Tibetan nomad communities. The *Christian Science Monitor* ran my articles and submitted them for a Pulitzer Prize. The articles formed the basis for a book I coauthored with my then husband, entitled *Chinese Awakenings: Life Stories from the Unofficial China*.

From Beijing, I moved on to Chicago and a new investigative challenge: probing the biggest African American gang in the region, the Gangster Disciples, and its leader, Larry Hoover. I wrote a series of articles that was one of three finalists for a Society of Professional Journalists prize. We moved to Washington, where I covered Congress during the Clinton impeachment proceedings as well as national security.

Then came the September 11 terrorist attacks, which I wrote about in a front-page story for the *Christian Science Monitor*. The next day, my editor called and again asked me to take on the Pentagon beat. I had turned down the job a few months earlier because I was working at home to care for my four young children, James, Sarah, Scott, and Kathryn.

After the terrorist strike, though, I felt I had no choice. I was a trained reporter, and I needed to do my job—not only in the public interest but to stay informed for my family. I said yes.

Starting with the U.S.-led overthrow of the Taliban regime in Afghanistan and the "shock and awe" invasion of Iraq in 2003, I covered the wars and embedded with dozens of U.S. military units. Almost immediately, I realized the public badly needed observers to translate the world of the military and hold it accountable. The challenge of reporting on the armed forces had parallels to news gathering in China. The military spoke a different language. It was a huge bureaucracy, extremely mistrustful of outsiders and especially reporters. And I'd sat in enough Pentagon briefings to know the Washington spin was nowhere near the realities of life of a combat soldier on the ground. I didn't just want Americans to know the facts about the wars; I wanted them to care. I felt obligated to get as close as I could to the taste, smell, and feel of combat.

I had other, more personal reasons for heading into the war zone. After many years during which my life revolved around giving birth, nursing my babies, and caring for my children at home, universal experiences of women, I wanted to see whether I could know and understand warfare, the most primal and universal proving ground of men. I struggled with leaving my children, even for a few weeks at a time. But as I watched members of the military leave their families time and again for a year or longer, I had to ask myself why my family and I should not make a sacrifice that paled in comparison. So I planned meticulously for my absences, timing trips when I could enlist the help of relatives, arranging backup care, and leaving detailed calendars of soccer schedules and music lessons. I believed that on some level my children would benefit from my example and experience. I knew how isolated and sheltered the public was from the costs of military intervention, and I wanted to educate my sons and daughters about the consequences of their country going to war. Still, saying goodbye was the most wrenching thing I had ever done. It was for them and other loved ones that I calibrated the risks I took. Intellectually, I grew numb to the danger I faced. I was not depressed or suicidal, but to a degree I stopped caring about my own life.

And then I met Jim Gant.

# CHAPTER 2

THE CALL CAME IN to Fort Bliss, Texas, at about 10:00 a.m. on October 29, 2009. Jim was ordered to board a U.S. Airways flight out of El Paso, Texas, destined for Fayetteville, North Carolina. From there he was to report to Fort Bragg, the home of the Army's Airborne and Special Operations Forces. His gear had already been shipped to the Middle East with an Iraq-bound armored brigade that was deploying within days from Bliss. But the brigade would leave one liaison officer short.

On November 2, after Jim arrived at Bragg, he was ushered into a video teleconference room. He took a seat, stared into a blank fifty-two-inch flat-panel television, and waited for Adm. Eric Thor Olson, commander of all the U.S. military's fifty-eight thousand Special Operations Forces, to appear on-screen. Jim's entire fighting life—the only life he'd ever known—came down to this conference call. He'd shed his blood with scores of men. But now he stood alone.

Jim's hair was too long. His sideburns were not regulation. Tattoos of Achilles on his right arm and the goddess of magic and war, Hecate, on his left poked out from his faded gray-green camouflage uniform. He'd been a Green Beret for nearly twenty years. Waiting for the four-star commander, Jim no longer cared if he ever ranked higher than major. A mentor long ago had told him, "You are the best soldier I've ever known, and the worst soldier I've ever known." Jim didn't argue.

In 1986, Jim was inspired to enlist in the Army by Robin Moore's classic Vietnam exposé *The Green Berets*—a book that was published as fiction because the Pentagon objected to Moore's embedding with Special Forces. He signed up straight out of Mayfield High School in Las Cruces, New Mexico. But his reasons for joining had little to do with his reasons for staying. The Green Beret tactics in Vietnam captivated him. No carpet-bombing from thirty thousand feet. They fought alongside Montagnard tribesmen against the Viet Cong, taking back land village by village. The Green Berets fought hand-to-hand, close-up, on the ground. And that kind of fighting was in Jim's blood.

Jim passed the Special Forces selection—which as many as 50 percent of candidates fail—the first time through in 1988. In the Darwinian training that followed, he was one of the youngest in his group and was repeatedly injured, but he held on to earn his Green Beret. He served as a communications sergeant in the 1990-91 Persian Gulf War and received training as an intelligence analyst. In 1996 he was commissioned as an Army infantry second lieutenant. He received superlative evaluations from officers above him and in 2001 returned to Special Forces as a captain.

Then came the long wars—deployments to Afghanistan in 2003 and 2004, and Iraq in 2006 and 2007—and with them multiple awards for valor:

**17 June 2003. Army Commendation Medal for valor.**
Captain James K. Gant, United States Army, displayed exceptional courage under fire while serving as the Detachment Commander for Operational Detachment Alpha (ODA) 316 during a double ambush in the Pech valley, Konar Province, Afghanistan.

Konar . . .

Jim grew up in New Mexico, but he believed he wasn't born until he landed in Konar. He and the six other men on his Special Forces team came of age in 2003 battling insurgent ambushes and conducting raids in the harrowing mountain passes of the Afghan province bordering Pakistan.

They used to swim naked in the Konar River, which was fed by melting glaciers in the Hindu Kush and coursed through a narrow valley bordered by sparse wheat fields. Huddled around fires under star-strewn skies brushed by the Milky Way, they listened to ethnic Pashtun tribesmen talk of the local ways of warfare and raiding and the honor they bestowed.

Jim learned to speak basic Pashto, the ancient Indo-Iranian language of some forty million Pashtuns, and developed a deep respect for their tribal code of ethics, Pashtunwali. He abided by the code to forge allies and shame opponents. To empower his allies he gave them guns and access to firepower, the ultimate status symbol for Pashtuns. The Pashtun tribes—traditional communities united by cultural, economic, and blood ties—dominated Afghan society. Pashtuns are honor-bound to show hospitality, or *melmastia*, to whoever comes into their territory needing protection—not only to friends but also to sworn enemies. But the hospitality only went so far. Jim learned which tribe he could trust with his life, and which tribe would invite him and his men to a midday meal, only to ambush them on the road out. Among Jim's most fearsome opponents were the tribesmen of Konar's Korengal valley, where dozens of Americans would later perish fighting.

Despite the duplicity, there was a mutual admiration between warriors that infused Jim's relations with even his worst enemies in Konar. "May your life be long, and may big wolves prowl at your door," went the Pashtun saying, meaning a man's honor grows with the ferocity of his enemies.

Afghans in the area referred to Jim's team as "the bearded ones"—a meaningful name, as many Muslims consider wearing beards a religious obligation—and called him "Commander Jim." Years later, after the Army transferred Jim to Iraq, the Afghans still remembered him.

In the sterile Fort Bragg conference room, Jim recalled the night that one of his sergeants swept an infant out of harm's way during an intense, thirty-second firefight inside a pitch-dark Afghan compound. He felt the exquisite burden and pride of having led men in combat, and wondered if he would ever lead again. And deep down, as he awaited his audience with the commander, he felt just a bit like an imposter. Because he knew his superiors had no idea who he really was. Because screaming in his mind was a demon that never left him: a desire to fight—not for his country, nor for a cause, nor even for his men, but for the pure sake of it. In Iraq, he'd been in combat so intense, so long, that he'd lost himself in it. Images flashed through his head of a Sunni fighter darting through a palm grove, and a military convoy on a deserted, sun-bathed road.

**11 December 2006. Silver Star.** Major James K. Gant distinguished himself by exceptional gallantry in combat actions against a determined and aggressive enemy of the United States. . . . MAJ Gant's masterful leadership and selfless courage under fire directly resulted in saving the lives of his men, a countless number of National Police QRF personnel and four Iraqi civilians, as well as 12 confirmed enemy kills and a much higher number of estimated enemy killed and wounded.

Just as sectarian bloodletting between Sunni and Shiite Muslims was pushing Iraq to the brink of civil war in the summer of 2006, Jim took over a nine-man U.S. military team advising an Iraqi National Police Quick Reaction Force (QRF) battalion. Attacks on U.S. forces by both Sunni insurgents and Shiite militias were escalating, and American casualties in Iraq were at an all-time high. In late November 2006, the vehicle Jim was riding in was hit by a massive improvised explosive device (IED) and caught on fire, trapping Jim inside. His Iraqi comrades pulled him out. Days later, after a brief hospital stay, Jim returned to the scene of the roadside attack. Using a loudspeaker, he taunted the local Sunni insurgents in an effort to draw them out. They took the bait. Soon afterward, on December 11, they struck again with full force.

Jim and his U.S. team plus 190 Iraqi police commandos in twenty-three vehicles were headed south from the town of Balad toward Baghdad through an area where insurgents had complete freedom of maneuver. What unfolded was a running gun battle stretched over five miles that required Jim and the patrol to push through three separate kill zones, all the while evacuating severely wounded commandos.

After the battle, the Iraqi police slaughtered a goat—swiftly slitting its throat in accordance with Islamic law—and covered Jim and his team with bloody crimson handprints to celebrate victory.

It was almost as good as Konar.

NOW SITTING IN FORT Bragg, Jim remembered that the Silver Star battle had been three years earlier . . . a lifetime. He knew why he'd been summoned. His career hung in the balance.

Only days earlier, on October 26, 2009, Jim had rocked the U.S.

military establishment by publishing a treatise on why it was destined to lose the war in Afghanistan. In the online paper "One Tribe at a Time," posted on a website popular in military circles, Jim exposed a gaping hole in U.S. strategy: the failure to systematically engage Afghanistan's powerful Pashtun tribes. The Pashtuns make up 40 percent of the population—including the majority of the Taliban insurgents—and occupy the volatile east and south. Jim argued that as the tribes go, so goes the rest of the nation. With the Pashtuns behind you, Afghanistan was won.

By the time Adm. Olson called Jim in, the paper had gone viral at the Pentagon, in the White House, and on Capitol Hill. Then it landed in the *New York Times*.

"The United States has killed tens of thousands of 'insurgents' in Afghanistan, but we are no closer to victory today than we were in 2002," Jim wrote.

Would the military establishment tolerate the criticism? Jim didn't care anymore. He had watched too many men die; he couldn't continue to keep quiet. Laying out his on-the-ground experience in black and white was the only honorable thing to do.

SIX HUNDRED MILES AWAY, Adm. Olson strode into a secure conference room at his MacDill Air Force Base headquarters in Tampa, Florida. Olson wore a crisply pressed uniform but chose not to flaunt the medals he'd earned as the highest-ranking Navy SEAL in U.S. history. Olson was the first SEAL to pin on four stars, and the first to head U.S. Special Operations Command (SOCOM). A meticulous man of few words, Olson studied the disheveled Army major on the screen in front of him. Jim chose to sit in a relaxed posture, seemingly unconcerned that Olson stratospherically outranked him.

As the debate over Afghan war strategy raged in Washington during the first year of the Obama administration, Olson had quietly staked out a position that was at odds with much of the top military brass commanding the war. The think-tankers lobbied for sending in as many as eighty thousand more conventional ground troops to wage a full-blown campaign of counterinsurgency (COIN) and nation building. In contrast, Olson's thinking was more closely aligned with that of advocates in the White House—led by Vice President Joseph Biden—who argued for a minimalist counterterrorism (CT) approach. It was a strategy with a lighter footprint that relied more heavily on the small teams of Special Operations Forces under Olson's command. Ironically, the COIN/CT debate was the perfect storm for Jim. Both sides could embrace his plan as buttressing their opposing strategies. But Jim didn't care about the debate. He just wanted to give Afghanistan back to the tribes, without shock and awe, corrupt governments, or boots kicking in doors in the middle of the night.

Afghanistan has long resisted having large numbers of foreign troops on its soil, from Alexander the Great's Macedonian army to the U.S.S.R., Olson observed. Testifying before a House subcommittee in June 2009, Olson even used a term many U.S. commanders shunned as the "*o*-word," noting that the Afghans perceived the U.S. presence as an occupation. Olson was also convinced that the U.S. military's understanding of Afghan culture—particularly the valley-by-valley, village-by-village distinctions that define Afghanistan's highly decentralized society—was woefully inadequate. "It's simply their culture to resist outsiders. . . . Afghanistan will require as small a footprint as we can get away with," he contended.

"We have to get beyond generalizations in Afghanistan into true, deep knowledge of tribal relationships, family histories, the nuances of the terrain and the weather and how that affects how business is done," Olson told the House panel. Only through such a robust understanding, he asserted, could the United States hope, over a long period of time, to convince the Afghans to side with us.

Olson dubbed this initiative to foster cultural knowledge within the U.S. military "Project Lawrence," a reference to the World War I British archeologist and military intelligence officer Maj. T. E. Lawrence. Lawrence "went native" and helped lead the 1916–18 Arab revolt against the declining Ottoman Empire, which had been weakened by a revolutionary movement of Young Turks and lost much of its territory in Europe and North Africa. The United States needs "Lawrence of Pakistan, Lawrence of Afghanistan, Lawrence of Colombia, Lawrence of . . . wherever it is," he said.

Olson naturally sought to nurture these advanced cultural skills inside the organization where they were already most concentrated, his

Special Operations Command, and in particular within the Green Beret community, the segment of the Army that specializes in working with indigenous peoples. So when Jim's forty-five-page "One Tribe at a Time" landed on Olson's Tampa desk, laying out a strategy for sending small, handpicked teams of Green Berets to empower and leverage Afghanistan's Pashtun tribes, Olson was intrigued.

The man who first brought Jim's paper to Olson's attention was the admiral's senior enlisted advisor, Command Sgt. Maj. Thomas Smith, a thirty-one-year veteran of Special Forces. Smith had found the paper fascinating and true to the Green Beret mission he'd long believed in. "Sir, this is exactly what we need to be doing," Smith told Olson. "We need this going on everywhere in Afghanistan, from tribe to tribe."

In the hours prior to the November 2 teleconference, Olson pored over Jim's military record while tasking Smith to call around and check out his bona fides. Olson had tasted his share of combat. In 1993, before he was named head of SEAL Team Six, the team that would be tasked with killing Osama bin Laden, Olson was assigned to the headquarters of a task force in Mogadishu, Somalia. There, on the video screens of an operations center, Olson watched as a mission that was supposed to last an hour turned into a bloody street fight after two U.S. Army Black Hawk helicopters were shot down. Olson volunteered to be a leader of the rescue force. He put on his body armor, grabbed his M4 carbine, borrowed some night vision goggles, and joined what turned out to be an all-night mission into the zone of the city where the Black Hawk helicopters were surrounded by Somali militia fighters. After the rescue force loaded the dead and wounded onto vehicles and daylight broke, Olson helped lead the battered patrol back through a daunting gauntlet of heavy fire known as the Mogadishu Mile. He was awarded the Silver Star. Because of his experiences there and in other places, Olson wanted to hear from men who had fought with Jim.

Olson then spoke with his next-door neighbor in Tampa, Gen. David Petraeus, who was in charge of all U.S. military forces in the Middle East and Central Asia as the head of U.S. Central Command (CENTCOM). Petraeus would later lead all U.S. and other foreign troops in Afghanistan, and afterward direct the Central Intelligence Agency.

Petraeus, too, had read "One Tribe at a Time" and was so impressed that he forwarded the paper to Gen. Stanley McChrystal, then the

NATO commander in Afghanistan, telling him that it should be required reading for all his subordinates. Petraeus remembered Jim from when they both had served in Iraq in 2007. Petraeus had singled Jim out there for his bravery in combat with the Iraqi police he advised. Although Petraeus's perspective differed from Olson's—Petraeus crafted the counterinsurgency doctrine and was its biggest evangelist—he agreed that the ability of officers such as Jim to mobilize indigenous forces was going to be critical to any winning strategy in Afghanistan.

Smith made a preliminary introduction via teleconference. The admiral remained silent.

Jim did what came naturally to an aggressive officer—one known to always carry triple the required ammunition for missions—and seized the initiative.

"Hey, sir, I appreciate you taking time to talk with me. The bottom line is: I will go anyplace and do anything you ask me to."

Olson had his Lawrence.

"Assemble your team. Send me the names of the men you want. Send me your training plan. The time you need to prepare is not important." Olson paused. There's an old Taliban saying: *you've got the watches, but we've got the time.* "What is important is that when you get on the ground, you do not fail."

# CHAPTER 3

ON NOVEMBER 29, 2009, three weeks after Jim got his "Lawrence of Afghanistan" mission, President Barack Obama summoned his top advisors to the Oval Office to give them their marching orders on the war. The meeting came after a protracted debate within the administration over Afghanistan strategy that summer and fall, as advisors grappled with how to fulfill the president's difficult-to-deliver campaign promises on what pundits had labeled "Obama's War."

Leading up to the Oval Office meeting, Obama and his advisors couldn't help but be mindful of historical parallels to another conflict: Vietnam.

Forty years earlier, in March 1969, President Richard Nixon had escalated the Vietnam War by authorizing B-52 Stratofortress sorties armed with thousands of tons of ordnance to carpet-bomb Cambodia. The secret bombing campaign later revealed as Operation Menu, which included missions lightheartedly code-named Breakfast, Lunch, and Dinner, was part of Nixon's covert war in "neutral" Cambodia. It targeted North Vietnamese army and Viet Cong guerrilla bases and sanctuaries in Cambodia and aimed to make good Nixon's 1968 campaign pledge to "have an honorable end to the war in Vietnam."

Like Obama, Nixon had inherited a war from a previous administration. The outgoing president, Lyndon Johnson, had not sought nor would have accepted the nomination for another four-year term, primarily because he and the best and the brightest of his advisors could not envision a way to deliver an honorable peace in Vietnam. To attain peace in 1969 would have required U.S. capitulation to communist North Vietnam and its Ho Chi Minh–inspired Viet Cong insurgency in the South. Whichever way you looked at it, Vietnam would be a definite mark in the loss column for the United States' war record. At the same time, an occupying superpower propping up the repressive government of South Vietnam was the furthest thing from honorable. Concealing disgrace by wrapping it in the ancient warrior code was good rhetoric, but in 1969 the only conceivable honor would come from turning the war around and doing what two previous presidents couldn't do: beat the insurgency.

The Nixon administration initially planned to accelerate a program that Secretary of Defense Melvin Laird called "Vietnamization." During preparations for Nixon's promised withdrawal, American soldiers would arm and train the South Vietnamese with the goal of eventually turning over the war effort to them. The security of South Vietnam would become the responsibility of loyalists of the U.S.-backed president Nguyen Van Thieu, a former general of the Army of the Republic of Vietnam (ARVN). If it could extract itself before the country fell, the United States would be spared "defeat."

But with increasing aggression from the Viet Cong and the North Vietnamese army, the People's Army of Vietnam (PAVN), just after Nixon's inauguration, the president felt compelled to respond. He had two unappealing choices. On February 9, 1969, the U.S. commander in Vietnam, Gen. Creighton W. Abrams, submitted a proposal to the White House. Why not bomb Viet Cong and PAVN outposts based across the border in neighboring Cambodia and signal to the enemy that the new guy was just not going to roll over and retreat? Less than two months into his term, Nixon had to decide whether to back his general and intensify bombing in Cambodia, thus expanding the battlespace for a war he'd vowed to quit, or stay the course and prepare for withdrawal while pursuing a truce. For a man who had once lost his composure after being beaten in an election and let loose with "You won't have Nixon to kick around anymore," the choice he made was perfectly in character.

Nixon secretly escalated the war and signed off on Operation Menu. The epic domestic and international debacle scarred the nation and cost thousands of Cambodian and Vietnamese lives in the process. By 1975, the United States had not only left Vietnam in dishonor but so destabilized Southeast Asia with its opening of the second front that Cambodia eventually fell to genocidal repressions of a communist North Vietnam offshoot, Pol Pot's Khmer Rouge. A war fought in the name of halting the domino theory, which holds that if one nation falls to communism, its neighbors will soon follow, ended up causing the very thing it was meant to deter—the expansion of communism.

Running for president in 2008, the forty-seven-year-old Obama, then a senator, knew the dark history of the Vietnam War. How could he not, running against "Hanoi Hilton" POW John McCain? McCain continually challenged Obama's determination on the wars, contrasting what he cast as his opponent's lack of commitment with his own pledge to stay the course and back U.S. troops. "We are going to win in Iraq and win in Afghanistan, and our troops will come home with victory and honor," McCain said in a November 3, 2008, election eve speech in Miami, Florida.

Facing the prospect of inheriting not one but two insurgency wars, Obama asserted during the 2008 presidential campaign that he would not make the mistakes previous wartime presidents (Johnson, Nixon, and George W. Bush) had made. Obama pledged that, as president, he'd shut down Bush's war in Iraq within sixteen months. Instead, Obama said, he would turn the nation's focus with renewed vigor back on the United States' primary foes—Osama bin Laden, Al Qaeda, the Taliban, and any nation or people that harbored terrorists. He'd put a full-court press on Afghanistan, kill bin Laden, and bring American soldiers back home as soon as he possibly could. That would be peace with honor.

"We will bring this war [in Iraq] to an end. We will focus our attention on Afghanistan," Obama repeated at one campaign stop after another.

But like Richard Nixon in 1969, the newly elected president now had to deliver on his rhetoric. The Bush administration had never come up with a clear strategy for Afghanistan. The rapid overthrow of the Taliban just months after the September 11, 2001, terrorist attacks and the influx of Afghan expatriates returning home gave rise to optimism. A handpicked, westernized Pashtun politician named Hamid Karzai was chosen as Afghanistan's interim president by a *loya jirga*, or assembly of delegates, in 2002. But soon complacency over a seeming victory led to the bungling of the war effort's critical last phase—stabilizing a new legitimate power and getting out. In the months and years to come, as the White House shifted its attention and resources to Iraq, stability operations in Afghanistan would be badly neglected.

In March 2003, Bush ordered the U.S. military invasion of Iraq and the toppling of dictator Saddam Hussein. Despite Bush's May Day 2003 "mission accomplished" speech, the administration again demonstrated the lack of a realistic plan for stabilizing the country. Instead, borrowing from the post–World War II model, the administration put in place an autocratic chief executive to oversee the occupation of Iraq until popular elections could be held. Bush chose a former assistant to Henry Kissinger at the State Department during the Nixon and Ford administrations, Paul "Jerry" Bremer, as head of the Coalition Provisional Authority (CPA)—essentially making him the governor of Iraq.

Bremer's disastrous edicts—to disband Iraq's military and forbid any former member of Hussein's Ba'ath Party to hold a public sector job—helped ignite another insurgency. Pentagon minds grappled with ideas for stanching the post-Saddam bloodshed to little avail. At last, after the carnage of four years of insurgency and sectarian strife, they turned to Petraeus, putting in charge a creative general with a bold strategy to turn things around.

As all major U.S. military resources flowed to Iraq, Afghanistan's Taliban had steadily resurged and Osama bin Laden remained at large. In a sobering report delivered to the White House in August 2009, the senior commander in Afghanistan, Gen. McChrystal, warned Obama that the military coalition had fought to a stalemate. The Taliban was seizing new territory. After exhaustive analysis and debate about the controversial report—inside and outside the government—Obama in November prepared for the Oval Office meeting with his top advisors.

Like Nixon in 1969, Obama faced a decision whether or not to escalate the war. He could dispatch tens of thousands more troops in a full-blown counterinsurgency campaign to try to defeat the Taliban, or he could wage a far more narrow counterterrorism strategy targeting Al Qaeda that relied heavily on aerial drone strikes with only small numbers of American boots on the ground. The first option risked entanglement in a protracted, labor-intensive, costly war that the American public was unlikely to stomach. The second raised the chances that Afghanistan would revert to civil war and suffer a Taliban comeback.

Obama's advisors were deeply divided between the two camps of counterinsurgency and counterterrorism.

Military commanders were pressing for a troop buildup in Afghanistan. Without significantly more troops, the campaign would "likely result in failure," said McChrystal's sixty-six-page report, which was leaked to the *Washington Post* in late September.

McChrystal and other top military advisors—chairman of the Joint Chiefs of Staff Adm. Michael Mullen, Defense Secretary Robert Gates, and Central Command chief Petraeus—lobbied for sending forty thousand fresh U.S. troops to the country for a robust counterinsurgency campaign. More troops were vital to bring enduring security and governance to the Afghan people and flush out the Taliban, they argued. The greatest challenge would be shoring up the government given serious doubts about the leadership of President Karzai. Karzai's administration had been accused of fraud in the 2009 presidential elections, and his family had been implicated in major corruption scandals as well as in Afghanistan's vast opium trade.

Another camp of Obama advisors, led by Vice President Joseph Biden, argued strenuously against a big troop influx. Instead, Biden called for focusing on targeted counterterrorism operations to degrade leaders of the Taliban and Al Qaeda along the Afghanistan-Pakistan border and elsewhere. Meanwhile, the strategy would boost training of Afghan security forces and reconciliation efforts to persuade Taliban fighters to give up their arms. Biden argued that the U.S. public lacked the political will for a drawn-out escalation of the Afghanistan war; instead, in a process similar to "Vietnamization," the Afghans would take over.

Both strategies had serious drawbacks. Obama knew that the public's patience with the war was wearing thin. Recent polls showed that most Americans had concluded that the war was not even worth fighting. And the public was solidly against sending any more troops to the landlocked country where U.S. forces had been fighting for more than eight years. Such sentiments were overwhelming within the president's Democratic Party, with seven in ten Democrats saying the war was not worth the cost, and fewer than one in five supporting a troop increase.

The cost of the war was another huge factor on the president's mind. With the country still emerging from the worst economic crisis since the Great Depression, the price tag of the wars in Iraq and Afghanistan was $1 trillion and counting. One report projected that another ten years in Afghanistan would double that cost. Obama wanted to engage in nation building in the United States, not in an impoverished Central Asian country whose economy was dominated by the opium poppy.

Instead of focusing on what the options lacked, Obama worked the problem pragmatically. If he did not commit to a troop surge, a promise he had made during his campaign and one he was reluctant to abandon, military hawks would accuse him of not giving them the opportunity to win the war. Petraeus had had great success with his COIN program in Iraq. To refuse to give the military the chance to apply the same principles in Afghanistan would deny the United States the chance to exit the war with honor. But Obama decided it would be a mistake to give the generals the expectation that he would be open to additional future troop commitments, concerned they would ask for more forces year after year, and Afghanistan would turn into as much of a quagmire as Vietnam. What Obama wanted was to give the military enough support to deal a fatal blow to the Taliban and get Afghanistan back on track, but no more.

It was a mild Washington Sunday, late in the afternoon, when Obama called the small circle of defense advisors into the Oval Office to tell them what he had decided. Present were Biden, National Security Advisor Jim Jones, Gates, Mullen, Petraeus, and a handful of others. The time for debate, he had made it clear, was over. He was essentially issuing an order.

The core goal of the strategy, Obama told them, was still to defeat Al Qaeda and deny the terrorist organization a safe haven in Afghanistan and Pakistan. But, he explained, the means to accomplish that goal in Afghanistan would be neither a full-fledged COIN strategy nor a light-footprint CT approach. It would be a combination of both, a surge of thirty thousand troops coupled with intensified, targeted raids. The intention was to weaken the Taliban and reduce its territory to the point where Afghan security forces could take over.

Then came the catch that surprised them all: they had eighteen months.

Obama told his advisors that the surge troops would flow in as rapidly as possible but would begin coming home in July 2011. This would not be an open-ended war, Obama stressed. There would be no blank check. Obama turned to Gates, Mullen, and Petraeus and asked them if they agreed with his plan. All voiced their support.

But the July 2011 deadline was troubling, particularly for Petraeus and many U.S. military commanders on the ground. The timeline was unreasonably short to anyone familiar with the history of counterinsurgencies, which historically had lasted on average more than a decade. Once Obama's clock became public knowledge, it would be hugely discouraging to Afghans who had sided with the U.S. military—and emboldening to the enemy.

During the meeting with the president, Petraeus mentally calculated how difficult it would be to make adequate headway by the time eighteen months rolled around. "I was concerned," he said later. "It was not about making progress faster; rather it was about making progress, period. The Taliban was very much on the march in 2008, 2009, and 2010, and we needed to halt their momentum." Key to doing that, Petraeus knew, would be to recruit a local Afghan security force. He thought of the plan outlined in Jim's paper "One Tribe at a Time." Small teams of Special Forces soldiers would eat, sleep, and live with tribesmen while training them and empowering them to defend their homes. In that way, the teams together with the tribes could cover large parts of rural Afghanistan and rapidly accelerate the traditional COIN timeline. It was all about relationships. If highly trained American soldiers could forge bonds of trust with the tribes, they could help the tribes regain control over their territory and push out the Taliban.

"It was hugely important . . . at a time when I was looking for ideas on Afghanistan," Petraeus recalled later. Others within the military had discussed the concept of grassroots security in Afghanistan, but Jim "was the first to write it down, in a very coherent fashion, very readable, very encouraging frankly . . . and there is enormous power in that," he said. Petraeus, McChrystal, and others made the paper a strategic catalyst for a bottom-up approach to security, according to several officers involved, using it to lay the groundwork as they began to formulate a plan for raising local forces nationwide.

Village by village and tribe by tribe, success could buy more time. Petraeus had pushed a grassroots security plan in Iraq that helped turn the tide of that war—and he intended to do the same thing in Afghanistan. The president wouldn't pull the plug on a winner—especially when the footprint and cost were so small and the payoff so large.

# CHAPTER 4

"BREAK LEFT," PETRAEUS ORDERED through a headset to the Black Hawk helicopter pilot.

"Hold on, Ann!" the wiry four-star general said, turning to me. I braced my feet against the floor and tugged on the shoulder belts strapping me into the seat beside him.

The chopper banked steeply, then leveled out and skimmed low over the rectangular tan rooftops and palm trees of downtown Baghdad. It was near dusk, and the war-torn city of seven million people, the second largest in the Arab world, sprawled out flat and bronze beneath us as far as the eye could see in the desert haze.

It was April 2007, just three months into the Bush administration's surge of more than 20,000 U.S. troops to bolster the 140,000 already in Iraq. Suicide bomb explosions and gunfire rocked Baghdad daily. I had been reporting on the war since the first day, embedded with the Army's 3rd Infantry Division as it spearheaded the invasion to topple Iraqi dictator Saddam Hussein, and this was the worst I'd seen it.

I'd flown into the Iraqi capital a few days earlier on a sleek C-17 Globemaster with Defense Secretary Robert Gates, covering his latest swing through the Middle East in my role as Pentagon correspondent for the *Washington Post*. In Baghdad, Gates was pressing Iraq's leaders to push through political compromises between rival Shiite, Sunni, and Kurdish factions. I was relieved to escape the security bubble that isolated Gates and his official entourage, which whisked reporters from five-star hotels to high-level meetings with just enough time in between to spoon-feed us a few quotes at the airport VIP lounge. Occasionally we were invited to have drinks with Gates at a hotel bar—off the record. He would roll up his shirtsleeves and regale us with stories of Cold War intrigue, giving a personal glimpse of the man behind the Pentagon briefing room podium. But the former CIA chief always played his cards close to his vest. The traveling press cocktail circuit was never my idea of reporting. Now I was headed back into the streets of Baghdad with ground troops, where I felt most comfortable. First, though, I would get Petraeus's view from three hundred feet.

I watched Petraeus gaze out the window onto the city, a maelstrom of violence and sectarian killing that he, as the new top U.S. commander in Iraq, had the mission of quelling. His angular face was taut. Five full U.S. combat brigades were flowing into Iraq, most of them bound for Baghdad. Thousands of troops were fanning out into dozens of tiny outposts in neighborhoods across the city in an effort to increase security for the Iraqi population—a key tenet of Petraeus's counterinsurgency plan.

Petraeus was largely responsible for the Army's new COIN doctrine, Field Manual 3-24, published in December 2006. He was involved at every stage from conceptualizing the strategy, to deciding on key content, to personally editing several chapters dozens of times. A core part of the doctrine was what Petraeus referred to as "counterinsurgency math"—namely, that victory against an insurgency required approximately one security force member for every forty to fifty people in the country. His Iraq strategy aimed to put enough U.S. troops into position in neighborhoods to protect Iraqi civilians and also strengthen the capabilities of Iraqi security forces and local governments. That, in turn, was intended to enable ordinary Iraqis to feel safe enough to provide intelligence on insurgents, or stand up to the enemy themselves.

That evening, Petraeus wanted to convince me that his plan was beginning to work—just in time for me to write a front-page story for the Sunday edition of the *Washington Post*.

It was part of Petraeus's pragmatic public agenda: speeding up the Baghdad clock by intensifying military operations on the ground, while slowing down the political clock in Washington by demonstrating enough progress to buy more time to try to win the war. It was essentially the same tactic he would later use in Afghanistan.

Petraeus was a master with the media, unlike many others who seemed intimidated by reporters. No other general officer I knew moved so fluidly from on the record to deep background to off the record—sometimes in a single sentence. One of his tactics—I learned the hard way—was to wax pedantic, eating up all of the reporter's time. Late that afternoon, in a conference room on the grounds of one of Hussein's old palaces, I had interviewed Petraeus for more than an hour on the fine points of his strategy. He answered my first question with a monologue that ran on for ten minutes. Finally I cut in and blurted out another question.

"Sir, I've got to interrupt you here—why have the Iraqi forces been so sluggish moving into Baghdad?"

One of Petraeus's military aides, sitting at the table beside him, burst out laughing.

"We're *very* glad you interrupted," he said.

"Yeah," Petraeus admitted with a grin. "I'm trying to run down the *Washington Post* clock!"

Still, I had to admire Petraeus's encyclopedic grasp of detail about Iraq. Brilliant and ambitious, Petraeus was already gaining a reputation in military circles as one of the most talented American generals of his generation. What was unusual about Petraeus's rise was that he was a true scholar—the kind normally shunned by the Army brass. A West Point graduate with a doctorate in international relations from Princeton, he had survived the Army's anti-intellectual culture with the support of powerful mentors. Now Petraeus faced the toughest challenge of his career—putting his cerebral war strategy into action, with countless American and Iraqi lives at risk.

As the interview wrapped up, Petraeus unexpectedly turned to me.

"How would you like to go for a ride?" he said.

"Sure," I answered, surprised.

"Let's go," he said. I followed him to the flight line, where his helicopter was waiting, the warm blast from its rotor blades whipping our clothes and hair.

We ducked and climbed through the open side door. Petraeus took a seat by the window, motioned to me to sit next to him, and handed me a headset with a pair of light green earphones and a microphone.

"You got me?" he asked through the mouthpiece, making sure I could hear him.

"Yep," I said, and gave him a thumbs-up.

We lifted off, and Petraeus began a running commentary as he studied the neighborhoods below, looking for signs of normalcy.

After one turn, he caught sight of a patch of relative calm. "He's actually watering the grass!" Petraeus said excitedly, peering down at a man tending a soccer field, with children playing nearby.

And so it went, all around the city. Telling the pilot to "break left" and "roll out," he scrutinized the landscape for even tiny improvements—a pile of picked-up garbage, an Iraqi police car out on patrol, a short line at one gas station—as if gathering mental ammunition that would help him deal with the next wave of carnage in Baghdad. An amusement park, its rides lit up, merited a full circle.

"We have certainly pulled neighborhoods back from the brink," Petraeus said.

But seconds later, the aircraft pivoted again, exposing boarded-up shops on a deserted, trash-strewn street. A bit farther, along the Tigris River, a hulking pile of twisted steel came into view—the remains of the Sarafiya Bridge, blown up ten days before our helicopter ride amid a series of spectacular and deadly suicide bombings.

"That's a setback," Petraeus said, his voice lower. "That breaks your heart."

Petraeus was right to point out that not all of Baghdad was a blazing inferno of bomb blasts, as shown on the nightly news. Overall, though, I knew from statistics that violence was spiking upward. Attacks were on the rise, including an upsurge in especially lethal bombings that were killing scores in Baghdad markets, bus stations, and mosques.

The massive bombings were so bad that Petraeus had ordered huge concrete walls built around neighborhoods—euphemistically calling them "gated communities"—to try to control population access and keep out car and truck bombs.

We flew over a west Baghdad district, and Petraeus pointed out a towering wall going up around it. "That's part of the 'concrete caterpillar,'" he told me.

U.S. troop casualties were also trending higher as American soldiers moved deeper into the city and insurgents assaulted their small outposts. Deaths among Iraqi forces were on the rise, too. While it was expected that such a large influx of forces into the war zone would initially bring more fighting and deaths—an argument I'd heard advanced again and again by U.S. commanders—at some point there had to be a turnaround and the killing had to subside. When would that happen? Petraeus wasn't sure.

"On a bad day, I actually fly Baghdad just to reassure myself that life still goes on," he told me, leaning back and propping his legs on the seat in front of him.

As darkness fell and some lights in the city came on, we passed over Baghdad's southern flank, where Petraeus had led the legendary "Screaming Eagles" of the 101st Airborne Division in the invasion of Iraq in 2003. After that, he served another full tour as the general in charge of training Iraqi forces. As he looked out the window, he said that he felt a personal sense of obligation to help the Iraqi people, because "General Colin Powell was right, it is Pottery Barn rules—if you break it, you own it."

But he told me that when he returned to Iraq this time, in February 2007, he was shocked at how far security had deteriorated. "The whole society is more fearful, more suspicious, more worried"—and more difficult to help, he admitted.

"I wouldn't be honest if I didn't say that this has an effect on all of us," he said. "And so every now and then we just get on the helicopter. . . . You go see some projects that you know have been built. . . . You see some police stations and you see people just sort of driving on, people getting on with their lives, and it sort of reassures you: 'Hey, these people are survivors.'"

A FEW DAYS LATER, I made my way with a foot patrol of 82nd Airborne Division paratroopers into Baghdad's Sadriya market and came face-to-face with some of those Iraqi survivors Petraeus spoke of.

As random gunshots rang out in the distance, I followed the soldiers down narrow streets jammed with shops and pedestrians in eastern Baghdad's Rusafa district. Rusafa was a center of markets frequented by Shiite residents and workers, and so was often targeted in car bombings by Sunni extremist groups.

The latest bombing, on April 18, had killed more than a hundred people at a bus depot beside Sadriya market as workers were crowding into vans and buses at the end of the day. Charred carcasses of vehicles were strewn at the blast site. I picked my way through the debris and met Sabah Abd, a thirty-three-year-old Iraqi man. Abd owned a fruit stand piled high with oranges just a few feet from one of Petraeus's concrete barriers that divided Sadriya market from the bus depot. I introduced myself in Arabic as an American reporter and asked a few simple questions.

Abd told me that when the bomb went off, it threw him to the ground, splattered with blood.

"The whole world shook," he told me.

But three days later Abd was back at his stand. "My wife said, 'If you go there, I will divorce you.' But what can I do? There is no work outside," he told me.

The American patrol advanced and met a group of Iraqi police who were moving through the neighborhood collecting scores of disfigured bodies to take to the morgue. I realized that despite the horrific scene, at least one thing was going right—U.S. troops were maintaining a close, relatively constant presence in the area. Shopkeepers knew the soldiers and were greeting them and inviting them to tea. As Petraeus dispatched troops to neighborhood outposts, American forces were learning more about the needs and concerns of Iraqi civilians and responding to them.

For too long I had watched the military move in the opposite direction. Since the 2003 invasion, I had traveled extensively across Iraq on multiple reporting trips and spent the better part of a year in the country. I had been shot at, rocketed, and mortared, and vehicles I was riding in had been hit by roadside bombs. I had patrolled in areas so dangerous that walking down the street was impossible—we had to use alleys and move directly from one house to another by climbing over their courtyard walls. In other areas, we had to run on patrol, darting across intersections in a crouch.

As the violence intensified in 2004, I observed U.S. troops progressively withdraw into large, garrison-like encampments. From there, they waged highly conventional, large-scale sweeps into Iraqi cities—kicking in doors, conducting searches, detaining young men, demanding information—and then leaving as quickly as they arrived. The "clearing operations" alienated and angered Iraqi civilians. No effective Iraqi forces filled the void left by the U.S. troops, and as a result, insurgents quickly resurfaced and terrorized any civilians who had cooperated with the Americans. So the vicious cycle repeated itself. Some insurgent-held towns had been "cleared" by different U.S. military units half a dozen times in just three or four years.

As U.S. military units fortified their bases, welded extra steel onto their vehicles, and added more Kevlar to their body armor, they were so distanced from ordinary Iraqis that it began to impact my ability to cover the war. On one trip to the insurgent-held town of Baqubah in eastern Diayala Province, the patrols I went out on kept hitting roadside bombs every day, forcing us to return to base. On another trip, I went north to Mosul, the second-largest city in Iraq with two million people, and discovered that the Army brigade stationed there had suspended patrols inside the town. I avoided asking units to depart from their missions to take me places—not wanting to be responsible for anyone getting hurt or killed—but this was an extreme case. So I convinced a military police element that was passing through Mosul to let me ride with them and make a few stops so I could interview Iraqi residents.

Against this backdrop, I arrived in late 2005 in Tal Afar, a northern Iraqi town of three hundred thousand people near the Syrian border, and was encouraged to find a much different military mind-set. Just a few months earlier, Tal Afar had been in the throes of sectarian fighting with a dug-in Sunni insurgency. Col. H. R. McMaster, a bold and ingenious commander, had systematically cleared the town and then moved his cavalry troops to live deep inside the neighborhoods among the people, greatly reducing violence. McMaster was no stranger to combat. He had earned a Silver Star in 1991 as a commander in a decisive Gulf War tank battle. In April 2005, I was riding in his Humvee south of Baghdad in an area known as the Triangle of Death when McMaster led our fight out of an ambush that killed the first soldier he lost from his regiment. What made the difference in Tal Afar, though, was McMaster's ability to outwit the enemy psychologically and politically.

Later in 2006, I watched as Col. Sean McFarland adopted Petraeus's counterinsurgency approach against a formidable insurgency in the town of Ramadi in the Sunni stronghold of Anbar Province. Anbar was the main base of the extremist Sunni insurgent group Al Qaeda in Iraq, and Ramadi was the most violent place I'd ever set foot in. You could count on coming under attack within about twenty-five minutes of leaving any U.S. base in that town. But McFarland and others pushed a plan to recruit local Sunni tribes to turn against the insurgents and join the local security forces in what became known as the "Anbar Awakening."

A short distance from Ramadi, I covered the work of a Special Forces team that was also effectively engaging local Sunni tribes around the Anbar town of Hit, where the population had earlier been completely estranged by conventional clearing operations.

"It's not about bulldozing Hit, driving through with a tank, with all the kids running away. . . . These insurgencies are defeated by personal relationships," said Master Sgt. Chris Heim, the Special Forces team sergeant, as we rode through the desert after meeting with a Sunni tribal leader. The real battles, he said, are unfolding "in a sheik's house, squatting in the desert eating with my right hand and smoking Turkish cigarettes and trying to influence tribes to rise up against an insurgency." That made perfect sense to me.

In 2007, the Sunni awakening in Anbar began to spread to other Sunni communities in Iraq, where Sunni tribesmen embattled by both Islamic extremists such as Al Qaeda in Iraq and Shiite militias were increasingly willing to switch sides and ally with U.S. forces.

Intrigued by the trend, I flew in the summer of 2007 to the eastern town of Baqubah, where for years I'd seen hard-core Sunni insurgents battling U.S. forces all along the town's canals and dense orange and palm groves. There, one night I walked with U.S. soldiers to a dim, abandoned house and met with a Sunni insurgent commander. He was sitting cross-legged, and—letting us know he was armed—lifted up his T-shirt, revealing a pistol stuck in his belt. It was a strange feeling to sit next to a man who just weeks earlier would have tried to kill all the Americans in that room, including me.

"We decided to cooperate with American forces and kick Al Qaeda out, and have our own country," said the twenty-one-year-old commander, who gave only his nom de guerre, Abu Lwat. "In the future, we want to have someone in the government," he said, holding his cigarette with a hand that was missing one finger. Abu Lwat was the first of many Sunni fighters I would meet in coming weeks, all of them risking their lives to defend their communities and aid the Americans.

As U.S. surge forces gained a foothold in Iraq's trouble spots and Sunni tribes and residents showed a willingness to switch sides, I got wind that Petraeus was moving to capitalize on the trend by launching a grassroots local security initiative. Petraeus, I learned from a subordinate commander, had pushed through a controversial program to allow the U.S. military to use commanders' funds, emergency payments, and other monies to pay local Iraqi men to protect their neighborhoods, formalizing the sporadic efforts by Iraqi tribesmen and villagers to defend their communities. Recruiting the security forces from their home turf was a major strategy shift—and a bold one.

In July 2007, I walked into an interview with Petraeus in his ornate, spacious office in a palace near the Baghdad International Airport. I had just arrived from a grimy outpost south of Baghdad and had not showered for at least a week. My clothes were dirty from a long foot patrol and streaked with white salt marks left behind after sweating in the Baghdad heat. My hair was pulled back in a ponytail. By that point, I just wanted answers—and Petraeus supplied them.

"The bottom-up piece is much farther along than any of us would have anticipated a few months back," Petraeus told me. "It's become the focus of a great deal of effort, as there is a sense that this can bear a lot of fruit."

Petraeus paused, choosing his words carefully. U.S. recruitment of grassroots forces was the most significant trend in Iraq since the beginning of the surge four months earlier, he said. It could propel the slow-moving national efforts to forge cooperation among Iraq's main religious sects and ethnic groups.

"This is a very, very important component of reconciliation because it's happening from the bottom up," he said.

It was page-one news—and, I sensed, a turning point in the war.

In coming months, the program spread more rapidly than Petraeus or anyone else had imagined. In a little more than a year, a local guard emerged of more than a hundred thousand men. They came to be known as the "Sons of Iraq." Petraeus's experiences in Iraq would profoundly shape how he thought about solutions for Afghanistan, a country with a largely rural population scattered across rugged and inhospitable terrain.

ULTIMATELY, PETRAEUS KNEW SUCCESS in counterinsurgency would rest on the shoulders of his men on the ground and their ability to grasp the mission.

Not long after our Black Hawk helicopter ride, Petraeus singled out one American officer for his exemplary work with the Iraqi police. Petraeus was well aware that in the end Iraqis would have to secure Baghdad and the rest of the country alone. But they badly needed skilled American mentors to help them get there.

"The story of U.S. Army Major Jim Gant and the Iraqi National Police Battalion that he advises is an important piece of the mosaic depicting the battle we're waging in Iraq," Petraeus told a group of reporters in early May 2007.

"A few days ago, Major Jim Gant received the Silver Star, the Army's third-highest award for valor in combat," Petraeus said. "He asked to be awarded the medal in front of his best friends—his fellow Iraqi brothers in arms—saying: 'We, both American and Iraqi, have to do this together. Neither of us can do it by ourselves.'"

Petraeus invited Jim and some of his Iraqi comrades to a meeting in his office in Baghdad's Green Zone and came away from it praising Jim as a tremendously effective advisor.

A week later, I was offered an opportunity to interview Jim. I took it.

# CHAPTER 5

RIDING ON THE HOOD of his Humvee, Jim cradled his M4 carbine and braced his boots on the bumper as the vehicle twisted through the trash-strewn streets of Baghdad's Sadr City. Beside him was his Iraqi comrade and interpreter Mohammed Alsheikh, nicknamed "Mack," who rarely left his side. Jim scanned left and tried to spot roadside bombs, while Mack scanned right and watched for gunmen.

It was early 2007, and Jim and the Iraqi police he advised were roaming a part of Baghdad that Petraeus's surge troops had not yet occupied. He and his small U.S. team had bonded with his Iraqi counterparts so closely that together they could go places that other U.S. units could not. The overcrowded Shiite slum of two million people in the middle of Baghdad rarely saw a daytime patrol by American forces. Apart from conducting night raids, they mainly stayed out of the area. Sadr City was ruled by the main Shiite militia, known as the Mahdi Army, or Jaish al Mahdi (JAM), which used the slum as a safe haven for staging attacks on Sunnis and U.S. forces across Baghdad. Sectarian killing was rife in Baghdad and other religiously mixed parts of Iraq, and the country was teetering on the brink of civil war. The Shiite militias had become as much of a threat to U.S. forces as the Sunni insurgents. But Jim had managed to gain safe passage through the area from Mahdi Army commanders, who ordered their fighters not to strike his patrols. So his Humvees, easily identified by the crimson Spartan helmets spray-painted on their doors, were able to roll through Sadr City exposed but unscathed.

A Shiite militiaman with an AK-47 slung over his shoulder waved the vehicle through one of the makeshift checkpoints that ringed Sadr City. The Humvee moved through dusty markets and streets that stank from sewage running through the gutters, and paused in front of a small hospital. One of Jim's Iraqi comrades went in to pick up a policeman who had been shot in a Baghdad raid a few days before. Sadr City was the only place the Shiite policeman could safely receive treatment. Shiites taken to Sunni hospitals in the city would often disappear, never to be seen again.

Jim had volunteered to go to Iraq in 2006. As a newly promoted major, he could no longer fight alongside the men he had led as a captain in 2003 and 2004 on his Special Forces team, Operational Detachment Alpha (ODA) 316, when they returned to combat in Afghanistan. It was how the system worked, but it was painful for Jim. The Army machine had to rotate officers through a limited number of command positions— but breaking up teams could hurt cohesion and morale. Jim felt incredibly guilty about not deploying with the men he'd taught to hunger for combat. The ache intensified in August 2005 when the comrades he loved most in the world faced their toughest battle in Afghanistan and one of them died fighting. Jim was proud of his men, brothers all, but devastated he was not by their side.

So Jim felt compelled to join the fray in Iraq, where the violence and U.S. casualties were escalating sharply. Having tested himself against Afghan insurgents, he wanted to find out what kind of enemy the Iraqis were. The American general in charge of training Iraqi forces, Brig. Gen. Dan Bolger, offered him a position leading a small U.S. military team advising Iraqi forces. Jim accepted immediately, and gave his team the call sign "Spartan." Over the next fourteen months, he would see the heaviest fighting of his life, sharpening his combat skills like never before. Yet the bloodshed and constant danger deeply scarred him psychologically. At the same time, his work with the Iraqis opened his eyes to the potential for comradeship and trust with indigenous forces that built on his Afghanistan experience and went beyond what he had ever imagined possible. In Iraq he led a non-American unit in battle for the first time, and he grew even more convinced of the power of bonding with and motivating a foreign force. But it didn't start out that way.

When Jim arrived in Baghdad in June 2006, the city was in the throes of the worst sectarian slaughter in its history. Warring between

Iraq's majority Shiite and minority Sunni religious factions had exploded following the February 2006 bombing by Sunni extremists of one of the Shiites' holiest shrines, the Al-Askari mosque in the contested city of Samarra. Thousands of Iraqi civilians were fleeing their homes as sectarian cleansing and executions took place in mixed neighborhoods. Exploiting the sectarian split, the Sunni extremist group Al Qaeda in Iraq was intensifying its attacks, and huge car bombs were exploding across Baghdad each week. Meanwhile, the U.S. military strategy in Iraq was still focused on the increasingly unrealistic goal of handing over territory to Iraq's fledgling security forces as quickly as possible. Iraq's newly installed prime minister, Nouri Al-Maliki, was predicting that Iraqi forces could take responsibility for security in the country in eighteen months. The top U.S. commander in Iraq, Gen. George Casey, called 2006 "the year of the police" and was pushing a plan to train Iraqi police and start putting them in charge of civil security by the end of the year. Jim was assigned to be part of that effort. But behind the façade of transition, concerns were growing within the U.S. military about the sectarianism and corruption of the Iraqi forces—in particular the National Police.

In July, Jim walked into a meeting in Baghdad's heavily guarded Green Zone with Col. Chipper Lewis, the newly appointed American senior combat advisor for the Iraqi National Police. Six-foot-five and blunt-spoken, Lewis had met Jim in pre-deployment training at Fort Bliss, Texas, and flew into Baghdad with him. A veteran infantry officer, Lewis knew the National Police would be a tough mission, and Jim was a man who could help. Looking Jim in the eye, he gave him his official assignment: he was to train, mentor, and advise the all-Shiite Iraqi National Police Quick Reaction Force Battalion. Then Lewis gave Jim the lowdown on the unit.

"It's a fucking Shiite militia," he said. "It's out of control."

The six-hundred-man battalion, Lewis explained, like many Iraqi units hastily recruited at the time, was formed out of a militia from Sadr City and legitimized by the government. Col. Taher Alasadi, a veteran Iraqi Army officer who grew up in Sadr City and served a year in jail in 2000 for bad-mouthing Iraqi president Saddam Hussein, led it. Called "commandos" by the Iraqis, the unit was to serve as a quick reaction force for the National Police, pursuing high-value targets in Baghdad and responding to emergencies anywhere in the country. The Baghdad-based unit was relatively autonomous. Unlike the two dozen other National Police battalions, it was not linked to the American military chain of command and had never before been assigned American advisors. It took orders directly from Nouri Al-Maliki, a longtime Shiite politician and dissident, as well as the minister of interior and the head of the Iraqi National Police, Gen. Adnan, whom Lewis was assigned to mentor.

Lewis paused.

"This is what we think is going on down there: government-sanctioned killings," Lewis said. The U.S. military was convinced that senior Iraqi officials were using the unit as an instrument of sectarian violence, to carry out politically driven executions of Sunni targets. The unit was going out on missions around Baghdad on short notice at all hours. Moreover, he said, Col. Taher appeared to be corrupt, taking pay for six hundred police while far fewer were on the job at any given time. "Everybody in the world is convinced that Colonel Taher and his organization are primarily responsible for all these killings at night," Lewis told Jim. "So don't go into this nonchalantly. Go in with your dukes up."

In reality, the situation was worse than Lewis let on. He did not tell Jim what his own marching orders were at the time. U.S. commanders, persuaded that the National Police were, in his words, "the root of all evil of sectarian violence in Iraq," had made Lewis an advisor in name only. His real job was to figure out how to dismantle the entire National Police apparatus. "My mission was basically 'make these guys go away,'" he recalled later.

But first Lewis wanted to see what Jim would turn up.

Initially, what Jim learned was disturbing. The police had the look and smell of an undisciplined militia. They lived in a small compound bordering the Green Zone known as Site One that was dirty and run-down, with no plumbing, electricity, or air-conditioning. It had a small detention center run by the National Police headquarters where beatings allegedly took place. The police wore mismatched uniforms and had bandanas wrapped around their heads. They fought each other, played cards, rolled dice, smoked, and dipped. "These motherfuckers were a gang. They were just a bunch of thugs the government turned into a military unit," Jim recalled.

During the first six weeks, Jim was unable to go on missions with the police because his U.S. military team had not yet arrived. So he focused on sizing up Col. Taher, the Iraqi commander, while monitoring his contacts. Taher, he discovered, had deep and extensive ties with members of the Mahdi Army, the main Shiite insurgent group led by Islamic cleric Muqtada al Sadr. But the relationship was not entirely what it appeared. Probing further, Jim learned that Taher's association with the Mahdi Army was familial and business-related, and involved no tactical dealings with the militia. Taher's parents lived in Sadr City, and he had to take his police there for medical care because the Sunni-run hospitals in Baghdad would not treat them. He also learned that Taher, while skimming off some government funds, was using the vast bulk of his payroll for the police. Jim spent hours drinking tea with Taher in his office and began to see his potential as a commander. Taher, he learned, was a seasoned officer, having graduated from the Baghdad Military Academy and spent fifteen years in Iraqi Special Forces and commando units, including fighting in the Iran-Iraq War. Pragmatic and secular in outlook, he drank alcohol and had a girlfriend, in violation of Islam. He was outraged by Sunni insurgent atrocities but was no radical Shiite. Jim decided Taher was someone he could influence.

"I am not here to stop you from doing your mission," Jim told him. "I am here to help. There will come a time when I will have men and guns and vehicles and ammo and I will fight beside you," he said. "You and I will become like brothers."

Meanwhile, Jim was given a nine-man U.S. military team and two Iraqi interpreters, and set out to train them. Jim spoke some Arabic, but he realized that grooming the right interpreter would be critical to his mission to try to mold the Iraqi police. He wanted an interpreter who could speak for him forcefully and clearly in any forum, and also one who would fight by his side, sparing a U.S. soldier from shouldering that risk. On the first day his team left the Baghdad base, headed for a firing range, Jim found his man. The patrol stalled in traffic along a stretch of Route Irish, one of the most heavily attacked roads in Baghdad. Jim dismounted from his Humvee and walked ahead a hundred yards down the hot, congested street to try to clear the way. He turned around and was surprised to see his twenty-four-year-old interpreter, Mack, a few steps behind him. Mack had no military training, and in fact in 2000 had left Baghdad to work at a potato chip factory in

Jordan to avoid mandatory service in the Iraqi Army. He knew an American soldier walking alone down that street made a ready target. Still, something made him follow Jim.

Mack had worked for the U.S. military since 2003, and he was aware from their first encounter that Jim was not like other American officers.

"I was outside in the Humvee smoking a cigarette and I saw a guy in DCUs [a desert camouflage uniform]. His uniform was plain. You could tell he didn't care what he looked like. He didn't look like an officer. He looked like someone who did a lot of things himself," Mack recalled.

Jim offered to help Mack and another interpreter, Joseph, as they moved with the U.S. team onto the Baghdad base next to the police compound. "You don't hear that very much from an officer," Mack said. "He didn't have to do that. He wanted to build a good relationship with the Iraqis."

For Jim, Mack's decision to follow him, unprompted, down a Baghdad road cratered from roadside bombs was the first indication he was the interpreter Jim was looking for. In coming days, he violated the rules and outfitted Mack with gear just like his own—a U.S. military uniform, M4 carbine, 9 mm pistol, body armor, helmet, and night vision goggles. Then he trained Mack and the rest of his team on close-quarters battle and other urban combat skills.

"You're going outside the wire with me. One day you will save my life," Jim told Mack.

No other U.S. military unit had ever given Mack a gun, let alone trained him. "Everyone was saying, 'Do not trust the terps [interpreters]. Don't give them weapons,'" Mack recalled. With Jim, he was treated no differently than a U.S. soldier, and looked so much like one that Iraqis started calling him "mister" when they went on patrol.

After two weeks of training, Jim and Mack walked into Taher's office at Site One. "The team is ready," Jim told Taher.

The U.S. team began going out on raids with the police, who were initially very wary of the Americans. At first Jim had little means of verifying who their targets were or whether they were legitimate. But after weeks of riding in the lead vehicle on missions and feeding Taher intelligence on roadside bombs and other threats, Jim slowly began receiving more information from Taher on who was who. He also learned that Taher would sometimes only pretend to go after targets, revealing a tension between the Iraqi commander and his political task-masters. Just as Jim's team was gaining some rapport with the Iraqi police, the battalion received a mission that would put their relationship to the ultimate test.

In mid-October 2006, a spate of revenge killings erupted forty-five miles north of Baghdad in the Tigris River city of Balad between primarily Shiite urban dwellers and Sunni tribesmen living in surrounding rural areas. It soon began to spiral out of control when Sunni insurgents and Shiite militia got involved. According to U.S. and Iraqi reports at the time, three Sunni men were killed, including a local insurgent leader. The Sunni insurgent group Al Qaeda in Iraq retaliated on October 13 by kidnapping a group of about a dozen Shiite laborers who went to work in date groves in an outlying village. The laborers were found beheaded and with holes drilled in their skulls. That prompted more Shiite killings of Sunni civilians. The Shiites then sent an SOS to the Mahdi Army in Baghdad, and by October 14, busloads of Mahdi gunmen were taking control of Balad's streets, while broadcasts from mosques warned all Sunnis to leave the city. At checkpoints, the Shiite militiamen were beating and killing suspected Sunnis, and scores of bodies began piling up at the Balad morgue.

Prime Minister Maliki's office called the National Police commander, then Gen. Hussein Alawadi, who ordered in Taher's quick reaction force to Balad on October 17. Lewis called Jim in to give him the mission.

"Jim, we are getting some pretty bad reports of sectarian violence up there. Here is the bottom line: we need you to get up there, find out what is going on, and fix it."

"Roger that, sir," Jim said.

About two hours later, Jim, Taher, and Mack rolled out in the lead vehicle of a patrol with the U.S. team and two hundred Iraqi police commandos. When they arrived at an Iraqi army base on the outskirts of Balad, they were ordered to retake several checkpoints across the city—some abandoned, others held by Shiite militiamen or Sunni insurgents.

The next morning they pushed into Balad, a deserted no-man's-land with an eerie, foreboding feeling about it. Feral dogs roamed the streets, and plastic trash bags were flying around in the wind. Not a person was in sight. They arrived at the first abandoned highway checkpoint, its machine gun position blown up by an RPG, and dropped off some police with guns. At the second checkpoint, they dropped off another group of police, but came under mortar fire from a nearby orchard. Helicopters spotted insurgents moving into a nearby school, but the police were afraid and refused to go after them, so Jim and Mack cleared the building. Then Mack got a radio call that insurgents were threatening police Jim had stationed at the first checkpoint, telling them to leave or they would be killed. Jim and Mack headed back there, but again the Iraqi police with them balked, cowering under the cover of a nearby bridge. As they got closer, insurgents ripped into the checkpoint with RPGs and PKC machine guns. Two of them ran into bushes by the side of the road. Jim and Mack chased them down, and their gunner, Capt. James Kim, mowed fighters down with his M240 machine gun. After the shooting stopped, they treated one of the police who had been shot in the leg. Then they saw an Iraqi man lying beside his car, which had been hit by one of the RPGs. His guts were spilling out, and Jim and his medic, Sgt. Robert "Doc" Minor, treated him. No one thought the man would live, but the police took him to a local hospital. Later they learned he had survived.

"That was our first day in Balad," Mack recalled.

Jim's American advisory team had proven it would fight to back up the Iraqi police, a major step in gaining their trust. The next day, when the Spartan vehicles arrived at one of the Balad checkpoints, the Iraqi police Jim had positioned there broke out in grateful cheers. Jim's U.S. teammates were disappointed at the skittishness of the police, but Jim urged them to be patient. As the fighting wore on in Balad, the police began to stand their ground—shooting back, transporting their wounded, and calling for reinforcements under fire. Other Iraqi units would bribe Sunni insurgents not to attack them on the road from Balad to Baghdad. Col. Taher refused to do so, braving the attacks because he knew the Americans would back him up.

"You let me make the decision to be honest," Taher told Jim in their makeshift headquarters in an abandoned fire station in Balad. "I do not have to be corrupt, because you are with me."

As the Shiite police helped restore order in Balad, initially they lashed out at some of their Sunni insurgent detainees, beating them badly. Word on the street was that the unit was a Shiite militia led by a "dirty American." Jim did not intervene at first to stop them, knowing that if he did he would lose his influence with the police. Instead, he brought the detainees food and water, and treated their wounds. Col. Taher saw him and began to do the same. The beatings lessened, and the police moved to curb the abuse of local Sunnis, for example by freeing Sunni civilians illegally detained in Balad by another Iraqi unit. The Shiite police commandos were still in essence a militia, but Taher's leadership, molded by Jim, was moderating their sectarian excesses and improving their competence.

"They are going to have to fight the war. We have to walk in their shoes. They are fighting in the streets for their families," Jim recalled.

The Balad mission, initially to last for only a few days, was extended to weeks and then months. Col. Taher decided to rotate out his force of about two hundred commandos every two weeks. Unwilling to allow the police to make the trip alone, Jim and his team would escort them from Balad to Baghdad and back, facing multiple ambushes in what became by far their most dangerous mission.

On November 24, Jim's Humvee was in the lead of several commando pickup trucks as they sped down a deserted, gritty highway north of Baghdad. Sunni insurgents detonated dozens of bombs along the road every week, and Jim knew the explosions would cut like butter through the flimsy, thin-skinned Iraqi trucks.

Jim shot out far ahead of the convoy, scanning for IEDs. He was circling back at high speed when he saw an IED thirty feet away—but there was no time to stop. His up-armored Humvee ran headlong into a 130 mm artillery shell rigged to detonate. The explosion flipped the vehicle over twice, as it tumbled down a slope off the side of the road. The gunner, Kim, was tossed out like a rag doll. When the Humvee stopped, it burst into flames.

Jim's head and body smashed into the windshield. His night vision goggles had been ripped off, and he could barely see, pinned by radios and other gear. He reached down and yanked on the door handle, but the door was bent and wouldn't budge. Ammunition inside started exploding like popcorn. The Humvee's halon fire suppression system engaged, sucking the oxygen out of the cabin. As the fire advanced deeper and deeper into the vehicle, Jim labored to breathe. He couldn't hear and could barely see. The flames started to singe him.

*I'm done. I'm fixing to burn to death*, Jim thought.

Jim tried to stand but was held back by his gun sling. He released the weapon, started to get up again, and passed out.

When Jim came to, he was on a medevac chopper headed for a military hospital in Balad. Only later did Jim hear from Mack how he survived.

Two Iraqi police had climbed onto the burning Humvee and—together with an officer named Maj. Fadil, who was in the seat behind Jim—pushed and pulled him unconscious from the wreckage. Jim had proven he would fight and die for his Iraqi comrades, and now they had risked their lives to save him.

Jim was soon released from the hospital and eleven days later, on December 6, he returned to the scene of the firefight. He wanted revenge. He wanted to stop insurgents from littering bombs along that strip of road to Baghdad. That meant luring the Sunni insurgents in the area, who were tightly linked with the Jabouri tribe, into a pitched battle. Jim's experiences in Afghanistan and his growing understanding of Arab tribal culture had taught him that to draw the tribesmen out, he had to attack their honor and the reputations of their women.

Climbing onto the roof of his new Humvee, next to the burned-out engine block of the old one, Jim shouted into a loudspeaker.

"*Salaam aleikum.* You did not kill any of my men. No one was even wounded. So we are all back, and I want to make sure you know it is me." Jim paused. "I am going to come back in six days. . . . Bring all your whore wives out to the side of the road. Bring your whore daughters and your whore mothers so that when I come back through, I can fuck them properly. Judging from the way you fight, I'm sure you guys really don't know how to fuck."

Jim then shot his entire magazine of twenty bullets into the air, waited a few minutes, and drove off.

On the morning of December 11, Jim woke up on his cot in the tiny concrete room that he shared with his nine-man U.S. team on the Iraqi base in Balad. The room was dank and reeked from the overflowing shitters around the corner. Jim pulled on his boots, walked down a dirty hallway, and stepped outside the one-story building, taking a deep breath as the first glints of sunlight hit the building. He crossed the gravel courtyard to his Humvee, opened the door, and pulled out a bottle of water and a razor, which he set on the hood. Splashing a handful of water on his face, he started to shave off three days of growth. As the blade scraped across his cheek, he couldn't help but think: *This might be the last time.*

That day, Jim and his team, together with about 190 Iraqi police commandos in twenty-three vehicles, were leaving Balad to return to Baghdad. News of their departure had been broadcast the day before on national television, giving Sunni insurgents in the area plenty of time to plan their attack. Jim had no doubt they would strike.

"Hey, bro." Mack, his interpreter walked over and greeted Jim. Mack looked like a biker with his goatee and his wavy black hair tied back in a desert camouflage do-rag. He paused to light a cigarette. "You ready for this?" Mack asked, a discerning look in his large brown eyes.

"Are you kidding me? Fuck yeah," Jim said. He rinsed his face and wiped it dry with his dirty uniform top. "Let's go get Colonel Taher."

They strapped on their pistols, went out the gate, and headed down a grimy street lined with two demolished hulks of Iraqi army vehicles. A mangy stray dog ran out growling at them, and Mack scared it away. After about four hundred yards they turned into the abandoned firehouse where Taher had his headquarters. Outside, a few police huddled around a fire that was heating water in a big blackened kettle. Jim and Mack walked into Taher's office.

"*Salaam aleikum*, good morning. Sit down," Col. Taher said, his voice calm but his large face creased with worry.

A policeman brought in a tray bearing several small glasses, each containing a heaping spoonful of sugar, along with a pot of Ceylon black tea. He poured the glasses full of steaming tea flavored with cardamom, and served one to each of the three men.

Jim stirred his tea, looked over at Taher, and spoke what was on both their minds.

"Sir, we know there are not many tactical options," he said. "There is only one road, and the entire area is crawling with Sunni insurgents."

Taher nodded.

"I have briefed my officers," Taher said. "If we get into a firefight, every man will move to the sound of the guns and shoot. I will keep communications, and accountability," he said.

"I will support you with everything I have," Jim said. He sensed Taher's concern, but also an underlying confidence. After some final preparations, he and Mack turned to leave. "Make no mistake, sir," Jim said. "We will be fighting today."

The sun was straight overhead a few hours later as Jim and his team pulled out the gate at the head of the long patrol of vehicles. Jim initially turned north in a tactical feint, then headed back south toward Baghdad. His commanders had promised him that two F-16 fighter jets would be overhead, but just as the patrol left he got word the fighters had been diverted.

At around noon, the patrol entered a stretch of road that ran between two palm groves. The stands of tall, graceful trees started about twenty feet back from the road on either side. Just as the entire patrol—which was spread out over about 800 yards—entered the groves, the crackle of AK-47 semiautomatic rifles burst out from the west side of the road. Iraqi police pickups at the rear of the convoy had come under fire.

"Peel off!" Jim, in the lead vehicle, yelled to his big, bespectacled driver, Sgt. Minor. Without a word, Doc swung the Humvee around and gunned it to the rear of the convoy. Eight to ten insurgents with rifles were lighting up the police vehicles from the palm groves.

"Fuck them up, Kimee!" Jim shouted. Kim laid into the attackers with his M240 machine gun. Short and stocky, the chain-smoking artillery officer had the one thing Jim needed in a gunner—a lead trigger finger.

Just then, insurgents unleashed a heavy barrage of machine-gun fire from either side of a long stretch of the road. Bullets were pinging off Jim's Humvee.

This was no harassment attack. It was intense, well planned, and coordinated. They had men, weapons, and equipment positioned close by. They were committed and dug in. Jim was getting his toe-to-toe throw-down fight—a complex ambush with machine guns, mortars, rocket-propelled grenades, and roadside bombs that would spread over five miles and three kill zones.

A radio call alerted Jim that two of the Iraqi commandos had been badly wounded. Col. Taher took charge, ordering his men to move the casualties to Jim for protection and treatment. One commando was shot through the leg, and another, a good-natured man named Ha'nee, had a life-threatening bullet wound to the face.

"Call a medevac," Jim radioed to Capt. Paulo Shakarian, the team intelligence officer, in vehicle two.

"Roger—got it," Shakarian said.

As the gunfire steadily increased, a vehicle carrying the wounded Iraqis pulled up near Jim's Humvee. Wearing a baseball cap because his helmet had been burned in the November 24 IED attack, Jim jumped out under fire. He pulled Han'ee over next to his Humvee, which offered some cover from the incoming rounds. Mack got out and started laying down suppressing fire.

Blood was pouring from Han'ee's face from a bullet that had hit just under his cheekbone, shattering his jaw and teeth and going out the other side. Jim knew Han'ee would die if he didn't stanch the bleeding. He was clearing the teeth from Han'ee's airway when bullets started kicking up dirt on the ground just behind and in front of them. It was too close to ignore.

Stopping treatment temporarily, Jim grabbed his radio. "Kimee, hold here and keep firing. I'm headed into the palm grove to the east."

As he and Mack moved in between the trees, Jim saw fighters darting back and forth. Mack threw a hand grenade from behind a small dirt berm, blowing two insurgents off their feet. Jim spotted a machine-gun position and saw a man take off running. He fired at the insurgent and clipped him.

A medevac chopper was inbound, so they ran back to the wounded. Jim managed to control Han'ee's bleeding and insert into his arm an IV with blood volumizer. Then Jim called on his husky teammate Sgt. 1st Class Jean-Paul "J.P." LeBrocq and Iraqi interpreter Joseph to set up the landing zone just down the road.

Jim noticed the insurgents start to shift their fire as about fifteen or twenty fighters maneuvered toward the likely landing zone. *Shit, here we go. They are going to target the medevac.*

LeBrocq threw a red smoke grenade to mark the landing zone. The chopper landed, and Jim and Mack carried the wounded police through a cloud of dust to the flight medic.

"Hey, man, you have about thirty seconds before this place starts getting hit with mortars!" Jim shouted to him.

"Oh, shit," the medic said, scrambling back onto the chopper. It took off, and indeed, about half a minute later a mortar landed in the road fifty yards from where the chopper had been.

At that point, the insurgents who had converged around the landing zone turned their fire toward Jim and the police commandos, now surrounded. Firing machine guns and AK-47 rifles and throwing grenades, the attackers closed within 75 yards of the convoy and opened up with everything they had.

The volume of fire made the ground shake; Jim could feel it inside his body. In his seventeen-year military career, he'd never experienced anything like it. He looked back down the road toward the police vehicles and saw tree limbs falling, shredded by bullets.

With rounds snapping beside their heads, Jim knew he had to do something to push them back. He grabbed Mack and put several commandos in overwatch positions behind a low dirt berm that lined the road, and again went after the insurgents in the palm groves. The members of his team covered one another as they moved deeper into the trees. Mack was Jim's closest comrade. They trained constantly together. On raids, Jim was often the first one to kick in the door, and Mack followed right after him. Mack was Jim's spotter during sniper operations; they dismantled roadside bombs as a team. Now they were fighting for their lives. They pushed forty yards into the palm grove, firing as they went, then lay prone, still shooting.

"Hey, Jim, I'm out of ammo!" Mack yelled with bullets flying overhead.

"Bro, what the fuck? Do you want me to give you mine?" Jim called back. He grinned and tossed Mack a magazine.

With the help of the commandos, who were laying down large volumes of rifle and machine-gun fire, they pushed the insurgents back. Then they returned to the road to try to move the stalled convoy—then spread out over about four hundred yards—out of the kill zone.

"*Yalla!*" Jim called in Arabic to one group of police after another. "Come on! Go, go, go!"

Just then Jim looked back and saw LeBrocq on a knee in the middle of the road shooting like mad. As the convoy prepared to move, LeBrocq disappeared on the other side of the road. Jim thought he had been shot, but a minute later LeBrocq's hefty figure reemerged carrying Joseph on his shoulder. Joseph had broken his ankle and LeBrocq was not going to leave him. LeBrocq carried Joseph about fifty yards back to his vehicle.

Still taking gunfire and RPGs from both sides of the road, the convoy started creeping forward. Suddenly Jim got a radio call from Shakarian in the lead vehicle.

"Hey, Jim, this is Shak. We have a possible IED in the road. What do you want me to do?"

"Hold on, brother. I'll be right there," Jim said. Then he told Doc, "Punch it! We have an IED up there." Doc sped forward while Jim told Col. Taher to instruct the rest of the convoy to hold.

Jim got to the front, and immediately saw the roadside bomb. It was a "drop and pop"—a 155 mm artillery shell lying in the road and command-wired to explode.

In a split second he calculated his options. The convoy was in the middle of a small marketplace. There was no way to get off the road to go around the IED. They could stop and fight, or they could try to turn around—but either way they would be trapped under enemy fire and likely all killed.

They had to push forward. But if the IED blew up on one of the Iraqi trucks, it would also be catastrophic—it could easily destroy two or three of their soft-skinned vehicles and cause a dozen casualties. That would leave them pinned down, too.

*This has to go off on my up-armor.*

"We have an IED here, and I am going to hit it," Jim said over the radio. To Mack he said, "Tell Colonel Taher to push his guys to the far left." Then he grabbed the gunner, Kim, by the legs and shouted to him to get down. Finally, he switched off the jammer on his Humvee that blocked certain frequencies of electronic signals used to set off IEDs.

"Doc, go straight at it!"

"Roger, sir," said Doc, not missing a beat. He spit a mouthful of tobacco juice into a Gatorade bottle and stepped on the gas, driving straight ahead.

"Push left!" Jim said, "I want it to go off on my side of the vehicle."

They inched forward. Nothing. Closer. Nothing. Then about ten yards away, the IED detonated in a massive explosion. Shrapnel tore into the door and windshield as the blast rocked the Humvee. The road disappeared in a cloud of dirt, rocks, and asphalt. But no one was hurt.

"We're good!" Jim radioed back.

When the dust cleared, he started pushing forward again.

They had advanced only a few hundred yards, when Jim spotted another bomb.

"Fuck me, there's another IED!" he said.

It was the same drill.

"Kimee, drop!"

Doc drove the Humvee closer. Jim turned the jammer off again.

*Boom!* Another 155 mm round blew up only five yards away. It knocked the Humvee sideways and pierced the engine block and radiator, which spewed out steam. Jim and his crew were dazed and their ears rang, but they kept moving.

Jim looked ahead.

*Shit!*

A third IED intended to halt the convoy lay in the road. This one was in a box. Same drill. They got right up next to it. Nothing. Jim cracked opened his up-armor door and looked down into the box. It was empty.

"It's a hoax!" he radioed.

Watching the convoy move on undeterred, the insurgents unleashed a barrage of PKC machine-gun fire, mortar rounds, and RPGs from buildings on either side of the road. They had concentrated their heavy firepower in this last portion of the kill zone.

"Everybody open up to the front!" Jim ordered.

For two or three minutes, it was full-up machine guns on machine guns. They punched through.

"Sir, we have a female seriously injured in a civilian vehicle," Mack said. A silver BMW had pulled into the convoy behind Jim's Humvee and been hit by shrapnel from the second IED.

Jim, Mack, and Col. Taher had the police pull security while they jumped out and ran back under sporadic fire to the BMW.

Jim opened the door and saw a man and a woman inside each holding a small child, all of them terrified and screaming. Blood pooled on the car floor. Jim moved to examine the woman's wounds, but she recoiled, horrified that an infidel man was reaching toward her.

"I've got to touch her," Jim told Col. Taher.

"He is my brother," Col. Taher told the woman in Arabic. "Let him touch you. He will help you."

The woman stared at him in fear, then nodded silently.

In less than a minute, Jim pushed her socks down, raised her dress to the knee and put two tourniquets on her bleeding legs. He bandaged the bottoms of her feet, deeply gashed by shrapnel.

Jim glanced over at the one-year-old baby girl on the woman's lap. In her face, he saw a younger version of his own daughter, Scout.

*We might all die today. But she's not going to.*

Jim picked up the little girl and told Mack to tell the family he would keep her with him until they reached safety.

Cradling the toddler, who wore a red sweater and tiny jeans, Jim climbed back into his Humvee. Kim was firing his machine gun off and on as they moved down the road. The little girl didn't cry or scream. Instead, in a moment that struck Jim as surreal, she started doing what kids her age do, pulling on Jim's nose and chewing on his radio antenna.

The insurgent fire petered out, the back of the ambush broken. After three miles, the patrol arrived at the U.S. military logistics base at Taji. Jim had the police set up a perimeter just outside the gate. An ambulance pulled out of the base, and soldiers picked up the wounded Iraqi woman, whose husband had driven along in their car with the patrol.

"Your wife is going to be okay," Jim told her husband as the ambulance drove off. Then Jim placed the little girl into her father's arms.

FROM TAJI, THE CONVOY made its way to the Iraqi police compound in downtown Baghdad. As they rolled in, all the commandos stationed there were yelling and singing. They had heard about the battle, and they mobbed Jim and their Iraqi comrades as they got out of the vehicles.

"Ai yi yi yi yi!" they cried. Jim joined in. It was their version of a war cry.

The commandos slit the throat of a goat, dipped their hands in its blood, and started plastering crimson handprints on Jim and his team and their vehicles. Jim felt at once strange and honored by the celebration. He knew from reading the Koran that the ritual came from the Islamic version of the story of Abraham's sacrifice—whenever someone risks his life for you, you slaughter a goat to celebrate. Everyone was hugging and kissing in the revelry, while the commandos danced around chanting and hoisting their rifles in the air. Jim was elated and proud of all his men, both Americans and Iraqis. They had fought well that day. They had won.

About that day, Jim said later: "I remember Mack lying in the palm grove by this tree with bullets flying over, a few feet away from me. The look on his face was one of excitement and calm. I remember thinking, this second, this moment in time is what I've trained for my whole life.

It was exhilaration. It was an opportunity for my men to show bravery and courage. In order to fight you have to expose your throat. You have to be willing to die. . . .

"There's not much to say after you've been through something like that. Neither one of us had any right to be alive, and there we were. . . .

"That feeling never left me."


THE BALAD MISSION AND December 11 battle solidified trust between the Spartans and the QRF battalion and gave them new stature in the eyes of senior commanders. They had helped quell the killing in Balad and turned the city back over to local forces.

The day after the firefight, Col. Lewis, Jim's commander, called him into his office on Forward Operating Base Prosperity, next door to Site One.

"Do you and your team need a couple of down days?" Lewis asked him.

"Fuck no, sir," Jim replied. "We need another Balad."

For Lewis, the battle had much broader meaning.

"If the whole idea was to create a functioning security force, so the United States military could leave Iraq, here was the first brilliant effort in proving that capability," Col. Lewis recalled. It bolstered his argument to the four-star commander Gen. Casey that the National Police should be re-formed, not dismantled—a position that ultimately prevailed.

The big firefight south of Balad also exacted a heavy cost on the Sunni insurgents and led to a sharp drop in the number of roadside bombs on that stretch of the road to Baghdad. Commanders rewarded the Iraqi police and their American mentors with public praise and several awards for valor. Jim was given the Iraqi National Police Medal of Honor, and later would receive the third-highest U.S. military combat decoration, the Silver Star, for "gallantry in action." Minor, Shakarian, and LeBrocq all received Army Commendation Medals with V for valor.

Commanders began to entrust the police with more sensitive missions. In December, National Police commander Gen. Hussein ordered the battalion to stay in Baghdad to carry out raids on a growing target list that now included some Shiite militia leaders. Shiite militias backed by Iran were growing more deadly, using a trademark weapon known as an explosively formed projectile (EFP) that could punch through the armor on an American tank. The National Police as a whole were under high scrutiny for sectarian abuses. Under pressure from the U.S. military, several of the police commanders were being fired. But Jim's relationship with Col. Taher had empowered the Iraqi commander to kill or capture fellow Shias and to refuse to execute targets for which intelligence was questionable. Iraqis joked that the QRF battalion had become the strongest Shiite militia of all—able to keep others in check. With Jim as an advisor it could call for backup from U.S. military troops and helicopters. With Col. Taher's Shiite connections, it could also roll through Sadr City unchallenged.

But the round-the-clock Baghdad raids and extreme hazards for his men only increased the pressure Jim felt to perform. He trained the team harder and worked out more. He spent three or four hours a day between missions poring over intelligence. To protect his men, he tracked and mapped by hand all IEDs, EFPs, sniper attacks, and complex ambushes reported by U.S. Army combat units on a laminated map of Baghdad with his own color-coding system. If the call came at two in the morning to conduct a raid in a neighborhood across Baghdad, Jim knew the best route to take. Every day he studied photographs and reports on all the IEDs found or exploded in Baghdad. Violating military rules, he refused to delay missions for hours to wait for military teams to clear the IEDs he spotted on the route. Instead, he disarmed them himself.

Again and again he approached the IEDs, knowing the slightest mistake could blow him to pieces. Booby traps, multiple explosives and detonation devices, hidden land mines—the risks and calculations were endless. Whenever he yanked an IED out of the ground, it was as though he was counting off another one of his cat lives.

The world beyond the war zone faded away as he felt himself losing touch with realities of life outside Iraq. The universe of the people he cared about shrank. He could no longer see the faces of his wife and children or remember what his house in Fayetteville, North Carolina, looked like. Time grew distorted, marked not by calendars and holidays but by missions and the little downtime he had in between. Maybe it was his mind's way of distancing himself from who he was and what he was doing. He started drinking much more—sometimes as much as a fifth of hard liquor a day. The alcohol made it even harder for him to sleep. He rested fitfully for two or three hours at a time; night and day blurred into one. The onslaught of killing and death collapsed his moral walls. Once in a while between missions, he began frequenting Baghdad whorehouses or sleeping with female soldiers. The only loyalty he felt was to his men. He got into fistfights, rode on the hood of his Humvee, and a few times went on night raids after drinking. Underlying it all was the inescapable fact that he could die any day, that each sip of whiskey could be his last.

"I sometimes feel like I do not want to wait for death," he wrote to his father. "I want to go out and find it. I want to be able to look myself in the mirror. I want no voices in my head telling me that I didn't go after the enemy every day. I have enough demons in my head without that. It is the battle I didn't fight that still echoes in my mind."

Jim had started hearing, faintly at first, what seemed to be voices inside his head when he was fighting in Afghanistan a few years earlier. They were like a "sixth sense," telling him what to do. Now those voices became louder, more incessant, and more vivid. They told him to work harder, to train more, to press his men every day. Over time, they began guiding him more actively.

*Slow down! . . . Slow down!* The voice seemed to come out of nowhere one day as his patrol approached an Iraqi army checkpoint north of Baghdad. Jim ordered his driver to slow and radioed the vehicles behind him to hold. Seconds later, a white sedan laden with explosives blew up just ahead.

Later on a raid, he heard it again: *Go! Get them! Push!* It sounded like his buddy Sgt. 1st Class Mark Read from ODA 316 in Afghanistan. Other old teammates spoke to him, too, as though they were watching over him.

Sometimes the voices were strange and unfamiliar. As he engaged them in mental conversation, he began to give them the names of ancient warriors, both mythical and real.

"Many, many centuries of warriors reside in my body," he wrote to his father. "This is my life. It is where I live daily. I may be going crazy. Hecate is the goddess of war. Although she has roamed the periphery of my life for many years, it is only recently that I have heard her voice and seen her body. I still have not seen her face."

The voices came to him when he was exhausted and on the way back from missions, telling him suddenly to stop and look for roadside bombs. He learned to always listen to them. The more receptive he was, the more he accepted them, the more they dominated his actions outside the wire of Site One, in the streets of Baghdad. It was to him almost like prayer. He believed that God spoke to people who opened their hearts, and that is how it was for him with the internal voices. They became his true north.

Baghdad at the time was a sea of carnage, with horrific suicide bombings killing scores of Iraqis every week. One day in November 2006, when Jim was in Baghdad between resupply runs to Balad, a huge explosion went off just outside Jim's base in central Baghdad. Body parts flew over the wall of Site One. Jim grabbed his medical bag and ran to the scene, armed only with his 9 mm pistol. As soon as he went out the front gate of Site One, he saw the mangled bodies of the dead and injured. A suicide bomber wearing an explosive belt had blown himself up next to scores of men waiting outside a recruitment center at the National Police headquarters. As the recruits scattered down a side street, another suicide bomber chased them and detonated himself. Jim decided to make his way down the side street, where there were fewer rescuers, looking for people who were still alive. He spotted a heavy man of about fifty whose leg had been blown off near the hip. The man was screaming. Jim wadded up his jacket, stuffed it in the leg cavity, and tied it up with a bandage. He moved farther down the street and saw another man walking in a daze, his face severely burned and his arm severed above the elbow. All Jim could do was try to stop the bleeding; he put on a tourniquet and kept going. Then, as he was bent over treating another man, he looked up. A crowd of Iraqi civilians had gathered and was starting to chant, irate. National Police on the scene shot a few rounds in the air. A shouting match started between the two groups, with Jim in the middle, the only American. He drew his pistol and looked around. None of his commandos was there. People were grabbing at him to get his help. As the crowd turned menacing, he ran back to the U.S. base, covered in blood.

Back on the base, a sergeant major looked at him disapprovingly.

"Sir, you are out of uniform. And you need to clean up," he said.

Jim walked numbly to a shower. He turned on the faucet, but no water came out. He went back to his room, opened a package of baby wipes, and tried to clean his hands. It seemed the blood would never come off. He slumped down against the wall.

*This is pure evil*, he thought.


IN MAY 2007, JIM was awarded his Silver Star. At his request, the ceremony was held at the ramshackle police compound at Site One. The U.S. team and the commandos stood in formation, while American and Iraqi generals and other dignitaries looked on. In the audience was then Brig. Gen. John Campbell, Jim's old battalion commander who had once chewed him out for challenging Army tactics as a young lieutenant. Campbell was spearheading efforts to integrate into the Shiite-led government thousands of Sunnis in Baghdad who had stood up to the insurgents and volunteered to defend their neighborhoods. Also there to shake Jim's hand was Lt. Gen. Ray Odierno, the three-star general in charge of day-to-day military operations in Iraq. Odierno had offered to pin on Jim's medal. But Jim asked his boss and friend Col. Lewis to do it instead. After the honors, Jim walked to a microphone and spoke of the December 11 battle.

"On that day, there were no Americans. There were no Iraqis, no whites, and no blacks. There were no Sunnis, Shias, or Christians. There was just a group of warriors working and fighting together. I would gladly and without hesitation lay my life down for all of them," he said.

For Jim personally, the award for valor held many meanings beyond showcasing the close bond between his American team and the Iraqi police. It relieved his frustration and guilt at not being able to return to Afghanistan to fight with ODA 316. It symbolized an achievement worthy, at last, of winning his father's approval.

"I wanted to make you proud of me," Jim wrote to his father on the day his Silver Star was pinned on. "I believe I have attained that. It was not easy. Your shadow was big."

It is hard to say whether Jim lost himself or found himself in his quest for greatness in combat. But there is no question that by the end of his time in Iraq he was possessed by war, captive to its horrors and addicted to the arena and roaring crowd.

In July 2007, after thirteen months in Iraq, Jim prepared to return to the United States. Fearless in the streets of Baghdad, he was terrified of going home. He had sacrificed everything at the altar of war. War was, by then, all he really knew. He could not imagine a world where the people he had loved most had become strangers, and where—unlike in Iraq—his enemies were not trying to kill him, making them much harder to find and impossible to destroy.

"I can no longer remember the person I once was," he wrote to his father.

> My perception of the world has taken an 180-degree turn. Everything I have always been taught, everything I believed in, has been turned upside down. . . . I have seen such evil that I don't think I can ever be free of the demons that will haunt me for the rest of my life. The smell of burned bodies. The smell of blood and death. So many innocent people killed.
>
> Killing another person, even in war, changes a person. Being in combat changes a person. Being isolated from the ones you love changes a person. When your life is in the hands of others it changes your life forever. . . .
>
> This is the most gruesome, violent environment you can imagine, even in your dreams. Yet, God forgive me, I love it.

Jim boarded a plane at the Baghdad International Airport that was packed with jubilant soldiers. A crew member noticed he was a major and moved him to first class. The plane hurtled down the uneven runway. The second its wheels lifted off, the cabin erupted in cheers.

Jim put his head in his hands and cried.

# CHAPTER 6

JIM LEFT HIS NOVEMBER 2009 video conference with Adm. Olson and burst out the main door of the towering tan U.S. Army Special Operations Command headquarters building at Fort Bragg into a chilly but clear afternoon.

"Ai yi yi yi yi!" he said, pulling off his beret and giving out his Iraqi war cry. He'd been waiting more than two and a half years since Iraq for another combat mission. Now he'd gotten new orders—for Afghanistan.

He strode past the manicured parade ground and through a stand of pine trees to his truck, jumped in, and gunned it down the drive. Pulling out the base main gate, he drove by the pawnshops, tattoo parlors, and honky-tonks on Yadkin Drive. But all he could see was a mud-brick village backed by snow-capped mountains. All he could think about was Olson's last instruction: *Assemble your team.*

Jim knew immediately four of the men he wanted for his tribal engagement team. Two he'd fought alongside—Capt. Dan McKone and Master Sgt. Tony Siriwardene. Two others he'd trained—Capt. Matt Golsteyn and Sgt. 1st Class Frank Benson. Jim knew each of them not only as soldiers but also as close friends. They were loyal to a fault. They were all warriors, and between them they had six awards for valor.

Dan McKone was a talented medic as well as the best gunner Jim had ever had. He would be the team medic and handle weapons. A former Peace Corps volunteer who had lived on four continents, Dan was worldly, mature, and coolheaded. He never bullshitted Jim about anything. With a master's degree in environmental systems, Dan was also in training to become a Civil Affairs officer and would oversee development projects. Dan was a proven fighter, a former Special Forces non-commissioned officer (NCO) with three awards for valor and five combat deployments under his belt, four of them in Afghanistan. He was culturally savvy, spoke Pashto, and had worked training Afghan soldiers. And from his time on Jim's team ODA 316, Dan knew Konar Province and the village of Mangwel—the people there remembered him.

Tony Siriwardene would be his team sergeant, responsible for the daily operation of the team. Tall and athletic, Tony was a native of Sri Lanka who had moved to Arlington, Virginia, at the age of eight. He graduated from Washington-Lee High School and enlisted in the Army as an engineer. In 2003, Tony was the second man assigned to ODA 316, and deployed with Jim twice to Afghanistan. There he proved himself the epitome of a soldier—always doing what was needed before he was asked. Tony was at Jim's side on every single mission. Tony was the only team member besides himself whom Jim allowed to disarm bombs. Once he ventured with Jim into the Korengal valley even though he was so sick he had an IV in his arm when he drove out the gate. "I will always have something for contact," he told Jim—and later that day earned his first award for valor in an enemy ambush.

Matt Golsteyn would be his second-in-command, focused on the team's strategy and operations. A 2002 West Point graduate and star baseball player from Winter Park, Florida, Matt was one of those rare officers who shone with both intellect and moral fortitude. He'd served with the 82nd Airborne Division in Iraq and then tried out for Special Forces, landing in 2008 in a group trained by Jim in a guerrilla warfare scenario in the backwoods of North Carolina. Matt turned out to be the best captain Jim had encountered among dozens of aspiring Special Forces team leaders, and the gutsiest. Better yet, he'd just returned from months of fighting with an Afghan commando battalion in some of the country's worst trouble spots. Matt knew how to navigate the clashing values of Western democracy and ancient Afghan precepts of justice—and make tough calls to save lives.

Frank Benson would serve as a medic and the team intelligence expert. Jim had known Frank for more than a decade, since Frank was seventeen years old and a newly minted infantry private. "Which of you sorry motherfuckers thinks you can whup my platoon sergeant's ass?"

Jim, then an infantry lieutenant, had demanded of a cluster of about twenty privates who'd just arrived at his Army base at Schofield Barracks in Hawaii. Frank was the only one who raised his hand. "Get your shit! Come with us," Jim said, grinning and pulling Frank out of the formation. Now an experienced Special Forces NCO, Frank had advanced intelligence training that would be vital to the tribal mission.

Three days later, all his men were on board.

As Jim pulled his dream team together on orders from the top, his phone began ringing with calls from subordinate levels of his Special Operations Forces chain of command. After all, Adm. Olson, who headed that chain of command, had paid Jim the ultimate tribute. "You are the first true 'Lawrence of Afghanistan' (as the ADM likes to put it) that we have found," Olson's top enlisted advisor, Command Sgt. Maj. Smith, wrote to Jim in an email hours after the November 2 video teleconference. Olson was working on a plan based on Jim's ideas to place tribal engagement teams in Afghanistan, and he intended for Jim to spearhead the mission and set an example for others. Jim was the bright, shiny object. The generals and colonels began swooping toward him like crows.

Brig. Gen. Michael Repass, the commander of the United States Army Special Forces Command (USASFC), who was in charge of all Green Berets, summoned Jim on November 6. Jim walked into Repass's large office suite wearing jeans, a T-shirt, and flip-flops. He had several days' stubble growing on his face—the beginnings of a beard for Afghanistan—and so by regulation could not wear his uniform.

Repass came around from behind his desk and invited Jim to sit down next to him at a table. The general took out a pen and pad.

"Major Gant, I've been tasked with supporting you any way that I can. What do you need?" Repass asked.

Jim pulled out the list of names for his team and a training plan. He began briefing Repass, who listened attentively and took notes. Jim had never experienced such deference from so many general officers. He was slightly amused. *They don't know me*, he thought. *If they are calling me, they must be desperate.*

At the end of their meeting, Repass turned to him.

"Jim, I want a T-shirt when you have them made up," the general said.

Jim smiled at the backhanded compliment. In the world of Special

Forces, team T-shirts, hats, and logos meant a lot. Repass wanted to be part of his mission.

"Roger that, sir," Jim said, and walked out.

Soon afterward, Jim was engaged by a general one step higher in his chain of command. The three-star general in charge of United States Army Special Operations Command (USASOC), Lt. Gen. John Mulholland, offered his full support for Jim and his team. Mulholland directed Col. David S. Maxwell, head of the command Strategic Initiatives Group, to work on getting Jim what he needed. Maxwell wrote to Jim with dramatic flourish about his mission. "To put things in historical terms . . . you are going to be the Lawrence and we have to find your Allenby in Afghanistan to provide you that top cover," Maxwell told Jim in an email on November 10. He was referring to the British general Edmund Allenby, who in 1917 backed the bright, eager intelligence officer, Maj. T. E. Lawrence, in working with the Arab tribes to oppose Turkish forces of the Ottoman Empire. Jim was thrilled. He had long been inspired by Lawrence's life story and his classic work, *Seven Pillars of Wisdom*, which he first read during the 1990 Gulf War and again while deployed in Iraq. He had grown to understand Lawrence's connection to the Arabs both emotionally and intellectually.

"You are getting your TET [tribal engagement team]," Maxwell wrote to Jim the next day. "You basically will be able to write your own ticket. ADM Olson wants you and your team to do some focused training to prepare you, but the longer-term goal is for you and your team to become the future Lawrences of Afghanistan. Bottom line is you are going to be on the cutting edge."

But as quickly as Jim had gotten his dream mission, forces within his Army chain of command tried to take it away. Jim was fully aware that he, a lowly major, had unleashed a rash of professional jealousy by winning such high-level praise. What he didn't realize was that two military hierarchies were about to battle over his fate—one in the United States and the other in Afghanistan.

One morning just two weeks after he spoke with Adm. Olson, Jim opened his email at his house in Fayetteville to find a terse and defensive message from Col. Mark Schwartz, operations director for the Combined Forces Special Operations Component Command–Afghanistan

(CFSOCC-A) in Kabul. The command was in charge of all Special Operations Forces in Afghanistan. Schwartz brusquely rebuffed the one-tribe-at-a-time idea and implied that his command was already conducting tribal engagement. In fact, a fledgling 2009 experiment in recruiting local police in the eastern province of Wardak had largely failed because it did not follow many of Jim's principles. Plans for a broader community defense initiative were under way but had only been attempted on a small scale, and had limited high-level U.S. and Afghan support. The initiative aimed to use volunteers like a neighborhood watch and did not involve arming or paying tribesmen—key elements of Jim's strategy.

"There is no intent to put you on a 'special team' conducting tribal engagement," Schwartz wrote. Instead, he said, Jim would be assigned as a "staff officer to the J35 Future Operations Directorate," putting Jim in the last place on earth he ever wished to be: behind a desk. Like many military officers, Jim shunned desk jobs as dead-end assignments.

"I understand that you were potentially putting together a select group of NCOs to accompany you to the headquarters. Now that you have a better understanding of the scope of your duties working within the J35, you realize you do not require a team of individuals to accompany you," Schwartz wrote.

Jim almost choked on his coffee. Who was this guy Schwartz? How could he override an order from Adm. Olson?

One reason Special Operations commanders in Afghanistan were so determined to confine Jim to a cubicle in Kabul was that the very public, four-star embrace of his strategy contained a damning subtext: an obvious Special Forces mission had been neglected at a time when U.S. commanders in Afghanistan were responsible for fighting the war. Jim's paper raised a number of other troubling questions. Why did Jim need a handpicked team? Were not all Special Forces trained to be "special" in their ability to work with indigenous fighters? But after eight years on the ground in Afghanistan, why did so few of them speak Pashto? The answer, Jim believed and wasn't afraid to voice, was that in the course of the Afghanistan and Iraq wars Special Forces teams had largely been tasked to conduct raids and other direct-action missions. He readily admitted that he, too, had been blinded by the "kill and capture" mentality during his early deployments as a captain. The Special Forces had lost sight of their core skill: integrating with local people and teaching them how to fight for themselves.

The Schwartz email also smacked of a turf war. Olson had plucked Jim out for the mission and—together with Petraeus and McChrystal—issued a "by name" request to change his orders from Iraq to Afghanistan. But as Jim soon learned, Olson's command, and those of Mulholland and Repass under it, were by law headquarters dedicated to providing forces and thus had limited or no authority over overseas combat operations. They were designed to train and equip forces to supply them to the overseas combatant commanders running the wars. Those commanders—mainly conventional military officers who were often risk-averse and wary of Special Operators—had the final say in employing the forces. So while Olson could create a team and try to influence how it was used, he often faced resistance from downrange.

Caught in the bureaucratic crossfire, Jim could only watch as the colonels and generals who had been behind him ran for cover.

"Your employment will be decided by the in-theater chain of command," Maxwell stated dryly. Privately, he advised Jim to be careful, warning that his conspicuousness could cause him to be "crushed under the weight of the senior leaders who are chasing the shiny thing."

Jim decided to take the matter directly to Olson. The admiral was no stranger to brawling, but he knew when to throw his punches and when to dance to the side. Olson realized that too much prominence could prove deadly for a military officer's career. Jim needed to assume a low profile until the opposition he faced from the military command in Afghanistan blew over. Olson assigned Jim to a new program in Washington, DC, designed to groom military personnel for long-term service in Afghanistan and Pakistan, known as the Afghanistan-Pakistan (AFPAK) Hands program. The downside was, instead of assembling his handpicked team and heading to Afghanistan, Jim would be cooling his heels studying Pashto for another six months.

At 8:00 p.m. on December 1, 2009, the same day Jim learned he was going to the AFPAK Hands program, President Obama stepped onto the stage at the packed Eisenhower Hall Theatre at West Point to address the nation on Afghanistan. At Fort Bragg, Jim watched on television as the president strode past a royal blue curtain and took the podium.

"Afghanistan is not lost, but for several years it has moved backward," he said, explaining how the Taliban reemerged while requests by U.S. commanders for more troops went unmet. "The security situation is more serious than . . . anticipated. In short: the status quo is not sustainable," he said.

"As your commander in chief, I owe you a mission that is clearly defined, and worthy of your service," Obama said, addressing the crowd of uniformed cadets in the audience. After a policy review, he said, he had determined that "it is in our vital national interest to send an additional thirty thousand U.S. troops to Afghanistan."

Jim nodded, feeling encouraged by the president's commitment to the war. But he didn't expect what came next. It was the precondition for the troop surge that had concerned the top U.S. military advisors handed the Afghanistan problem in the Oval Office just two days before—one they might have preferred to keep quiet.

"After eighteen months, our troops will begin to come home," Obama pledged.

*Fuck me!* Jim thought. *Our commander in chief just told the Taliban how long they have to wait us out!*

In his view, Obama's public announcement of a deadline to begin the troop withdrawal sent exactly the wrong message—to U.S. troops, to the American and Afghan people, and, most important, to the enemy.

Obama had just set the clock on the war ticking, but Jim would have to wait another half year before he got on the ground.

# CHAPTER 7

WHEN I MET JIM, one of the first things I noticed was his startle reflex.

Soon after I stepped into his Crystal City, Virginia, apartment in January 2010, the sharp clack of a door down the hall sent him momentarily into a crouch, his arms poised at his sides. I ignored the combat-zone impulse, which didn't surprise me.

Anyone who's had repetitive exposure to incoming fire is hypervigilant. When I returned from covering the invasion of Iraq in 2003, I imagined that I saw charred bodies lying on the side of the road in my manicured Bethesda neighborhood. When the annual Fourth of July fireworks exploded at the country club across the street, I curled up in a ball and pressed my hands against my ears.

"Would you like a drink?" Jim asked as he reached for a glass, filled it with ice, and poured in Johnnie Walker Black and a splash of Coke.

"Why not?" I said.

I glanced across the living room and noticed a few empty scotch bottles standing on a corner table. He'd only lived in the apartment a few days. Self-medication. Some used alcohol, others pills, but they all did it. Many in the military—especially in Special Forces—saw any admission of post-traumatic stress disorder (PTSD) or psychological problems as career ending. No wonder the Army's suicide rate was soaring.

Funny thing is, I had brought Jim one more bottle—fuel for the flames. It was Early Times, the rotgut Kentucky whiskey his dad mailed to him in Afghanistan because it was sold in plastic containers. As a reporter, I picked up on that kind of detail—personal preferences, habits, likes and dislikes. I did my homework before interviews. It was something I had learned years earlier, when I met former *New York Times* publisher Arthur Ochs Sulzberger in his Paris apartment. "Remember every bit of information. It will be useful one day," he advised me, after showing me a wall full of signed black-and-white photographs of famous people he'd interviewed. So when I called Jim's dad in Denton, Texas, to ask him about his son for the commentary I was writing about Jim's one-tribe-at-a-time strategy, he mentioned Early Times. What he didn't tell me, though, was that Early Times made Jim crazy.

Jim smiled when I handed him the bottle, his glance at me making me feel ever so slightly uncomfortable. I took a sip of my drink. He tried to nurse his, but it was soon gone. He poured another.

A few weeks before we met I'd finished my article, an opinion piece for the Outlook section of the *Washington Post*. Jim's sophisticated insights on the Pashtun tribes, coupled with his rough demeanor, had intrigued me. "I am not a very nice guy," Jim told me when I interviewed him by phone. "I lead men in combat. I am not a Harvard guy. You don't want me on your think tank." I needed photographs to accompany the piece, so I met Jim at his apartment that chilly evening to go over images of his Special Forces team in Konar Province in 2003. He took out a stack of photos and offered to tell me about them.

"Afghanistan was still the Wild Wild West. There weren't a lot of U.S. forces in the area," he began, showing me a picture of the Konar River valley. "This region was of strategic importance because of where it sits along the border with Pakistan."

I knew Jim didn't trust reporters. In 2007 his commander in Baghdad had handed him a three-by-five card with my name on it and told him to call me in Washington for an interview about his Silver Star battle. Jim's command apparently knew I had a solid reputation—I'd gained the confidence of a lot of wary military officers—and so picked me to become the only reporter he would speak with. Ten minutes into the conversation, Jim was telling me his beliefs about God and warfare. We talked for well over an hour. I was impressed by Jim's deep understanding of the culture of the Shiite Iraqi police he fought with. He was appreciative that, a day later, I took the time to interview his close Iraqi comrade and interpreter, Mack.

As with Iraq, I knew the intensity of Jim's approach to Afghanistan had caught the eye of senior commanders. A couple of days before I met Jim, Petraeus weighed in on Jim's one-tribe-at-a-time strategy in an email. "Ann—on the record: Major Jim Gant's paper is very impressive—so impressive, in fact, that I shared it widely."

*Petraeus wants this out*, I thought. *He's pushing this.* I later learned that the politically savvy Petraeus considered Jim the "perfect counterinsurgent" to infiltrate the halls of power in Washington and Kabul and launch a local security program on the ground.

But as I sat listening to Jim talk about Konar, one thing was certain: Jim Gant was no David Petraeus.

"I was born in the Konar," Jim went on, looking at another scene of the harsh yet beautiful landscape. "I became a man there."

Afghanistan, he had discovered as a thirty-five-year-old Army captain, was a nation of ancient, warring tribes. The largest tribe was the Pashtuns, who make up about 40 percent of Afghanistan's population and inhabit the east and south. The Pashtuns live by the tribal code of Pashtunwali, which requires a Pashtun man to protect his honor by fighting for what is his—his village, his valley, his tribe. Jim learned about Pashtunwali at the knee of a sixty-eight-year-old tribal chief, Malik Noor Afzhal, whom he respectfully nicknamed "Sitting Bull."

Jim showed me a photograph of Noor Afzhal, imposing in a silver-gray turban, and recounted the day he had met the venerable tribal leader in his village of Mangwel in Konar. He had learned of a land dispute involving Noor Afzhal's Mohmand tribe and made a gut decision to ally with the powerful elder.

"I looked him right in the eye—after knowing him for about four hours—and told him that we would do whatever he needed us to do," Jim said. "I took a lot of heat for that."

Noor Afzhal and his men bonded with Jim and his team because they were themselves a small tribe united by a similar set of beliefs, he said.

"All of our values and ethos are those of a warrior. The tribesmen knew that we would fight next to them, that we would die. They knew that we would kill for them," he said.

"Tell me more about Sitting Bull," I asked.

"Ah, Sitting Bull," Jim said. "He was like my own father. We would sit together and speak Pashtunglish. . . ." His voice trailed off. "I am hoping that someday, when they write the story of the war, Sitting Bull will be included, because he was a great man."

Jim took another drink of his scotch and turned to look at me.

"Now," he said. "You tell me about Afghanistan."

I thought for a minute.

"I think I'll show you instead," I replied.

I did a quick search of the *Post* website for articles and video I'd taken the previous summer while embedded with a company of U.S. Marines in Helmand Province, the most violent place in Afghanistan. I hit PLAY.

On the screen, a sweaty Marine platoon sergeant and his men were crouching in grass in a tree line, when suddenly we came under Taliban fire at close range.

"Fire on it!" the sergeant screamed to his men. "Where's my fucking SAW gunner?"

Gunfire exploded in footage I'd taken of several close-up firefights with the Marines. It always made my pulse quicken to watch it.

Jim, kneeling beside me, stared at the screen. He was silent, and then he said: "Ann, you have seen more combat than a lot of men I know. You have credibility with me." He paused again before asking, "Would you like to go to dinner?"

"Sure," I said with a smile.

We walked to a nearby Spanish tapas restaurant, talking nonstop. The room was so dim that we could barely read the menu, and the food was terrible, but neither of us cared.

As we picked at the dishes, Jim began telling me in detail about his inner battles—the violent nightmares, the visions, the demons that mauled him whenever he tried to sleep. I listened attentively. I thought of all the times, in dusty outposts and lonely barracks, soldiers traumatized by combat had broken down while talking with me. Sometimes I put down my notebook and simply hugged them or held their hands. I was no longer a reporter; I was just another human being. Still, I had always kept my own trauma to myself. Not that night.

Something about Jim made me want to tell him—the first person ever—about the aching depression I had suffered after returning from the war zone.

"One morning on a walk, I had an epiphany," I told him. "I realized that I had just been born. I was an infant. I had to give myself time to learn and explore. I started crawling out of the depression," I said. "Before long, I was back in combat."

Jim reached across the table and took my hand.

"Ann, I am certain that in past lives, you too have fought many battles as a warrior—one far fiercer than I."

For the first time all evening, he seemed calm.

# CHAPTER 8

SOON AFTER HE ARRIVED in Washington on a bitterly cold day in January 2010, Jim began a five-month Pashto course. The curriculum was part of the AFPAK Hands program. Military officers enrolled in the program could study several different languages, including the Pakistani dialect Urdu and the standard Afghan tongue Dari, which is derived from Persian and used for official government business and media reports.

But Jim chose to concentrate on Pashto, the language of the Pashtun tribes, spoken by about fourteen million ethnic Pashtuns in Afghanistan and another twenty-four million across the border in Pakistan. The Pashtuns are the largest tribal society in the world. They derive their power by dominating a large swath of territory in eastern Afghanistan and western Pakistan that they call Pashtunistan. Jim understood that no Afghan ruler or foreign invader had ever held power in Afghanistan without backing from the Pashtun tribes, and that is one reason the country has not had strong central governments. As hundreds of years of history have proven, as the Pashtuns go, so goes Afghanistan.

Jim was advised by military superiors to stay under the radar in Washington, but the buzz over his tribal engagement strategy made that hard. After arriving in DC, Jim got a flurry of invitations to talk about the tribes in Afghanistan and explain how he believed they could be leveraged to help secure the country. Still, Jim initially held off. A self-taught man, he wanted first to read everything he could get his hands on to better understand the Pashtuns. Jim reached out to some of the leading tribal experts in the world, including David Ronfeldt and William "Mac"

McCallister. He devoured books on Afghanistan and tribalism—from *The Pathans* by Sir Olaf Caroe, a British administrator who governed a Pashtun territory in the 1940s, to the *Muqaddimah* by the fourteenth-century Tunisian Muslim historian Ibn Khaldun. Often he found himself reading four or five books at a time. What he learned fascinated him and reinforced what he had intuitively sensed during his 2003 and 2004 deployments in Afghanistan.

Historically, Jim found, the Pashtun identity has been so strong that in large parts of Afghanistan "Pashtun" was synonymous with "Afghan." Belonging to a tribe, being a peer, not only protects the individual but provides a man with a deep sense of identity. Abdul Wali Khan, a learned urban Pashtun politician and leader of Pakistan's opposition party in the 1970s, was once questioned about his loyalty to his country. His response: "I have been a Pakistani for thirty years, a Muslim for fourteen hundred years, and a Pashtun for five thousand years." Pashtuns have little doubt about their identity as Muslims, believing their earliest ancestor, Qais Abdur Rashid, was directly mentored by the prophet Muhammed. Given such a deep shared lineage and connection to particular patches of land, no matter how barren or remote, tribal life fulfills every one of the five strata of psychologist Abraham Maslow's hierarchy of needs. The tribe provides food, security, love and belonging, self-esteem, and an understanding of one's place in the world. Simply put, Jim believed that the eastern Afghan tribes offered Pashtuns everything that a person could want.

Especially revealing to Jim were the writings of Ibn Khaldun, one of the earliest thinkers to analyze the Islamic peoples living independently of central authority in the region's mountains, steppes, and deserts independent of central authority. Ibn Khaldun focused mainly on marginal groups such as the Bedouin nomads of North Africa and the tribes of Arabia. But modern scholars—in particular Boston University anthropology professor Thomas Barfield—found Ibn Khaldun's model of social organization equally relevant for the Pashtun tribes of Afghanistan. Most compelling to Jim was Barfield's application of Ibn Khaldon's theory in his 2010 book *Afghanistan: A Cultural and Political History*.

In the *Muqaddimah*, completed in 1377, Ibn Khaldun laid out simply and brilliantly the primal political divide of peoples such as the Afghans. Long before modern democracy, communism, or socialism, Ibn Khaldun proposed that these peoples fell into just two categories—tribal and urban. He then put forward the notion that most civil unrest in the country was the result of these two diametrically opposed cultures. Dogma did not divide men on any level in nearly the manner that their chosen way of life did.

Ibn Khaldun described the tribal world as a "desert civilization," one founded on what environmentalists today would romantically call "sustainable living." Tribes, whether settled farmers or nomadic herders, are autonomous. They live off the land, growing the crops they need to eat and raising the livestock necessary to sustain the group. Because a majority of Afghanistan's tribes are rooted in rugged terrain or isolated valleys where individuals alone would perish, kinship is vital. This isolation also prevents the tribes from being easily conquered.

As tribal economies are based on subsistence, not surplus, there is little value in growing more food or slaughtering more livestock than the group can live on. Material personal possessions, beyond the necessity of weapons to defend the tribe and protect its natural resources, are few. One man's home is just like another's. A chief eats roughly the same food in the same quantities as any other peer. With resources scarce in the marginal areas that Afghanistan's tribes call home, wealth is equated not with money but with land and livestock.

What little wealth there is in a tribe is invested in relationships. After the community has been fed, whatever surplus remains is traded in town or village bazaars. The extra food or goods brought back from the marketplace are then used for hospitality and communal feasts, not personal consumption. Gift givers are highly esteemed, and power is reflected in a man's ability to live simply, not in his ability to hoard.

Tribal life is based on honor. An honorable man brings tribute to his tribe, a venal man humiliation. Similarly, an assault on one peer is an assault on the entire tribe. Justice is meted out not by a policing arm of a centralized governing body but by the entire community. If a man is murdered by a member of another tribe, blood revenge or reparations must be enforced. If a tribal warrior is called to battle, he fights bravely and will not retreat unless the tribal elders determine it is best to fight another day. Retreating to save oneself, though, is the most shameful act imaginable.

The tribe is led by elders, those peers with the most experience. But in essence a tribe is egalitarian, run by consensus. A chief does not have dictatorial power. He does not live in a palace. He cannot give orders and expect blind compliance. Instead, he must earn and constantly re-earn trust. If his leadership is lacking, there is no shortage of other ambitious peers capable of taking his position. While a chief with the respect and support of a majority of tribe members can strongly influence the direction of a tribe, he is expected to listen to and refute the arguments of detractors in order to get his decisions approved. The wisdom of the crowd counterbalances charismatic dominance.

In stark contrast to this tribal world or "desert civilization," Ibn Khaldun wrote, is the world of the urban or "sedentary civilization." Cities in Afghanistan have greater population density, more wealth, greater social inequality, and a political order that is more hierarchical. In the urban centers, divisions of wealth and class are far stronger, social bonds weaker. Individuals take care of their nuclear family first, putting it before the larger kinship group. Governments exert more authority over the population through bureaucracies and taxation.

When it came to warfare, the contrast between the sedentary civilizations and the tribes was most striking, Ibn Khaldun observed. For city peoples, a man's honor was defined differently than in a tribe. His position in the social hierarchy was determined by his net material worth. War was outsourced to specialists and was not a skill valued in a society driven by its surplus economy. But the self-sustaining tribes living in mountainous lands were adept at deadly conflict. By necessity, they defended their homes from neighboring threats and were well versed in the strategies and tactics of war. Their warrior's ethos was built on death before dishonor. Tribes viewed the city people as weak and venal. Urban dwellers looked down upon the tribes as primitive and uncivilized. When the two cultures collided on a battlefield, however, one prevailed over and over again. Historically, Ibn Khaldun found that when the central urban powers became vulnerable in places such as Afghanistan, with flagging economies raising the ire of unpaid mercenaries and a widening segment of the rural population, desert civilizations rode herd. Camel-riding Bedouins and Turkish nomadic horsemen attracted by the wealth of the cities attacked and conquered. The autocrats were cast out and glory and wealth came to desert empires. Then a process by which the desert people were seduced by the luxuries available in the sedentary world began. The allure of better food, better clothes, and power transformed and corrupted tribesmen into the urban life. Within a few generations, the process would begin anew. Thus in Afghanistan social status at times overlaid and trumped ethnic affiliation—a Pashtun living in Kabul had far more in common with his Tajik neighbor in the city than he did with a Pashtun living in a qalat in the Konar River valley.

The divide between these two worlds, tribal and settled, remains strong in Afghanistan today. Most important, the geographic isolation of the tribes in the mountains and large expanses of desert hinders the central government from asserting control over the tribal people.

Jim fell hard for the desert civilization code and its ethos of Pashtunwali in 2003, while living with the Mohmand tribe and fighting the Taliban alongside them in Konar Province. He related to their warrior creed as parallel to the life he'd embraced himself as a Green Beret and one he preached to lead his small band of men into battle. It resonated with the ancient laws abided by the obedient three hundred Spartans at Thermopylae in 480 BCE. Honor, strength, and loyalty were not empty platitudes to Afghanistan's tribes; they were as important to tribal members as were water and wheat. As important as they were to Jim. As important as Jim assumed they were to the U.S. military establishment.

In 2010, as Jim prepared to return to Afghanistan, he increasingly realized that the only way to stabilize the country was to empower the desert civilizations, the Pashtuns still living in the rugged lands bordering and inside Pakistan. It was the pursuit of this honor, through physical courage and battling a common enemy, that Jim believed would allow him to become close to the Pashtuns. To ally with these proud fighters, to befriend them and help them recover their economies while also giving them the power to defend themselves, would not only take the fight to the Taliban but also draw disgruntled Taliban foot soldiers back to their villages, back to their tribes, secure and confident in their ability to come home and make a life for themselves. They could retreat with honor.

Jim was convinced that the reintegration of the Taliban fighters was a key component of any plan for long-term stability and could happen through the tribal network. Taliban foot soldiers continued to fight after years of hardship and immense disadvantage on the battlefield. Why? One reason was that battling the Americans was a way to earn their stripes as warriors. If they could fight against the Taliban and alongside fellow tribesmen, would they? The answer, Jim knew, was yes.

Jim viewed the Taliban's top leaders—Islamic extremists such as the one-eyed Mullah Omar—as championing a dogmatic, tyrannical movement that by its very nature threatened to dismantle the millennia-old rule by tribal elders. If the U.S. military were to convincingly help village elders take back their clans, defend their honor and traditions, and return their tribes to the authority of these egalitarian peer councils, the Taliban would be hollowed out and ultimately destroyed. The men who left the villages to join the Taliban in the turmoil of the civil wars would come back and take their rightful places inside their tribes. With no foot soldiers, the Taliban would lose power. The best way to empower the rough-hewn tribes, Jim believed, was with small teams of Special Forces such as his ODA 316, living among them one warrior to another. Once one tribe was secure, the team would leave and knock on the qalats of the tribe next door and start all over. It required little manpower or money, but could help Afghanistan begin to change from a war-torn terrorist haven to a more stable U.S. ally.

This was heady stuff. As Jim spent hours in Washington, DC, memorizing Pashto vocabulary, practicing the ornate Arabic-style script, and reading about Afghan culture, the speaking invitations continued to pour in. After Jim's "One Tribe at a Time" hit in-boxes at the Pentagon, in Congress, and at the White House, everyone from four-star generals to policy wonks wanted to hear from him directly. Jim's ideas were fresh and adaptable to both counterinsurgency and counterterrorism proponents. After Obama announced his Afghanistan surge, policy makers were continuing to wrangle over specific initiatives that could help turn the tide of the war. Recruiting Afghanistan's tribesmen and villagers to defend their own communities, as Jim had argued, was one of those initiatives. In Washington, policy reports were constantly circulated by think tanks as a form of not-so-subtle political warfare, and Jim's paper was no different. So Jim soon found himself in a suit and tie—completely out of his element—debating strategy as he gave a series of talks on his tribal engagement plan. Jim's friend Steven Pressfield, the author who had edited and published "One Tribe at a Time," was on the speaking tour with him.

At the think tanks, there was a mixture of support and skepticism. Critics argued that tribal engagement could sap power from the Afghan government and its security forces, and risked creating local militias.

Others said a strategy focused on tribes was too narrow, and Jim agreed that in some areas it would be necessary to work with other types of local leaders, as years of war had weakened the tribal structure in some parts of the country, such as the south. But there was broad agreement that leveraging the tribes was important and had been overlooked. Perhaps the most serious criticism was that the U.S. military lacked the pool of talent needed to carry out the strategy on a wide scale. "I just don't see how the United States can back a strategy that is predicated upon it being implemented by geniuses," said Andrew Exum, a former Army Ranger and advisor at the Center for a New American Security, a prominent Washington defense think tank.

Jim and Pressfield were also asked to speak at the Washington-area military schools and bases. There, they found the audiences of young officers hugely enthusiastic—like Jim, they all seemed hungry for a worthy mission. At Quantico and Annapolis, hundreds of Marines and Naval midshipmen turned out. Jim's last stop was at Mahan Hall at the Naval Academy.

It was nine at night when Jim spoke. A crowd of midshipmen filled the audience, taking time away from their required studies. Jim was tired. But his real self came out. He looked into the young men's eyes, and he made them members of his team.

> I don't want you for eight months. I don't want you for eighteen
> months. I want you committed for as long as it takes. You will
> not be just good. Good doesn't count for shit. You will be
> great. I will make you great.
>   When I send you out with the tribes, you will wear what
> they wear, eat what they eat, sleep where they sleep and live
> and die not as an American but as an Afghan. If you get home
> once in eighteen months, you'll be lucky—and I'll give you
> only enough time to kiss your wife before you turn around and
> come back. How long will we stay? Till we win. Till that tribe
> is standing on its own two feet and doesn't need us anymore.
> You are there for them, not for yourself and not for the United
> States. Your agenda is their agenda. You can't lie; they'll see
> through it. You can't bullshit; they'll know it before you speak.
> You must be truer than you've ever thought you could be and

braver than you've ever dreamed. Whatever you imagine your limits are, I will push you past them, and when I'm gone you'll push yourself beyond even those. I won't give you a medal, I won't give you a warm bunk, I won't give you a hot meal unless it helps the tribe you're working with. You'll eat rice and beans, wear Afghan clothes, and carry an AK-47. You won't see an American uniform, helmet, or body armor. And if you die there, I'll die alongside you.

When Jim finished, the middies swarmed the stage. "Where do I sign up, sir?" they asked him. "How soon can we leave?"

Yet despite the surge of professional recognition, the high-level accolades for his plan, and his new mission in Afghanistan, Jim felt drained by the bright lights, the uncomfortable suits and ties, the speeches and political games. The war of words so intrinsic to Washington was not the kind of combat Jim thrived on. At the same time, his mind and soul had never recovered from Iraq. But now he had to steel himself to kill and die once again. He felt himself slipping into a precarious, downhill slide mentally and emotionally. He just wanted to get back in the mountains with a couple hundred guns, some men like him, a bundle of dry kindling for a fire, and a couple of sacks of beans and rice. He wanted to get away from the empty sparring of Washington and into the heart of the Taliban insurgency.

# CHAPTER 9

A COOL MIST WAFTED over the long rows of white marble gravestones at Arlington Memorial Cemetery and disappeared into a line of trees, their tall trunks and bare branches like silent sentries silhouetted against a wintry sky.

On the south side of the cemetery, Jim and I gathered with a small group of mourners at the columbarium, where cremated remains are interred. It was January 22, 2010, and within days a major snowfall and blizzard would hit Washington, DC, burying the cemetery in several feet of snow. But that afternoon, the only hint of the cold front was a chill wind that cut through my navy coatdress and Jim's thin green jacket as we stood listening to the eulogy for his driver from Iraq, Sgt. 1st Class Jean-Paul LeBrocq.

I could tell that Jim, standing next to me, was deep in thought.

Jim's mind rushed from Arlington to the Iraqi desert, where for six months in 2006 and 2007 he and LeBrocq had lived and fought together in the heat and dust. They had come to know each other in battle. They had walked the same path. He recalled LeBrocq, steady at the wheel of his Humvee, when an explosion rocked the vehicle and shrouded them in debris. He saw him kneeling in the middle of that road to Baghdad on December 11 with insurgent rounds cracking over his head. He remembered him speaking fondly of his wife, Teresa. Jim glanced over at the soldier's teenage son, Bryan, his pale face somber and brave beyond his years. And he thought of Liam, just a toddler, conceived after LeBrocq returned from Iraq and before he was diagnosed with a fatal cancer. LeBrocq died just before Christmas.

Across the grassy field, an Army rifle party, wearing white gloves, dark navy uniforms, and caps, fired a volley in a final salute to LeBrocq. As the echo of the rounds dissipated into the haunting and strangely peaceful place that is Arlington, I looked over and saw a single tear running down Jim's cheek. I took hold of his arm, and he pulled my hand to his chest and gripped it tightly. He did not let go.

Jim had been back from Iraq for two and a half years. But he had never really made it home. Now he was bracing himself for Afghanistan and the inevitable mental and emotional strain of another long combat tour. I understood. It was my job, and I felt it deeply, too.

I had seen my share of combat covering the wars in Iraq and Afghanistan since 2001 for the *Christian Science Monitor* and the *Washington Post*. When I first ventured into a war zone in March 2003, I rode with the force that spearheaded the invasion of Iraq all the way to Baghdad. I put on an Army-issued chemical suit and climbed into a Humvee in Kuwait with the 3rd Infantry Division. Along the way, we braved sandstorms, chemical weapons alarms, and suicidal enemy attacks, and left a trail of death and destruction.

Two of the journalist colleagues I set out with were killed in a rocket strike. I was threatened with a rifle at point-blank range by a spooked U.S. soldier. With everyone else, I lost much of whatever innocence I had left. My sense of time and space grew severely distorted, and my faith was challenged. With every bloody mile, the soldiers and I felt more betrayed by the Washington officials who had ordered the invasion, resentful that they couldn't fathom what they'd asked of the troops, and that the weapons of mass destruction that were the justification for the war didn't seem to exist. I flew out of Iraq in 2003 in a C-130 plane full of dead U.S. service members in gray body bags. I had to step carefully over the rows of body bags in the plane's belly to make it to my seat in the cockpit. I landed in Kuwait after weeks in the desert, took my first sip of cold water, and felt the unfamiliar firmness of asphalt under my boots. In a surreal twist, after weeks of sleeping in and on military vehicles, a well-intentioned hotel attendant gave me a room in the presidential suite at the Kuwait Hilton Resort. All I could do was heat up an MRE packaged military meal, wash off a little of the dust that caked my face, hands, and body, and try to sleep. All night a yellow light glowed in the hallway, and I slipped in and out of a dream that I was in another sandstorm on the outskirts of Baghdad. I struggled to keep the rest of the world and all of its sensations at bay while I gradually reentered life a changed person.

I came home to suburban Bethesda, Maryland, in April and found myself in a grassy backyard surrounded by blooming daffodils and my four wonderful children. But I had only partly returned. Every day I lived and breathed the war as I covered the news. Senses honed by danger and hardship added a strange new dimension to ordinary life at home. Hikes along the Potomac felt at times like combat patrols, and I fell asleep in my clothes on the couch, just as in Iraq, where I slept anywhere I could find a place, only taking off my boots. I was living in a twilight zone between two separate worlds. But I had no one to talk with about it, particularly since my husband was opposed to my war coverage.

Jim held on to my arm as we left LeBrocq's resting place with the rest of the mourners, walking between the gravestones on the damp grass. The cemetery was a gut-wrenching place for both Jim and me, in different ways. He had fought with men buried there. For my part, after years of writing about the war dead and receiving reports on every single military casualty, I recognized far too many names on the headstones and knew their stories. And every time I wrote about one of the fallen, I would eventually hear from the family. It could be days later, or months. But it always happened, because they could never know enough. The questions they asked me were as surprising as they were heartbreaking: "Was he cold when he died? Was he alone?"

Jim and I shared an unspoken bond through our acquaintance with the sacrifice of men like LeBrocq. The sadness ran through us, eased ever so slightly by the poignant notes of "Taps" at the end of LeBrocq's ceremony. The wind whipped around us as we walked out the gate of the burial grounds and down tree-lined Memorial Drive to the Arlington Cemetery Metro station, where we waited for a train. Jim was shivering in his thin jacket, plaid shirt, and mismatched tie, and his hands were freezing. His eyes were sallow and he looked drained.

A train rumbled into the station and screeched to a stop. We got on. Jim looked out the window as the train accelerated past the cemetery, a grim landscape he passed every day on the way to and from language school.

"I don't want to be buried there," he said. "At Arlington, there is no differentiation between warriors and soldiers, between fighters and cowards."

We stopped at a coffee shop in Crystal City to warm up a little before I went back to work at the *Post*. He asked me again whether I would have dinner with him. I said yes.

Then I smiled and added: "Listening, although we are already in agreement," referring to a Pashto proverb he taught me. He laughed.

Unexpectedly and mysteriously for us both, over the next few days we fell headlong in love.

It was one of the most emotionally intense weeks of our lives. For me, falling in love with Jim felt like stepping into a Humvee for a night raid in Baghdad. It was tense and exhilarating, there was no turning back, and I never knew what lay around the next corner. Jim compared it to boarding a helicopter together and lifting off.

A few days into it, he wrote me a note asking me to marry him. I laughed it off.

Neither of us was looking for a relationship. I had been separated from my husband for a year, and Jim was also separated and headed for divorce within a few months. Outwardly we were an unlikely couple. We came from starkly different cultural, social, and professional worlds, ones frequently at odds. He was a self-taught graduate of New Mexico State University; I was a Harvard-educated professor's daughter. He had roots in Texas and considered George Bush a capable wartime leader; I grew up in Seattle and reported for a liberal East Coast newspaper. He was a member of a secretive military unit; my job for years was to ferret out and expose secrets about the military. Peel back the externals, though, and we were more alike than it seemed—adventurous, intellectually curious, passionate about our work, and rebels of sorts.

Our relationship crossed so many lines and broke so many unspoken rules that if we had hesitated, we might have faltered.

"Don't look down," he would warn me. "Just jump. I will catch you."

We needed each other.

I was making the leap from a long-troubled marriage and anguishing over its impact on my three younger children, who were living with me. Suffering from my own combat trauma, I needed a comrade to cry and laugh with. I thought about the wars every day and cared deeply about the people involved. "You have two families now," one of my military friends told me.

Jim was hurting, too, personally and professionally. He felt incredible pressure over his mission—the mission of his lifetime. He needed to steel himself to kill again, to go back to that place in his mind, to become that other person. He had to be ready to give his life for everything he believed in. Part of him expected—wanted, even—to go to Afghanistan and die in battle, to prove beyond a doubt he was the warrior he said he was. But he was living so hard he risked never making it there.

Drinking all day, he was barely eating. One afternoon he told me he wanted to show me one of his favorite movie scenes. It was a montage of Nicolas Cage as his character drank himself to death in the movie *Leaving Las Vegas*. Sleep for Jim came only in spurts. Nightmares made him cry out and sit up in bed. I spent as much time with him as I could, telling my children I had to help a very sick friend.

Some nights Jim got into bar fights in the worst parts of DC. Once I arrived at his apartment and found his fists cut up and blood all over the sink and floor. I took a deep breath and cleaned it off. On the bathroom counter, he had a bowl full of multicolored pills that turned out to be speed, Valium, and a lot of painkillers such as Percocet. I later found out he was addicted to Percocet and had used cocaine. He was also cutting himself. He would slit the throat of one of the figures tattooed on his arms and let the blood stream out onto the floor. One day he told me he had a present for me. I opened up a black-and-white checkered Afghan scarf and found wrapped inside four plain wooden hearts. He had written the words "I love you Meena" on them with his blood. *Meena* is the Pashto word for "everything" or "love." It was his name for me.

Jim scared a lot of people. I might have been afraid of him, too, had I not seen what combat could do to men. Instead, his behavior seemed oddly familiar. I felt as though I had known him for a very long time. On top of that, what made all his craziness less frightening was that he was open about it—at least more open than most.

Days and nights blurred together as we spent hour upon hour talking in his apartment, oblivious to the world outside. One night I shared one of my favorite pastimes with him by reciting poetry. As he listened to "The Love Song of J. Alfred Prufrock" by T. S. Eliot, he wept at its beauty, and because no one had ever done that for him before.

"I feel as though I am deep underwater," he told me one day as I lay in his arms in his room. "I can look up and see a little bit of light, but I can't reach it. I can't breathe."

Often he contrasted his own darkness with my light. He was drawn to me and my inherent optimism and happiness.

"Your light hurts me," he said, his blue eyes cold and ancient. "I have many shields and weapons, but your light pierces straight through them."

He was cutting through my armor, too, making his way into my most heavily guarded rooms. We both knew that our time was limited, and that we might not ever fall in love that way again. Just to have known it, though, was enough. One day as he embraced me in the kitchen of his apartment, I told him so.

"It just doesn't matter anymore," I said.

He understood and hugged me harder. "No," he said, "nothing else matters."

The deeper we fell in love, the harder it was to be apart.

In February, after a huge blizzard brought Washington to a standstill, Jim trudged through the two-foot-deep snow to my house at night and tapped on my bedroom window. At first he didn't come through the front door out of consideration for my children. I pushed the window up and he crawled through, shaking with cold. We stayed together until sunrise, when he left to go lift weights and study Pashto.

Jim began to imagine a future, and he started pulling out of his self-destructive state. He was coming back to life, and doing it before my eyes.

"You make me want to be a better person," he told me.

Without me ever asking, he cut back sharply on his drinking and quit abusing painkillers—virtually cold turkey. Every night for about a week, he stayed with me and I held him as he sweated and shook through withdrawal from the drugs. One night I cooked him some chicken and dumplings, and he ate bowl after bowl of it.

"It is like blood," he said of the rich soup, asking me to make him some more.

Finally one night he stopped shaking. He had made it through. The next morning he woke up and smiled at me, his face glowing in a way I had never seen before.

"You have saved my life," he said.

ONE AFTERNOON IN APRIL 2010, as his deployment approached, Jim asked me to meet him at Rick's Tattoos on the second floor of a redbrick building in Arlington, Virginia. Established in 1970, the tattoo studio had cartoon blondes in swimsuits painted on the window and boasted some talented artists. Jim's tattoos during his years at war were never frivolous—each one spoke something powerful about his identity, as did the one he would get that day.

Jim greeted me with a kiss and pulled out a piece of paper with four words in Afghan lettering written by his Pashto teacher, a young man named Najmal. After speaking with the artist, he invited me to sit and watch as she inscribed on his body words that described the cornerstone of what he stood for and was about to do.

I looked as he winced while the artist moved her tattoo machine around his right wrist, writing in black, indelible ink the Pashto words *ghairat* and *nang*, meaning "individual honor" and "collective honor." On the top of his left wrist she traced the word *namoos*, referring to those things a man has—women, land, and guns—that he must protect. Underneath, she tattooed our Pashto name for each other—*meena*.

We had started talking about me joining him in Afghanistan. Jim was excited to take me to Mangwel to meet Noor Afzhal and the tribe, to make that forever part of our life together. The country, the war, the sacrifice, the suffering—both of us felt passionate about it. We were convinced that a deep understanding of the Pashtun tribes and their needs was critical to any positive outcome of the war. We wanted badly to help the tribal people. And we were willing to risk everything to put his strategy to the test on the ground. Only by giving it our all would we find out how far he could succeed.

As most would view it, I crossed over to the dark side professionally by becoming involved with Jim, and he with me. I saw it differently, particularly because he is so open about his own failings and those of the U.S. military and Special Forces. If anything, through being close to Jim I have gained a far more unvarnished view of the military and its flaws, having seen the institution from the inside. We talked about writing a book that would chronicle his efforts and work with Noor Afzhal and the tribes, so that others could draw lessons from it. We had no idea then how we would pull it off, but we knew we wanted to try.

Jim believed he could recruit one Pashtun tribe after another in eastern Afghanistan, and that this just might be enough to achieve a tipping point in the war. But even as he packed his duffel bags and memorized Pashto sayings, he was troubled by doubts about whether the American political and military leadership would back him up and stay the course. One morning he woke up in my bed ashen-faced. In his nightmare, he had been walking down a narrow road past a cemetery in an Afghan village, one that may have been Mangwel or was similar to Mangwel. It was night, and the qalats beside the road were burning, their walls crumbling. He looked beyond them and realized that the entire village was flaming, a huge orange glow. It was deserted.

"It was very clear it was a village that had been standing, and I had caused its destruction," he told me. Jim always trusted his dreams and visions, and this one was a horrible premonition.

But it was too late. Afghanistan was deteriorating, his mission was under way, and Jim believed that it was the only one with any chance of success. He was willing to bet his life on it.

So he prepared to deploy as he always had. He took the Spartan shield that he wore around his neck and placed it in my hand. "This will keep you safe," he said. In its place, I gave him an inch-high golden lighthouse on a leather cord. It was a replica of the tall brick Cape Hatteras light station that beams over dangerous shoals off the North Carolina coast—I said it would guide him to the shore.

Many soldiers—perhaps to protect their loved ones—choose never to speak of the possibility of their death, almost as if making it taboo would prevent it from happening. Jim embraced the idea as the essence of who he was. At first it was hard for me to hear him talk about it, but I gradually came to understand and accept it. Jim had a very real and tangible acceptance of his own death as the consequence of his decision to be a warrior. Thousands of times in his head, he had played through the possibility of giving his life on the battlefield and watching his own burial. On some days he wished to die fighting because he believed it was all he was good at, and because he found it difficult to envision any future for himself beyond that. He also had an urge to seal his legacy by proving once and for all that he would bravely lay down his life when the time came. Over and over, he imagined his death in a special place he created in his mind, a white hallway lined with the people who loved him, with a door at the end.

So Jim packed his gear and wrote his goodbye letters, and when he boarded the plane at Dulles Airport on June 16, 2010, in his mind he was already dead.

# CHAPTER 10

A FEW DAYS AFTER Jim landed in the Afghan capital in June 2010, Gen. David Petraeus walked into his well-appointed office at the Kabul headquarters of NATO's International Security Assistance Force (ISAF), assuming his new position as the top U.S. commander in Afghanistan.

In an unexpected switch, Petraeus had agreed almost overnight to replace Gen. Stanley McChrystal and take the reins of what by some measures is the longest U.S. war. President Obama had abruptly fired McChrystal on June 23 following the publication of a *Rolling Stone* article that attributed remarks ridiculing key Obama administration officials to McChrystal and his staff. Then the president asked Petraeus to take charge of the faltering campaign.

Petraeus had a slew of reasons to turn Obama down. He had often said that the war in Afghanistan was a far more intractable challenge than the fight he'd been widely credited with winning in Iraq. The Afghan post was a step down from Petraeus's job as head of U.S. Central Command, where he oversaw the entire Middle East and Central Asia, including the Iraq and Afghanistan wars. On a personal level, Petraeus had been deployed for more than five and a half years since the September 11 terrorist attacks. And he had recently recovered from early-stage cancer, diagnosed in 2009.

At his Senate confirmation hearing, Petraeus looked tired. Nevertheless, his testimony left no doubt that he wanted to put more time on Obama's clock.

"It is important to note the president's reminder in recent days that

July 2011 will mark the beginning of a process, not the date when the U.S. heads for the exits and turns out the lights," Petraeus told the senators. The drawdown will be "responsible" and "based on conditions on the ground," Petraeus emphasized.

What Petraeus didn't reveal during the hearing was a key behind-the-scenes initiative he was pushing with Afghan president Karzai to create village defense forces—inspired in part by Jim's ideas. Petraeus considered the plan a "game changer." According to his own "counterinsurgency math," Afghanistan had less than half the number of security forces it needed to cover the population of about thirty million. He wanted to use Special Forces teams to recruit thousands of rural Afghan tribesmen to help fill the gap. After his confirmation, he called Karzai twice to discuss the plan, but reached no agreement.

In early July 2010, Petraeus walked into the presidential palace in Kabul for his first face-to-face meeting with Karzai as the U.S. commander in Afghanistan. With only twelve months left on Obama's clock, it was now or never. If Petraeus didn't get Afghanistan's president to sign off on the village defense program, it would be a major setback for his counterinsurgency strategy.

But from the other side of the desk, Karzai had reasons to put up stiff resistance to the United States' local security initiative. Karzai knew how delicate his hold on the country was. He was nervous about arming Afghanistan's most seasoned warriors. In his view, the isolated tribesmen had little concept of Afghanistan as a nation-state seeking to scratch out a prosperous niche in an interconnected global economy. If anything, the tribes in the hills and deserts had contempt for the goings-on in Kabul and viewed the urban world as base and dishonorable. And the Taliban drew its rank and file from the same local tribes that Petraeus sought to arm and empower. The Taliban's *jihad* was not just against the infidels; it was against westernized, educated, and pragmatic men such as himself.

Still, Karzai was above all a survivor, a master of discerning the winds of power and capturing them to his advantage. He referred to the Taliban as a "great evil" one day and offered to negotiate a place for them in his government the day after. Karzai played Afghanistan's powerful factions and ethnic groups off one another by means of titles and tribute in the form of political appointments and cash payoffs. He'd even tapped the powerful Northern Alliance Tajik responsible for his arrest and interrogation in 1994, Mohammed Fahim, as defense minister. Fahim later became the seniormost of Afghanistan's two vice presidents. Karzai made his enemies allies.

In the early July 2010 meeting with Petraeus, Karzai was acutely aware that his relationship with the United States, critical to his selection as Afghanistan's president after the fall of the Taliban in 2001, was as complex as his relationship with his fellow Afghans. He genuflected to the Stars and Stripes one day and damned them the next. But by far the greatest source of his power and money was represented by the highly decorated four-star officer sitting in his office. To say no to Petraeus would alienate his most important benefactor and could halt the flow of the aid that kept him in power. And if the money stopped, his allies would revert to threats.

So Karzai didn't say no—but he didn't say yes, either. He told Petraeus that he and other senior Afghan officials were concerned that the program to arm local tribesmen would result in the creation of militias that would threaten the government's power. No program would be approved without a direct tie-in to the central government, he demanded.

In response, Petraeus launched intensive negotiations that over the next few weeks led to a series of critical compromises with the Karzai regime. Most crucially, the United States granted the Karzai government control over which of Afghanistan's roughly four hundred districts would be earmarked for the program. The initial plan was to recruit ten thousand local police, to be trained and mentored by twelve-man Special Forces teams. That number was later increased to thirty thousand, to be recruited by 2015. The local security effort was to be coupled with economic development projects and the building of grassroots governance, an effort the U.S. military called village stability operations (VSO). But in order to get the program approved and off the ground, Petraeus agreed to have the Afghan Ministry of Interior put in charge of the vetting, pay, and weapons distribution for the forces, which were formally named Afghan Local Police (ALP). In time, the bureaucratic and rampantly corrupt government control would hamper the responsiveness of the Special Forces teams to local needs. Ethnic politics would delay approvals for police for months in different regions. Meanwhile, the U.S. military bureaucracy would present its own challenges for carrying out the strategy.

Petraeus instructed Jim's new chain of command—the men who oversaw all Special Operations Forces in Afghanistan—to make the local defense mission their top priority and to expand the program as rapidly as possible. Petraeus turned in particular to Brig. Gen. Austin "Scott" Miller and Col. Donald C. Bolduc.

Bolduc oversaw day-to-day Special Operations missions as the commander of the Combined Joint Special Operations Task Force–Afghanistan (CJSOTF-A), based at Bagram Air Field. Bolduc was a veteran Special Forces officer who first deployed to Afghanistan in 2001 with a team sent to link up with and advise Karzai, who was just returning to the country from overseas. Bolduc already had three tours in Afghanistan under his belt and had been awarded two Purple Hearts and two awards for valor. A wiry fitness fanatic, Bolduc was serious and aggressive as a commander. He had a knack for articulating a crystal-clear intent as a military leader and seeing that everyone below him carried it out. Bolduc knew Jim well. He had been one of Jim's battalion commanders when Jim was a captain and had consistently rated Jim highly in evaluations. Bolduc and his staff drew on Jim's ideas as they wrote a detailed, step-by-step plan for implementing VSO. Petraeus considered Bolduc's work masterly.

One level above Bolduc, Miller was head of the Combined Forces Special Operations Component Command–Afghanistan (CFSOCC-A), based in Kabul. Tan, with close-cropped silver hair and a tennis player's build, Miller had spent most of his career in the military's secretive Joint Special Operations Command (JSOC). In Kabul, Miller was instrumental in negotiations to gain the support of senior Afghan ministry officials for the local police program. He believed the tie-in of the program to the government was vital, as was the development portion. Petraeus and others praised Miller's efforts for helping to expand the program to a national scale.

Miller had sat down with Jim when they were both studying at the Defense Language Institute in Rosslyn, Virginia, in the spring of 2010. Miller was taking a six-week course in Dari, the dialect of Persian spoken by non-Pashtun peoples in Kabul and other parts of northern, central, and western Afghanistan. The two officers respected each other's combat experience. As a captain, Miller commanded elite Delta Force troops as they fought off waves of Somali gunmen during the infamous

1993 battle surrounding two downed U.S. military Black Hawk helicopters in Mogadishu. Miller was awarded a Bronze Star for valor for his bravery in the battle. Yet Miller grew up in a force that specialized in man-hunting raids and hostage rescue—a military tribe distinctly different from Jim's.

In Washington, Miller had warned Jim to "lie a little bit low." "There are some out there who fault you for your access, your article," Miller wrote Jim in an email. "Just trust me, it's the nature of the beast and something we all deal with."

Once Jim arrived in Kabul and was sent to work for Miller, he was put on guard right off the bat. Miller's first instruction to Jim after he stepped off the airplane was to shave his beard and get into uniform—a signal he would be working close to the flagpole. Jim sensed the one-star general was slightly wary of him and his direct line to Petraeus, Olson, and Gen. James Mattis, a Marine Corps general who led a division in the 2003 invasion of Iraq and who had been promoted in July 2010 to take over Central Command after Petraeus went to Afghanistan.

Still, Miller initially instructed Jim to travel the country and gain an overview of how the preliminary village defense effort championed by Petraeus was unfolding. The program was in its infant phases at a several locations, mainly in the south, where U.S. forces were concentrating their main effort. There, Jim would uncover some serious problems for Petraeus, Miller, Bolduc, and other senior commanders to address.

Jim headed south in July 2010, just as Obama's surge forces were flowing into southern Afghanistan in a bid to reverse the Taliban's growing momentum. The troop influx brought fresh fighting, and violence in Afghanistan was at its highest since the start of the war. Jim scanned the windswept, barren southern Afghan desert as the mine-resistant vehicle he was riding in sped down a bomb-pockmarked road that fed into the troubled Argandab valley, then into the village of Adehera, just outside Kandahar.

The hulking armored vehicle rolled past several mud-brick compounds and stopped outside a defensive position at the edge of the village. A young Special Forces team leader, Capt. Joe Quinn, came out to greet Jim, his face weary from stress. Quinn's eleven-man team was under siege, having difficulty recruiting Afghans for a local security force. Making things worse, Jim found, the team and others he visited were hampered on several fronts by an overly rigid and risk-averse U.S. military bureaucracy.

In "One Tribe at a Time," Jim argued that to be successful, the military chain of command had to give the Special Forces teams working with the tribal forces unusual trust and latitude. They needed to be able to hit targets quickly, gaining approval if necessary through a single radio call instead of having to abide by the standard glacial rules of engagement. They had to fight side by side with the tribal forces, not in segregated teams. They had to be free from burdensome reporting requirements and have greater leeway to spend money to benefit the local communities. "A strategy of tribal engagement will require a complete paradigm shift at the highest levels of our military organization—and the ability to push these changes down to group/brigade and battalion commanders," he wrote.

But Jim found that Quinn and other team leaders faced lengthy delays of days or longer in approvals to conduct operations—a laborious process that then had three dozen people involved. When one team gained intelligence that a key insurgent was just a mile away, the team leader had to decide whether to risk going after the target without approval, which would be a serious legal violation, or let the insurgent go. Another issue was that the "battlespace owners"—conventional military units that had responsibility for the areas within which the Special Forces teams were operating—were often unsupportive or even hostile to the teams.

The teams were further hamstrung by a requirement that they conduct all missions with an Afghan partner force—but they had not received permission to operate with the very Afghan Local Police they had recruited and trained. How could they build esprit de corps with their Afghan allies if they weren't allowed to launch missions with them or back them up in combat? Team commanders were also asked to write scores of different reports informing various superiors of their activities. Simply obtaining funds needed to operate was an onerous process. It required two team members to fly to Bagram, Kandahar, Jalalabad, or another large U.S. military base and spend on average two weeks doing paperwork and gaining signatures to draw money from a myriad of different sources.

One particularly maddening problem to Jim was the military's imposition of uniform and grooming standards on the teams. How were they supposed to blend in with the Afghans? The military required the team members to wear U.S. flag patches Velcroed on the shoulders of their traditional Afghan clothing. Team members also had to shave their beards if they went on leave. No facial hair was an absurd requirement for soldiers operating in an honor-based culture where beards are one of the most important symbols of manhood for tribesmen. Jim was wrestling with the problem himself.

Apart from the frustrating tactical issues, Jim had a strategic concern based on his research on Afghan tribes. He questioned the decision to put most of the initial teams in southern Afghanistan, where tribal organizations were generally far weaker than in the east. That decision was made in part because the south, where violence was highest, was the focus of the overall NATO campaign. In contrast, at the time there were only three teams working on the local defense plan in the east. Jim believed that was a huge missed opportunity, because the stronger tribal structure in the east would allow for greater security gains with the investment of a relatively small number of Special Forces soldiers. Jim's recommendation for a greater focus on the east proved especially relevant when the Taliban and other insurgent groups steadily increased their attacks in eastern Afghanistan in an effort to open up a new front.

On a personal level, Jim's reconnaissance visits brought home to him what a daunting task it was, even for the Special Forces, for teams to live in such austere conditions all the while abiding with the alien Afghan tribal culture. Getting them in wasn't enough. They had to spend long periods to win the tribe's trust. The average six- to eight-month Special Forces deployments were too short, and rarely were teams sent back to the same locations to enable them to build on existing relationships. And after nearly a decade of fighting in Afghanistan and Iraq, the Special Forces as a whole were suffering heavily from the psychological and physical strains of waging two simultaneous wars and back-to-back rotations. Jim understood combat fatigue.

In the village outside Kandahar, Jim discovered that Quinn's team was in shock from the recent loss of its team sergeant. The team was particularly young, and five of the members had gone straight into Special Forces without prior military experience. It was a recent program aimed at expanding the ranks of Green Berets more quickly. The young recruits were referred to as "18 X-rays"—in Jim's day they were "SF babies." The team sergeant—who was a father figure to the younger men—had been literally blown to pieces before their eyes when an IED he was disarming detonated. Team leader Quinn was required to radio a situation report to higher while the other members collected the sergeant's body parts and placed them in Ziploc bags. Later, Quinn confessed to Jim that he was anguished that he hadn't helped them in the gruesome task. The event tore the team apart. Quinn finally broke down as he told Jim about the horrific scene.

With these and other observations, Jim got on a helicopter back to the main U.S. military base in Kandahar. Lugging all his gear and wearing a dirty uniform, with several weeks of growth on his face, he was walking through the Special Forces camp when he unexpectedly ran into Adm. Olson—the SOCOM commander who had given Jim his "assemble your team" mission. Olson was delighted to see Jim in-country and stepped away from his entourage to greet him warmly and pat him on the back. "Are the men prepared for the mission?" Olson asked. Jim said he believed so.

Accompanying Olson a short distance behind was Miller. Miller shot a stern look at Jim, noticing he had allowed his beard to grow in the field, and ordered him coolly: "Jim, you need to shave." He then walked away. Jim bridled at the comment but did as he was told, and soon returned to his headquarters at the military base in Kabul.

A few days later, Jim attended a staff meeting in which Miller's and Bolduc's senior enlisted officers, Command Sgt. Maj. Ledford H. Stigall and Command Sgt. Maj. Jeff Wright, were discussing combat stress and how to help returning troops. Afterward, Jim, fresh from his encounter with the young Quinn in Adehera, pulled Stigall aside. He confided in him and candidly told him about his own struggles and those of the soldiers he had led. "Everyone I know in Special Forces who has seen a lot of combat drinks too much, has anger issues and PTSD, and needs help, including myself," Jim said with typical bluntness. He knew that the majority of Special Forces teams in the country had members who were drinking, smoking hashish, or taking other illicit drugs—not to mention their widespread use of illegal steroids. At one tiny outpost, he had to intervene to stop a drunken brawl in which a Special Forces team member had smashed computer equipment and threatened to kill some conventional Army troops.

As one of the main jobs of noncommissioned officers is to take care of soldiers, Jim was outraged when he found out the next morning that Stigall had apparently reported what he said to Miller. Jim was immediately called on the carpet—summoned to a meeting with the CJSOTF-A deputy commander, Col. Wade Murdock. Murdock told Jim that he was "an alcoholic, womanizing, mentally unstable maverick." He then read him the riot act. "You can go home now. You can marginalize yourself on the staff. Or, if you step out of line in any shape, form, or fashion, you are going to be fired—and we will be watching every move you make."

As had happened so often in the past, Jim had felt obligated to tell his commanders the truth on the ground as he saw it—only to learn they didn't want to hear it, at least not from him. Unfazed, he brushed the criticism off, stayed in Bagram, and set about writing his first assessment of the local security strategy. He laid out the problems that were hampering the teams on the ground, and called for a greater focus on eastern Afghanistan. He then sent the assessment directly to Olson, Mattis, and Petraeus, who had asked for his input. Petraeus wrote back a long email a few hours later telling Jim he agreed with virtually all his recommendations and asking him to help carry them out. Shortly beforehand, Petraeus had walked into a video teleconference with Miller and other Special Operations Forces personnel. Without saying where the recommendations came from, Petraeus ticked off Jim's proposals one by one and instructed Miller and the other officers to implement several of them, according to a Special Forces officer who attended. Miller, realizing what had happened, was furious.

Petraeus moved to protect Jim, writing him an email stating that he had ordered Jim to report to him directly—"sua sponte"—of his own accord.

Soon afterward, Bolduc—who knew Jim's strength was not as a staff officer but as a tactical leader on the ground—moved to put Jim under his immediate command. "I've got Jim," he told Miller in a brief phone call. Then Bolduc assigned Jim to be the advisor for the village defense strategy in eastern Afghanistan and let him grow his beard. Finally Jim was returning to where the Pashtun tribes were strongest, where he had first learned of the power of the tribes in Konar Province in 2003.

# CHAPTER 11

IT WAS PITCH DARK on the night of May 24, 2003, as Capt. Jim Gant and his Special Forces A-team crossed a footbridge over the river that coursed through the rugged Pech valley of Konar Province, one of the deadliest regions for American troops in Afghanistan. The team's target was an Afghan insurgent named Dr. Naimetullah, a financier of roadside bombs. Jim expected Naimetullah would not go out without a fight, and the new team had rehearsed exhaustively for the raid. One of the biggest challenges was the treacherous terrain deep in the Pech where Naimetullah lived, in the village of Dag.

To reach the village, the team had to navigate up the valley on a long, winding dirt road so narrow that in places its edges crumbled from the weight of their Humvees. Mountains dotted with scrub brush rose up on either side, seeming to close in on them the higher they went. The soldiers scanned the hillsides through night vision goggles that made the landscape glow in fuzzy shades of green and white, creating an eerie sense of detachment. They turned off the lights of their Humvees to make it harder for Afghan insurgents to spot them, but the engines' growl and grinding of wheels on the road left no doubt an American patrol was approaching. Several Toyota Hilux pickup trucks followed with an Afghan Counterterrorism Pursuit Team (CTPT) and their CIA handlers. All Jim could hope for was momentary surprise.

A short way off from the village they stopped their vehicles and got out. Below they could hear the rushing Pech River. Without speaking—they had rehearsed the operation several times—Jim and his men walked swiftly up the road and over the footbridge. The Afghans and CIA para-militaries took up position behind them, encircling the target area to create an outer perimeter and prevent anyone from escaping. Jim and his assault element turned and picked their way down a path that ran between the wall of the doctor's compound and a steep fifty-foot drop off to the rushing Pech River. They lined up with their shoulders pressed against the wall, and an Afghan police officer with them called the doctor out. The only response, from behind the door, was muffled voices and the distinctive metal clacking as men inside racked bullets in their AK-47 rifles.

"Breach it!" Jim ordered.

Staff Sgt. Scott Gross, a young ox of a man from Indiana, breached the outer courtyard door with a heavy crowbar known as a "hooligan tool." Staff Sgt. Tony Siriwardene pivoted right through the doorway, followed by Sgt. 1st Class Mark Read and Jim. They immediately came under fire at close range from men inside.

Tony hadn't even rounded the first corner when heavy gunfire started coming in their direction. As he turned into the compound, he saw muzzle flashes from inside a room that stood about thirty feet directly in front of them in the courtyard's interior. He fired back and moved along the courtyard wall as bullets hit the wall and pelted him with pieces of dirt.

Mark turned left in the door and saw a shadowy figure in a vest with an AK-47 run past the window in the inner room. He opened fire on the figure. *You dumb motherfucker. You have fucked with the wrong guys*, Mark thought.

Tony and Jim were both returning fire on the right. They needed backup.

"Bravo team in!" Jim yelled. A second group of his men rushed in the door under fire, including Scott, the breacher, and Staff Sgt. Dan McKone, the former Peace Corps volunteer turned Special Forces medic.

As the gun battle raged, a group of Afghan women and children ran into the front courtyard—apparently pushed out by the male shooters to give them time to escape. Caught in the line of fire, the women scattered, leaving an infant lying on the ground. As bullets from an automatic rifle shot across the courtyard, Scott scooped up the baby and moved it to safety in an adjacent room.

Soon after that, the shooting stopped. Dan spotted two Afghan women lying wounded in the courtyard, and rushed over to begin treating them. Other women and children were wailing around him, and Dan was operating in the dark. Reserved and meticulous, Dan had joined the Green Berets to be a medic, not a machine gunner. He moved methodically to assess their wounds. He saw that one of the women had a serious leg wound, and quickly applied a tourniquet. Then he turned to another and applied a pressure bandage and some antibiotics. Finally, he slipped an IV into the arm of the woman with a severe leg wound. "They will be all right," he reassured the family members, speaking through the team interpreter.

Meanwhile, Jim got a report from Special Forces Chief Warrant Officer 3 Brian Halstead, who had gathered intelligence for the raid and was the overall ground force commander, that an AC-130 gunship was circling overhead. The pilot had spotted some other fighters moving toward an adjoining qalat next door. Jim, Tony, and Mark looked over the wall into the blackness of the neighboring compound. Jim couldn't see anything. Were the fighters inside? He couldn't tell. Tony found a ladder and dropped it over the wall. Jim climbed down it.

*I'm about to get shot*, Jim thought as he stepped down the ladder. He reached the bottom, and covered Tony and Mark as they followed him down. They began searching the rooms, found four men huddled in one, and detained them. Halstead then directed soldiers in scouring the compound for intelligence and other sensitive items and confiscating a computer, wads of cash, and opium.

The gunfight marked the first close-quarters battle for Jim's brand-new team, ODA 316. They accomplished the mission. Dr. Naimetullah was dead. Sometime during the chaotic night he was killed or fatally wounded and his body surfaced downstream in the Pech River a few days later. All the other males in his family had been captured. The Americans had all lived to fight another day. When they finally reached their tiny camp near the town of Asadabad, the capital of Konar Province, at about four o'clock the next morning, their emotions poured out.

"With that proximity and the volume of shooting, we were amazed no one got shot," said Tony, the tall Sri Lankan who was Jim's right-hand man. "How did we walk away from that?"

Just six weeks before, Jim and his men had arrived in Afghanistan having been given no idea where they were going in the country, no intelligence reports, no current maps, no briefings on the people or the culture. Standing in the darkness on the tarmac at Bagram Air Field north of Kabul, about to board a Chinook helicopter for Asadabad, the intelligence officer of Jim's battalion walked over and gave him a piece of paper with a handwritten mission: "Kill or capture ACMs [anti-coalition members]."

"Roger that," Jim replied.

The terse mission statement summed up the U.S. military strategy for Afghanistan—or, more accurately, the lack of a strategy. It had been less than two years since the Al Qaeda terrorist network had killed thousands of Americans in attacks on U.S. soil on September 11, 2001. The Bush administration had swiftly retaliated, leading a punitive invasion in October to overthrow the Taliban regime for giving Al Qaeda safe haven in Afghanistan. The U.S.-led coalition, spearheaded by CIA operatives and Special Forces teams, ousted the Taliban from power in December 2001. But the Taliban and Al Qaeda were far from eradicated. As the U.S. military turned its attention to Iraq, launching a full-scale invasion in March 2003, the Taliban regrouped and made a steady comeback. But the U.S. strategy in Afghanistan failed to evolve. It remained simplistic and driven by revenge.

"The American people wanted blood, and we were that instrument of destruction," Jim recalled. "There was no plan."

At the time, the pure kill-and-capture mission suited Jim, a new ODA commander itching for his first taste of combat. Jim relished the opportunity to lead men in battle all the more because it had been so long in coming.

In the seventeen years since he left for Army basic training, Jim had worked to hone his skills, first as an enlisted Green Beret and then as an officer leading small combat units. He tested himself on the Army's shooting ranges and training fields and immersed himself in the study of military tactics and history. He earned the respect of commanders and the love of his men. Through all of this he emerged with a distinctive creed and identity—one that revolved around the Spartan ethos and ancient call of men to arms: he wanted to prove himself as a warrior, and Afghanistan was a place to fight.

In August 2002, Jim had been assigned to the 3rd Special Forces Group, which was heavily engaged in the Afghanistan conflict. The Afghan war and the looming showdown with Iraq had dramatically increased the demand for Special Forces teams, and the 3rd Group was reactivating some teams eliminated in the 1990s because of a shortage of manpower. In January 2003, Jim was given command of one of the new teams.

"Congratulations, Capt. Gant. You are the first member of ODA 316," the battalion commander, who was in charge of more than a dozen ODAs, told him. The team was scheduled to deploy to Afghanistan in April with the rest of the 1st Battalion, 3rd Special Forces Group.

"I knew I had a lot of work to do," Jim recalled. Because the team was starting from scratch, it had to beg and borrow and steal everything from existing units—including gear and personnel. What it got were hand-me-downs and cast-offs, and some inexperienced new Green Berets. "We had no team room, no equipment, nothing. And we were hitting the box in ninety days," Jim said. The "box" was Afghanistan.

Two days later, Tony Siriwardene, a friend of Jim's from Arabic-language school, walked into the company area. Having learned Jim was the team leader, Tony signed on right away. A staff sergeant, Tony had just graduated from the Special Forces qualification course and had never deployed before. Next came Dan, who had lived all over the world but also never deployed to a combat zone. The third brand-new Green Beret on the team was Scott, a muscular farm boy who was a communications sergeant. Two other soldiers were assigned to Jim after other teams released them: Sgt. 1st Class Chuck Burroughs, also trained in communications, and Staff Sgt. Brian Macey, whose specialty was weapons. So when Jim got word that he was about to get another apparent reject soldier, he balked. "Quit giving me everyone's fucking trash!" he complained to the company sergeant major. Outside, he got into a fight with the soldier's team sergeant, who was offended by the accusation. Still, just hours before they left to deploy, he got one more seeming outcast: Sgt. 1st Class Mark Read, a boisterous medic from the backwoods of Georgia who had just been kicked off the HALO team, an ODA specially trained in high-altitude low-opening parachute jumps. An ODA is supposed to have twelve men, but Jim's team would deploy to Afghanistan with only seven.

Training was an urgent priority for the new team, given the short amount of time before they left for the war zone. But Jim had a problem: the team had no weapons. So he obtained "rubber duckies"—nonfiring

M16s made of rubber—and used engineer tape to mark off some "rooms" for training in close-quarters battle.

Other equipment was also missing or in poor condition. The team received broken radios, Global Positioning System devices that did not work, malfunctioning machine guns, and outmoded night vision goggles. Team members had to buy their own chest racks for carrying ammunition and other supplies. In a particularly egregious logistical shortcoming, the military failed to give the team tourniquets or blood-clotting agents, so Dan bought them with his own money. They also had a mishmash of maps—the old Russian ones were accurate but illegible, while the American maps had the wrong grid coordinates.

In April 2003, the seven members of ODA 316 flew to Bagram Air Field and soon afterward boarded a Chinook helicopter for Asadabad. They touched down at a landing zone near the banks of the Konar River and dragged their bags into a tiny, dusty compound. The Asadabad camp, known as a "mission support site," was austere. Jim and his men lived in a single large room with mud walls and dirt floors covered with straw mats. Another Special Forces team and command element were also stationed at the camp, along with a small CIA contingent and their Afghan partner force, known as Counterterrorism Pursuit Teams (CTPT). Jim and the head of the CIA paramilitary unit in Asadabad sat down over a beer one night and figured out how they could cooperate and thereby skirt operational restrictions each side faced. The CIA could go after terrorist targets, for example, but not after insurgents who were laying bombs in the road outside the base. So when the CIA had information on a target they could not pursue, they had Jim do it for them. But for all intents and purposes, the team was on its own, with $30,000 a month in funds and very little guidance from above.

Immediately Jim conducted a short train-up in Asadabad and then launched into an aggressive pace of operations. With almost no information on where they were or who was who, they began scraping together any intelligence they could get from Afghans on the ground. Their first mission, eight days after the team landed in-country, involved a six-mile patrol up steep mountain paths in the freezing cold and wind to locate a village, Adwal, that was of interest to commanders. On their next combat foray, they were ambushed near the village of Pashad by two enemy fighters firing rocket-propelled grenades. Together, the missions set the tone for the deployment: the terrain and people would make this a tough fight—harder than Jim had expected. In their first ninety days operating in Asadabad, they spent eighty-nine in missions outside the wire of the camp. Their objective was making contact, either with the enemy or with friendly Afghans. Certain rules applied to every mission: Jim always rode in the lead vehicle, the patrol always carried triple the required ammunition, and his men were authorized to shoot whenever they felt necessary—these became known on the team as "Gantisms." Jim kept telling his men they would have a day to rest. It never came. "You can rest when you die" became his motto—written on a sticker on the back of his helmet.

With so few men on his team, Jim had to make sure every body counted, and he badly needed another gunner—one who would kill without hesitation. The least likely choice was Dan. Unlike the other team members, Dan hadn't gone to Afghanistan to hunt insurgents. A hardworking and dedicated medic from the northern California coast, Dan once asked to stop a patrol to help injured Afghans by the side of the road. He spent hours treating the parade of Afghans who showed up at the base—farmers who had blown up their feet stepping on old land mines, or children burned by cooking oil. But something told Jim to test out Dan on his vehicle's M240 machine gun.

At first Dan was apprehensive about the responsibility. Not for long. After a few days with Dan up on the gun, the team rolled in from a mission and Jim jumped on the top of the Humvee to pull the M240 off for cleaning. Dan stopped him.

"What the fuck?" Jim asked.

"Jim, that's my gun," Dan said point-blank.

Jim looked at Dan and backed off. He had his gunner.

From then on, when Dan wasn't treating patients he spent most of his time on the camp cleaning and zeroing his machine gun, and upgrading the turret. He figured out a way to get more than 800 rounds of 7.62 mm ammunition linked and ready to go in the gun, with more than 3,500 rounds in the turret and another 4,000 rounds in the Humvee. The total of more than 8,000 rounds far exceeded the roughly 1,800 carried by other units—following the "Gantism" on maximizing ammunition. Other members of the team proved themselves beyond Gant's initial expectation—medic Mark Read turned out to be highly aggressive in combat and communications sergeant Chuck Burroughs mastered every job on the team and was the best shot of all the men.

In between missions, the team trained hard by day and at night would drink the cheap Kentucky whisky Early Times. At the Asadabad camp, rain poured in through a large hole in the collapsed roof, creating a stream of water that flowed through the sleeping quarters and crude operations center where Jim planned missions and held team meetings. "We sweated our balls off. We didn't sleep for three months. Our house was a fucking swamp," Mark recalled. Sanitation was poor and at first they got horribly sick with vomiting and diarrhea from eating Afghan food. "We all took turns going down for three or four days," Tony said. The camp had no Internet and the only phone was a satellite phone that they used to call home once a week.

"All we had was each other," Jim said.

As the team bonded and fine-tuned their combat skills, Jim realized he needed to somehow give his men a more meaningful mission than "kill or capture." So he gathered them together one night in the operations center and told them the story of the three hundred Spartan warriors at the epic battle of Thermopylae in the fifth century BCE. Shoulder to shoulder in a narrow gorge, the Spartans held out to the last man against an onslaught by thousands of Persians. Jim's expectation, he said, was for his men to fight and show bravery together with the same warrior ethos that defined the Spartans. He showed them the Greek letter lambda he had had tattooed on his left forearm shortly before they left Fort Bragg. The lambda stood for Laconia, the Spartan homeland in ancient Greece, and appeared on the shields that Spartan warriors wielded in an impenetrable phalanx. Together they were invincible, he told them. "That was the mission, and they all accepted that. It was what we all needed," he said. The next day, Jim turned to his men as they put on their body armor and readied to roll out the gate and hit another target. "A Spartan never asks how many, just where are they," he said. It became a way of life for them.

The team later adopted the call sign "Spartan." Several members got lambda tattoos identical to Jim's on their left forearms—even Dan, the most skeptical of the group, who got his last. They began to wear unauthorized shoulder patches with the lambda. Tony and Mark made the first set of woven cloth patches, which were round and three inches in diameter with a black lambda on a gray background. The patches had to be earned. The significance of the patches grew over time, both for Jim and his men, many of whom coveted them as a symbol of belonging to a unique group of warriors.

Jim's leadership, his philosophy, and the team's rigorous training created what Tony called "combat chemistry." "What it comes down to is, we wanted to try to win the war," he said. "People like us can identify other people like us. It's just energy," he said. "We were better trained than teams that had been together three or four years. I am the first guy going in the door, I will take the round, as long as we win the battle," he said. "In Konar, we were in our zone."

What Jim didn't fully realize until years later, though, was that by proving they were aggressive fighters, he and his men were also taking the first and perhaps single most important step toward gaining entry to the warring culture of Afghanistan's Pashtuns. Indeed, in many ways Jim's assimilation into the culture started naturally, because the stark values and dichotomies of the Pashtun people so closely mirrored his own. He would offer them food with one hand and a knife with the other; they would invite him to lay down his arms and eat with them, then attack him on the road out.

On one April mission in Konar, Jim was told there was trouble in the village of Mangwel, a twenty-mile drive southeast from Asadabad down a rough dirt road. The team rolled warily into the village, their vehicles bristling with guns, only to be greeted by a mild-mannered man with soft brown eyes who introduced himself using some broken English as Dr. Akbar. Dr. Akbar lived next to the road and, unbeknownst to Jim, had been selected by the powerful and fierce old tribal chief, Malik Noor Afzhal, as the emissary to the first American soldiers to set foot in the village since the war had begun eighteen months before. In fact, through word of mouth, the Mohmand tribesmen in Mangwel had learned the Americans were headed their way and had enough time to squirrel away their weapons in the nearby hillside. Everyone in the village, including Dr. Akbar, was afraid. But Noor Afzhal decided the doctor's house was the safest spot to meet with them.

"Will you drink tea?" Dr. Akbar asked Jim, who agreed and followed him into the house.

Nearby, Noor Afzhal was determined to show these foreign soldiers the pride and hospitality of his Pashtun people. "I am not one bit scared," he told his younger brother. "The only person I am scared of is up there," he said, pointing toward the sky. So he put on his best silver turban and a black vest. Then, a prayer on his lips, he strode into Dr. Akbar's greeting room, his head held high.

"*Salaam aleikum*, peace be upon you," Noor Afzhal said to Jim, placing his hand on his heart.

"And peace be upon you," Jim answered. He immediately noticed the aura of respect the broad-chested, *malik*, or chief, commanded.

Noor Afzhal sat down cross-legged on a red carpet facing the bearded American.

"I am Commander Jim, and these are my men," Jim said.

After an exchange of pleasantries and another serving of tea, Noor Afzhal turned to Jim and looked him in the eye.

"Why are you Americans here in Afghanistan?" he asked.

"Our country was attacked. We came here to fight the Taliban and others responsible for this," Jim replied. Then he pulled out a laptop and showed Noor Afzhal video footage of the Twin Towers crumbling to the ground on September 11. "My men and I are warriors. But we are not here to fight you," Jim said. "We want to help you."

Noor Afzhal was visibly moved. He was silent for a moment, and took a sip of tea. Then he spoke again to the young American. "If you can come all the way to Afghanistan from the United States to help us, then why should I not help you?" Noor Afzhal said. "We don't want the Taliban here."

In a conversation that lasted about two hours, the tribal chief told Jim of his people's bitterness over Taliban rule. After the Taliban took power in Afghanistan in 1994 and extended their control to Konar Province in 1996, the movement's strict form of sharia (Islamic law) and justice began to clash with the less dogmatic ways of the Pashtuns. The Taliban's religious enforcers forbade villagers to play music, listen to the radio, or smoke. They kept a list of who was not going to the mosque five times a day to pray, and carried out draconian punishments such as cutting off the hands of thieves. Over time, they grew increasingly corrupt and predatory. Not long ago, Noor Afzhal recounted, the Taliban caught a group of local thieves who were breaking into houses in Mangwel and nearby villages and stealing money, jewelry, and other valuables. Taliban officials were about to cut the men's hands off, Noor Afzhal said, until he and other tribal leaders bribed the officials with 100,000 Pakistani rupees and got the men released.

"After about two years of the Taliban, we got tired of them," Noor Afzhal told Jim. "They forced everyone to follow their strict rules, and if you resisted, they would torture you and put you in jail," he said.

Jim nodded. He was listening to Noor Afzhal, but what struck him more than the tribal chief's stories was his confident tone and the way every other Afghan in the room was hanging on his words.

"When the Americans came to Afghanistan, the Taliban came here in the middle of the night and knocked on everyone's door and said, 'Be ready, we will make holy war tomorrow,' " Noor Afzhal said. "But in the morning they ran off."

"The cowards!" Jim laughed.

A slight smile crossed Noor Afzhal's lips.

"At first, everyone was afraid you Americans would be just like the Russians," he said. "There were a lot of rumors that because you were infidels you would destroy our country and kill people," he said. "But you didn't burn villages like the Russians did."

"And we will not, I give you my word," Jim said.

"But until today, not one American has come to speak to me," Noor Afzhal said, his words trailing off in an implicit question.

"I am not like other Americans," Jim answered. "We heard there was some kind of trouble here."

Noor Afzhal stroked his beard, surprised at how much the officer seemed to know. "You are right," he said. He hesitated for a moment. The Taliban was one thing—but this was a tribal matter. Still, he went on.

"For many years, we have had a blood feud with some highland people near here," Noor Afzhal said. "They killed three of our people, and we killed three of theirs. Now they are cutting wood on our land. I have a paper signed by the king of Afghanistan showing that it is our land. I can show it to you."

"I believe you," Jim said.

"We are preparing to take the land back," Noor Afzhal said. "The problem is, I do not have enough fighters. I have eight," he said.

"No," Jim said without missing a beat. "You have sixteen," he said, referring to his eight U.S. soldiers.

Noor Afzhal's eyes lit up. As the conversation progressed, he bent over toward Jim and told him quietly that he had eighty fighters within the village and could rally eight hundred from his Mohmand tribe. As the tribal elder's trust in Jim grew, he was willing to hint at how many weapons he had and even to boast about his power.

In return, Jim agreed to throw the full force of his team and the bombs and rockets he had at his disposal behind the tribe.

In a gesture of Pashtun generosity, Noor Afzhal insisted on serving the Americans dinner, but lacked enough food. So families all around the village brought dishes for their guests. Noor Afzhal asked several tribesmen to take their rifles and guard the house as they ate. In some other villages, Jim would have confiscated the weapons for security. But that day in Mangwel something about Noor Afzhal made him feel protected, and Jim was beginning to understand the importance of firearms to Pashtun male honor. Jim made no move to take away the tribesmen's rifles, telling them he had a gun in his house at home, too. "I trusted him right at the start because he was brave and he stayed with us. He trusted us, so I trusted him," recalled Noor Afzhal. From that night on, Jim's team and the tribesmen would be allies.

In the weeks and months that followed, Jim and Noor Afzhal traded stories about fighting the Taliban and Russians, and where the best ambush locations were. Jim learned much from Noor Afzhal about the Pashtun codes of honor and revenge. He backed up Noor Afzhal in the land dispute between the lowland clan of the Mohmand tribe in Mangwel and the highland clan of the Sarkan valley. Soon afterward, the highlanders backed down and the dispute was resolved peacefully. The team helped refurbish a well, build a clinic and girls' school, and provide children with clothing and school supplies. Dan, Tony, Mark, and other team members were practically adopted by village families. Noor Afzhal took Jim and his men to the ruins of his ancestral home high in the valley behind Mangwel, and they shot each other's rifles there. They began staying overnight at the malik's house. It was the first true relationship Jim had with an Afghan tribesman.

Jim and his team at times patrolled together with the Afghans. He began to learn Pashto. One day his interpreter, Khalid Dost, gave him a set of Afghan clothes, and he began to wear them around his camp. Khalid had worked with Special Forces units since 2002, but he was struck by

Jim's rapport with the tribe. "Jim was the only team leader who really meant business," he said.

One sweltering day Jim and his men stopped on patrol near the banks of the Konar River. Dan, the gunner, pulled guard at the vehicles. Leaving their rifles with local Afghans in a gesture of extreme trust, Jim, Tony, Mark, and Scott stripped down, wrapped head scarves around their waists, and jumped into the flowing water for a swim. As they laughed and splashed in the cool water, Jim was overwhelmed by the feeling that he belonged in Konar.

Elders from other tribes heard of Jim and also began to approach him. A group of elders from the Korengal valley, where the bomb financier Dr. Naimetullah was from, had met Jim the day after the raid when they pleaded with him to return opium the team had confiscated from the compound. With all the males in the family either killed or captured, the lucrative opium was all that remained to support the women and children in the family, they said. Jim weighed the risks—he knew the opium might make it into insurgent hands—but then agreed to the elders' request. He believed it was more important and ultimately safer for his team for him to show respect for the elders' need to protect the women of the family—the central tenet of Pashtunwali known as namoos.

Those same elders in June 2003 asked Jim to travel deep into the Korengal, where so many U.S. troops would later meet their fate, to help them resolve a tribal dispute over wood. They invited his team to have lunch with them and guaranteed their safe passage. Jim and his men had to leave their own vehicles across the river and get into the trucks of the Korengalese to ride up the narrow, twisting dirt road to their village. Jim and Chief Halstead held a jirga with tribal elders, and then they all sat down to a generous lunch of flatbread, seasoned rice, and stewed chicken. But Jim found the tribal dispute too complicated and murky, and no ready solution was found. The elders began arguing and the tone changed. As they left, one of Jim's Korengalese hosts warned him: "We have guaranteed your safety here. But do not come back through the Pech River valley past Watapor," he said.

"I am here for a mission," Jim replied. "I will go wherever the enemy is."

"If you pass Watapor," the elder repeated, "we will attack. As a matter of fact, we may attack you today."

"So," Jim said, "we will fight."

The tribesmen took the Americans back to the footbridge and they departed. Jim decided to move far ahead in the lead vehicle in an "eagle" formation, a tactic that helped give him time and flexibility to react to enemy contact. Suddenly a blast went off near the second vehicle, with shrapnel peppering Tony's arm. After Mark checked him out, they pushed forward. But insurgents opened up on the patrol from the left ridgeline with machine guns, rifles, and rocket-propelled grenades. With rounds impacting nearby, Scott returned fire on the .50-caliber machine gun. Jim pushed his vehicle back into the kill zone, and Dan hammered some suspicious dwellings with the M240 gun. "We slugged it out until they stopped firing," Dan recalled. Dan was proving to be the best gunner Jim ever had. "We never ran out of ammo, and Dan never let 'em breathe, not for a second," Jim recalled. The team headed back to Asadabad.

Constantly attuned to the danger, Jim honed what he called his "sixth sense" and learned to listen to his instincts about when and where the insurgents were likely to strike. One day in mid-July, after returning from a trip to Kabul, he intended to train but was tasked to go on a mission planned by another team. He expressed reservations but was directed to treat it as a "training mission"—a concept that to Jim was an oxymoron. That day, when the patrol was moving slowly and towing a broken-down vehicle, a strike by a massive IED made of antitank mines seriously injured three soldiers on ODA 316, including two assigned to the team after it reached Afghanistan. One of them, Staff Sgt. Luke Murray, a civil affairs soldier, lost a leg. Another, a psychological operations soldier named Jonathan Wines, lost the full use of one of his arms. A third, Staff Sgt. Brian Macey, sustained a traumatic brain injury. They evacuated the wounded. Then back at the camp in Asadabad, Jim began explaining to Sgt. Maj. Robert "Buzz" DeGroff what had happened. A comrade from Desert Storm, Buzz made no effort to hide his dislike for officers in general, but he had observed Jim training and caring for his men. As Jim spoke, Buzz saw him start to break down. Buzz kicked the table that was between them out of the way, wrapped his arms around Jim, and let him cry. "I felt completely responsible," Jim said. He always considered it his highest priority to protect his men, and he blamed himself for not spotting the IED. But he also swore he would avenge the casualties.

In coming weeks, ODA 316 would pursue all four Afghan insurgents involved in the IED strike. The financier of the attack was an Afghan named Kharni Jan Dahd, who owned a mine and factory that produced marble and other stone. Because of that, he was able to purchase explosives. Jim, together with Chief Halstead, who was in charge of gathering intelligence, lured the Afghan, nicknamed "Marble" Jan Dahd, to the Asadabad camp in what was known as a "Trojan horse" raid. As a veteran Special Forces noncommissioned officer, Halstead had played a key role in Jim's early development by acting as a sounding board and serving as the overall ground force commander during key missions, allowing Jim to lead the assault element. Once Kharni Jan Dahd arrived with his bodyguards, Halstead had tea with him while Jim and his team prepared to detain them. In an agreed-upon signal, Halstead picked up a can of Coke, triggering the detention operation, after which Jim, his men, and another Special Forces team left to raid Kharni Jan Dahd's compound. They confiscated explosives and bomb-making materials, and Kharni Jan Dahd was sent to the U.S. military detention center at Guantánamo Bay, Cuba, for three years.

Another insurgent named Norullah, who oversaw the IED operation, was a member of the Afghan Border Police (ABP). He worked under Haji Jan Dahd, the Safi tribal leader and former mujahideen fighter, who was then commander of the ABP in Konar Province. Jim and his men tracked Norullah to the border brigade outpost at Barabat, across the river from Haji Jan Dahd's large house. The police at the outpost denied knowing where Norullah was, irritating Jim. "Do you have any weapons here other than your AKs [AK-47s]?" Jim asked them. They shook their heads. "If I find anything other than AKs, I am blowing this place up," he told them. So Jim and Tony began searching the compound, and in the first room they pulled up the floor and found piles of mortars, rockets, and a heavy machine gun. They demolished the outpost and the munitions in it, setting off a mushroom cloud. At the time, it seemed like an appropriate retribution for the uncooperative police, but years later both Jim and Haji Jan Dahd would, only half jokingly, bemoan the destruction of the weapons. The next day, Jim's team found and detained Norullah.

Meanwhile, Jim was spending more and more time in Mangwel with Noor Afzhal. Noor Afzhal promised to protect Jim as he would his own son, and he was true to his word. The tribal intelligence network had more than once tipped off Jim and his men to danger, or given them critical information to capture a target. Noor Afzhal also faced growing threats, as insurgents accused him of spreading Christianity in the area by associating with Americans. The two men would talk late into the night under the stars on the roof of Noor Afzhal's mud-walled Afghan home, with the smell of smoke rising from wood fires and dogs barking in the village paths below. Jim began to think that no matter how many insurgents he killed or captured or how many weapons he blew up, the only place where he was making a difference in Konar was in Mangwel, and that was because of his relationship with Noor Afzhal and his tribesmen. Jim and the team called Noor Afzhal "Sitting Bull" after the great nineteenth-century Sioux warrior and tribal chief. Jim felt that, like Sitting Bull, Noor Afzhal was a proven fighter, but he had been chosen to lead primarily due to his strength of character, charisma, and concern for the safety of his people. From then on, Noor Afzhal would be known to many Americans and Afghans alike as Sitting Bull.

In August 2003, Jim and his team gave Noor Afzhal a 12-gauge shotgun with his name engraved on it and ammunition. Then Jim said goodbye to his Afghan father, promising to return one day.

OVER THE NEXT SEVERAL years, the fighting in Afghanistan escalated sharply, taking a toll on the men of ODA 316 as well as Noor Afzhal and his tribe in Mangwel.

In October 2003, ODA 316 flew home—but not for long. Jim got a rare opportunity to stay on as the leader of the team for another rotation into Afghanistan in January 2004, this time to the southern province of Helmand, a Taliban stronghold and opium-growing area where virtually no other U.S. forces were operating. There, ODA 316 bonded even more closely. One chilly February day a new weapons sergeant arrived for the team. ODA 316 was so tightly knit that newcomers had a hard time winning acceptance, and the team had already run off a string of them. Sgt. Christopher M. Falkel was different. The twenty-one-year-old from Littleton, Colorado, had a baby face and sometimes looked even younger than he was. With a slight build, he appeared no more than a boy next to others on the team. Still, he quickly won over the team's veterans with his hard work and terrific sense of humor. Jim, wanting to bring Chris up to speed with the rest of his battle-hardened men, was deliberately tough on Chris. He made him ride in the back of his Humvee for the first month. Finally one night Tony came to Jim.

"Chris is good to go," Tony said. "He's ready for a team patch."

Two days later, Jim placed a lambda patch on Chris's shoulder, a ritual he would repeat many times in coming years with men he led in combat. Soon afterward, Chris was made a gunner, along with Dan and Scott.

In 2004, the team again returned to Fort Bragg. Jim was relieved to have brought all his men home alive. He handed over command of ODA 316 to another captain. In a final gathering with his teammates and their families at his home in Fayetteville, he reflected on the Afghanistan mission as he saw it at the time.

"We will never win in Afghanistan," he told the team. "But know—now and always—that does not matter. That is an irrelevant fact. It gives us a place to go and fight, it gives us a place to go and be warriors. That's it.

"Think back to who you were two years ago, and some of the fears you felt before you were seasoned and proved yourself under fire. Never forget that person," he went on.

"My biggest battles are in front of me . . . as it is with you. Your next great triumph will be celebrated without my presence. And your first huge defeat, I will not be there, either."

If any of them were wounded, he said, he would stay close to them. And if any died, he promised to look after their families "until I cross the river of death and we are together again."

He closed with a poem from Steven Pressfield's novel *The Virtues of War* about the sarissa, the long, tough wooden spear that Alexander's Macedonian army used so effectively in their phalanx formations.

> *The sarissas' song is a sad song.*
> *He pipes it soft and low.*
> *I would ply a gentler trade, says he.*
> *But war is all I know.*

In the summer of 2004, Jim was selected to serve for the next two years on a special team conducting a classified mission. Sitting at the gate waiting for a flight in the Denver airport a year later in August 2005, he got a call from Afghanistan. It was Buzz, his old sergeant major. His voice was tense. ODA 316 was pinned down by Taliban insurgents in a remote valley in the Mari Ghar region of Zabul Province in a series of gunfights that had already raged over two days.

More than seven thousand miles away, high in the dusty Mari Ghar valley, Dan was crouched next to his vehicle pumping out rounds from an M249 light machine gun toward insurgents ensconced behind large granite boulders several hundred feet above. Dan, Scott, Tony, Chris, and other members of ODA 316 were fighting their way through the fifth heavy ambush of a two-day patrol. Their higher command had earlier failed to resupply them with machine-gun ammunition. Worse, it had inexplicably ordered them to circle back through a hostile village that had already been cleared. That gave the Taliban insurgents ahead an extra hour to prepare the biggest attack of all in a U-shaped ambush location that Afghan mujahideen used against Russian troops in the 1980s.

*Why the fuck did you have us turn around?* Dan thought when he heard the order.

Soon afterward, the team and the Afghan army soldiers with them were hit by a massive barrage of machine-gun fire and rocket-propelled grenades from the high ground. The Afghan soldiers stopped their trucks on the dirt road in front of the three U.S. Humvees and ran off, blocking the Americans in the kill zone.

Bullets were hitting the vehicles and the ground all around them, and set a bag of ammunition on fire in the lead vehicle. Dan grabbed one of the Afghan soldiers cowering behind him with a rocket-propelled grenade and told him to fire it. Air Force A-10 jets flew over and lit up the insurgent positions with Gatling guns. But the boulders were impervious, and the enemy onslaught continued.

Then Dan heard a call over the radio. It was the team sergeant, Master Sgt. Al Lapene.

"Chris is spark," Al said.

*Shit!* Dan thought. His head reeled. He felt as though he'd been punched in the stomach. *Spark* was the code word for "dead."

"Is there anything I can do?" Dan radioed back.

"No," Al said. Chris had been shot in the head in his gunner's turret and killed instantly.

Blood was everywhere in Chris's vehicle. Al was having a hard time getting Chris's body down from the turret. Scott was the only soldier firing a heavy weapon at the time. Meanwhile, the enemy fire intensified. Dan saw the Humvee twenty yards ahead of his hit by a burst of bullets that struck the gunner's ammunition, setting it on fire. The gunner, Sgt.

1st Class Cliff Roundtree, had to scramble out of the turret down behind his vehicle, temporarily leaving his gun to escape the flames.

"We have to fucking move out of this kill zone!" Dan yelled to Cliff, taking cover behind the hood of his Humvee as he returned fire.

Al got in the turret where Chris had been, and the team pushed out of the kill zone toward a small village. They went into the village shooting but found it was empty, and took cover there, as darkness was arriving soon. The soldiers set up a hasty security perimeter and began furiously reloading ammunition. After they took up position, Dan asked Al: "Where is Chris?"

"He's in the back of the vehicle, under a poncho," Al said.

Dan winced. Chris was the youngest member of the team. They were all close to him.

*I'll take care of him*, Dan thought. *I don't want anyone else looking at him.* Dan steeled himself, took a black canvas body bag out of his vehicle, and walked over to the vehicle where Chris was lying. He slowly lifted the poncho off Chris, exposing his blood-smeared, pale face.

"God, brother," Dan whispered, tears welling up in his eyes. He placed his hand softly on Chris's cheek. He checked for any signs of life, just to be sure. Then he looked at Chris's wound. He was filled with an indescribable sadness. He had seen many dead bodies before, but never the lifeless form of a person he was so close to. With others he was able to distance himself, and depersonalize them. With Chris, that was impossible.

*This is unreal. This can't be*, he thought.

After a minute he pulled himself away, mechanically going through the steps of preserving his comrade's body. He knew there was more fighting ahead, and he tried to stifle his tears. He had to stay strong for the team, and mentally prepare for the attacks to come. Dan spread out the body bag, and carefully lifted Chris onto it. Then he zipped up the bag, tears streaming down his face. He secured the body bag in the backseat of the vehicle. No helicopter was coming to evacuate Chris. They had to drive him out.

The ordeal for ODA 316 was not yet over. After a tense night on guard at the village, they hit another ambush, taking heavy fire from insurgents in nearby orchards who shot at them with rocket-propelled grenades and machine guns. The team fired back with rifles, machine guns, and an antitank weapon. The vehicle Dan was driving lost all brake and steering fluid after a round severed a line, and the Humvee ahead of his was disabled. Dan had to ram it repeatedly to get it out of the kill zone. Finally, the patrol limped back to their firebase. In fifty-six hours of battle, the team had proved its mettle like never before. For their bravery, Scott and Tony earned Silver Star medals and Dan a Bronze Star with V for valor. Chris was given a Silver Star posthumously. But the team didn't recover from the loss of Chris.

"The team was never the same after that," Dan said. "It couldn't be."

Anger simmered over what Dan and the rest of the men perceived as faulty decisions by their command, including a lack of resupply. Had it not been for Jim's training, Dan said, the casualties might have been far worse.

"The only reason we didn't go black [dangerously low or empty] on ammo was it was our SOP [standard operating procedure] to carry massive amounts of ammunition," Dan said. "It was the fight Jim always talked about, and it came true. Jim was there in spirit."

Jim was proud of his old team but devastated by Chris's death and tormented that he had not been fighting alongside his men as they faced their biggest battle. After all, he had forged and molded the team—Chris died with a Spartan lambda tattoo on his left arm. As the Afghan war dragged on and the casualties increased, Jim began racking his brain for a better strategy. He knew that Taliban insurgents were making gains nationwide, including in Konar Province, the home of Noor Afzhal and his tribe in Mangwel. Would Noor Afzhal survive and hold the line against the Taliban until Jim returned? He wasn't sure.

# CHAPTER 12

MALIK NOOR AFZHAL PUT on the silver-gray turban reserved for important gatherings, picked up his best walking stick, and shut the heavy wooden door to his mud-walled home. Then he stepped into an old yellow taxi with his son Azmat and rode down a bumpy dirt path out of the village of Mangwel in Konar Province south toward the city of Jalalabad.

The seventy-five-year-old elder of the Mohmand tribe looked out the window as the car wound up a steep hillside above the Konar River, which shimmered as it coursed past fields of ripening wheat and corn. Mountains green with vegetation rose abruptly on either side of the narrow river valley.

The car descended again to the riverbank and slowed as it met with a herd of sheep coming up the road. A skinny boy wearing a skullcap and sandals shooed the animals out of the way with a stick. A woman covered from head to foot in a flowing blue burkha walked by, followed by a girl in an embroidered violet dress carrying a jug of water on her head.

They continued for miles through the territory of the Mohmands, a proud and unified Pashtun tribe whose lands extended far south into Nangarhar Province and east into Pakistan. Passing through one Mohmand farming village after another, Noor Afzhal spotted familiar faces of fellow tribal elders. But this day he did not stop to chat and drink green tea.

The car crossed over the river on a partially collapsed concrete bridge called Zire Baba that dipped in the middle, and headed down the main road to Jalalabad. Carts piled with ripe watermelons lined the way, and boys waved down travelers, seeking contributions for their village mosques. An hour later fields gave way to market towns and the thronged city of Jalalabad came into view.

It was August 22, 2010, and Noor Afzhal was ignoring the warnings of other maliks and elders by making the dangerous trip to the main U.S. base in Jalalabad to meet with an American military officer. Taliban insurgents had more than doubled their attacks in Konar in the past year and were moving with greater impunity from the valleys where they took sanctuary through lowland villages such as Mangwel. The Taliban commanders had little regard for tribal leaders and regularly vowed to kill them and their people for cooperating with the Americans. Noor Afzhal and his sons often received death threats. That day, he worried that the call he had received from an Afghan contact in Jalalabad about the meeting might have been a trap laid by his enemies, and braced for an ambush along the road. Would the American officer be there? Noor Afzhal could hardly believe he would be. But he would not turn back. He considered the meeting a matter of personal honor. The American he remembered as a young Army Special Forces captain named Jim, whom he had protected as a son, had promised seven years earlier to return to Afghanistan to see him. Noor Afzhal had to take the risk.

At the base in Jalalabad, Jim pulled on a light blue Afghan tunic, or *khamis*, loose pants, and a cylindrical white cap, and rehearsed again in Pashto what he would say to Noor Afzhal. For Jim, the reunion was one of the most anticipated moments of his life. He had worked for years to return to Konar and make good on his pledge to see the tribal leader again, haunted by the prospect that Noor Afzhal would die before he could reach him. Now that fear was gone, and he felt a deep sense of relief and gratitude that bordered on the spiritual. It was as though the moment was a gift from God, or the gods, for all his sacrifices—allowing him to keep the last promise he would ever have to make as a warrior.

But Jim had another purpose. He wanted to ask Noor Afzhal, one of the few people in the world he trusted, for his wisdom on the tribal strategy. Would it work? Would it help the people? Jim believed it would, but at this decisive moment he wanted to hear from Noor Afzhal. Another critical question was whether Noor Afzhal would support testing the strategy in Mangwel. Nagging at Jim, too, despite the high-level support from Petraeus on down, was the question of American staying power. Would the U.S. military enlist the tribes, only to abandon them before the war was won? He wrestled one last time with whether to go ahead, and decided that Noor Afzhal would make the decision.

Jim waited outside a secret meeting room at Jalalabad Air Field (JAF) as a car pulled up and Noor Afzhal stepped out. Jim approached and kissed Noor Afzhal's hand, and the two men embraced for a long time.

"I have told people that this day, this moment, after I saw you again, I could die a happy man," Jim said. "I told you I would see you again. Do you remember?" he asked.

"*Ho.* Yes," Noor Afzhal replied, beaming. "What took you so long?"

Jim laughed and grasped the older man's hand more tightly. They spoke for a minute about each other's families, and Noor Afzhal asked about Jim's father, who years before had sent him a knife and asked him to look after his son. Noor Afzhal had promised the father that not one hair on Jim's head would be harmed while he was with the tribe.

"You are my father," Jim said.

Noor Afzhal looked at him, his expression both kind and wise.

"From the bottom of my heart, you are my son," he replied. "And you have returned."

Holding hands, as is the custom among Afghan men, they walked to the meeting room and sat down. Noor Afzhal's son Azmat put a *tsera*, a garland of red paper flowers, a traditional gift for pilgrims returning from Mecca, around Jim's neck.

As they drank tea, they spoke of their years apart. Jim studied Noor Afzhal and saw that his health was weak. Although the elder man stood tall, as he always had, he walked with difficulty. In fact, Noor Afzhal had a serious untreated heart ailment, Jim later learned. Noor Afzhal, too, saw how the intervening years of warfare had taken a heavy toll on Jim, who appeared far older than the cocky Green Beret who had rolled into his village seven years before. At the same time, Noor Afzhal noticed something more serious about Jim.

Jim showed Noor Afzhal a copy of "One Tribe at a Time," with photographs of the tribe and Jim's Special Forces team in Mangwel. Then he spread out a large map of eastern Afghanistan and began to explain his plan for raising tribal forces.

"Mohmand territory is here, and this is where Safi territory starts," Jim said, tracing his finger along the map.

"Yes, the Safi lands are across the river and from Asadabad to the south and into the Pech valley," Noor Afzhal said. Like most rural Afghans, he had rarely seen a map, but looked on with interest as Jim described the locations of different tribes.

"Here is the Pakistani border," Jim said.

Then he outlined in broad strokes his strategy to train, arm, and pay tribesmen to defend their villages against the Taliban. He would begin in Konar and from there expand south. As security increased, so would funds and resources for development. Jim turned to Noor Afzhal.

"Do you think this plan can succeed?"

Running his hand through his long white beard, Noor Afzhal was silent for a minute.

"The tribes have always fought to protect their territory, and they will again. But the Taliban is a problem," Noor Afzhal said.

"How strong is the Taliban in your area?" Jim asked.

"They are everywhere now," Noor Afzhal said soberly. The insurgents had steadily gained strength in his area in recent years, he said, and had infiltrated the local police and district government. They conducted limited but lethal attacks—mainly aimed at U.S. troops—and moved through the tribal territory unhindered. The tribe had a tacit understanding with the Taliban: they could transit the Mohmand area if they kept the fighting out of there. U.S. military units had further weakened the tribe, he explained, by repeatedly sweeping through Mangwel and other villages and taking away their weapons—a shameful event for the Pashtun men.

"We can do something about that," Jim said. "I would like to start by bringing a team of Americans to live with your people in Mangwel, and give your men guns and training. What do you think?"

Noor Afzhal looked down, deep in thought. His people had suffered greatly in recent decades. During the Soviet occupation of the 1980s, Russian aircraft had bombed Mangwel, causing almost the entire population of the village to flee to Pakistan. The Soviet withdrawal was followed in the early 1990s by the lawlessness of mujahideen rule. After that had come the zealous and oppressive Taliban. Noor Afzhal did not want a Taliban comeback. But to ask the tribe to invite Americans into their village? That would make Mangwel a Taliban target. Some of the tribal elders were sure to oppose it, including ones who were already maneuvering behind his back to succeed him.

"I must go back and talk with my people," Noor Afzhal said.

"Of course, I understand," Jim said. The tribes had their own traditional form of democracy. All important decisions had to be deliberated in a gathering of elders, or jirga. Jim believed that Noor Afzhal had the stature to sway a final decision, but he was not sure.

"There is one more thing I must tell you before you speak with the tribe," Jim said. "The U.S. military will not be here forever. Most American troops will begin to leave next year. My commanders tell me the Special Forces will stay for five or ten years, not more. But this is an opportunity."

They continued speaking for an hour, and then it came time for Noor Afzhal to leave. But before the elder man left, Jim felt the need to unburden himself to the Afghan father who had seen a side of him—the essence of who he was as a warrior—that his own father never had.

"Father, I have killed enough," Jim said simply. "I am here to help."

Noor Afzhal looked into Jim's eyes. "You are not just a fighter anymore," he told Jim. "You are *nangyalee*."

In a verse by the seventeenth-century Pashtun poet Khushal Khan Khattak, *nangyalee* refers to a brave man who also has honor and who never gives up. "A brave man has only two options in the world, to fight to the death or secure victory," went the famous line.

But Noor Afzhal explained the term to Jim as meaning a champion who is both brutal and compassionate. "*Nangyalee* is a warrior who rides a white horse, and when he sees someone who cannot protect themselves, he rides there with his men and fights for them," he said. Jim was deeply moved.

NOOR AFZHAL RETURNED TO Mangwel, and the next day called a meeting of elders at his house. In strode Malik Mir Salaam, one of the tallest men in the village, his very long gray beard a sign of seniority. Following him were two other elders, Malik Qayum and Malik Angur. They sat cross-legged in a circle and were joined by Noor Afzhal's younger brother, Dost Mohammed Khan, an educator who had spent years living away from Mangwel in the Afghan capital, Kabul, and other provinces. Noor Afzhal's sons Asif and Azmat sat at one end of the greeting room, while other young male relatives served tea.

Noor Afzhal opened with an appeal.

"You have known me for many years. I have always made wise decisions for the tribe. I have always been fair and just, even when it came to punishing my own family," he said.

The other maliks in the room murmured and nodded.

"Since the first Americans came to Mangwel in 2003, I have sided with them. There have been problems, I know, but our tribe has stayed united. Every day, though, I see the people becoming more afraid of the Taliban. In Jalalabad, I learned of an opportunity that will make us stronger." Noor Afzhal outlined Jim's plan. "Now I will listen to you," he said.

Immediately Malik Mir Salaam spoke up in dissent.

"If the Americans come to Mangwel, the Taliban will fight them here. Our village will be hit by rockets and mortars!" Mir Salaam said in a booming voice. "How can you claim this will help our people?"

"I agree, the Taliban will retaliate," said Noor Afzhal's brother, the most educated and urbanized of the group. "And surely we can't do this without a piece of paper from the government?" he said.

"Yes, the Taliban may try to attack us," Noor Afzhal agreed. "But we have always protected our village and we always will. We will have our guns back, and the American soldiers will have machine guns and mortars," he said. "Men in the village will have jobs with pay. They will not have to work in Jalalabad or Kabul far from their families."

"But how can we trust anything the Americans tell us? Didn't they help our fighters kick out the Russians and then leave?" Dost Mohammed said.

At that, Noor Afzhal's face grew flushed with anger. "Trust? All of you remember Commander Jim. He built our school and clinic. He is the only American we can trust, the only one to keep his promises to us!" he said. "We will welcome him to live in Mangwel!"

The room fell silent. No one said another word.

Just as he had when he first met Jim in 2003, Noor Afzhal followed his instincts and used the force of his will to bring the tribe along with him.

In coming days, Jim received word from Noor Afzhal that the people of Mangwel were ready and willing to supply fighters for a tribal force. Jim was excited, and traveled to Mangwel once more to make sure the tribe accepted the plan for a U.S. team to embed in the village. It was a rare opening, the one Jim was looking for.

But once again, the U.S. military chain of command had other plans.

# CHAPTER 13

JIM'S VOICE SOUNDED TENSE over the satellite phone line from the U.S. base in Jalalabad. I got up from my desk in the noisy fifth-floor newsroom of the *Washington Post* and walked to an empty conference room where I could speak privately. It was September 2010, less than a month after Jim's euphoric reunion with Noor Afzhal.

"I just got word that the battlespace owner is opposed to sending a team to live with the tribe in Mangwel. It turns out the conventional Army guys now get to decide where Special Forces teams operate," he said.

In Afghanistan, the U.S. military had one chain of command for Special Operations Forces—the elite troops such as Green Berets and Navy SEALs—and another for conventional forces such as Army and Marine Corps infantry units. The conventional force commanders were the "battlespace owners," they controlled the ground. Afghanistan was divided and subdivided into geographical sectors known as "battlespace," and commanders at different levels approved where military units were assigned and how they operated within those sectors. They tended to be very territorial.

"The battlespace owners want to send us up north to a place called Marawara," Jim went on.

"Marawara was where the 101st Airborne took heavy casualties this summer," I said as I paced back and forth across the room. Soldiers from the 101st Airborne Division had fought for several days in June 2010 to secure the Marawara valley, one of the most contested parts of Konar Province. But when they left two weeks later the valley quickly fell back into insurgent hands.

"Launching this in Marawara would be a disaster," Jim said. "The population there is divided between several different tribes. The insurgents are hunkered down and intimidating the people. The local leaders won't even meet with us without preconditions, such as demanding they be given heavy weapons and vehicles. We would have to fight our way in." Marawara stood in sharp contrast to Mangwel in Khas Kunar District, where the Mohmand tribe was relatively unified and anti-Taliban, and a degree of trust existed because of Jim's prior ties with the village.

"Starting in Marawara makes no sense," I agreed. "It risks short-circuiting the whole effort. That can't happen."

We knew there was only one option. Jim would have to somehow persuade his higher-ups to change their minds. I would help. I believed in his mission, and I didn't want to see it or him put at greater risk by some ill-founded command decision.

In the four months since Jim had deployed, we had talked almost every day about his work with the tribes. I supported him however I could. I acted as a sounding board, brainstormed ideas, and helped articulate his recommendations in several assessments, including the first one that he sent in August directly to Petraeus, Central Command chief Gen. Mattis, and Special Operations Command head Adm. Olson. I couldn't help but imagine, reading their responses, how shocked they would be if they learned of my behind-the-scenes role.

That role would quickly expand in September, when Jim and I decided that the time was right for me to make my first foray to join him in Afghanistan. Ever since the start of our relationship, when Jim spoke of bringing me to meet Noor Afzhal and his tribe, we had looked forward excitedly to this moment. We both knew instinctively that it was what we wanted, and the decision to execute was an easy one. Neither of us was a stranger to the war zone and its risks—we knew them all too well. The challenge was one of overcoming bureaucratic obstacles and restrictions that kept us apart given my status as a reporter and his as a Special Forces officer. I pressed for weeks to get official permission to report on Jim and other Special Forces personnel working on the local security initiative, but I was told there was a blackout on such coverage. However, I was able to gain approval for book-related interviews with U.S. military units located at the same bases in Bagram and eastern Afghanistan where Jim was working. For my initial trip—a reconnaissance of sorts—I decided to take a relatively brief, three-week leave from the *Washington Post*.

On a sunny afternoon in mid-September, I landed at the main U.S. military base in Bagram, Afghanistan, to carry out a series of officially sanctioned book interviews. Little did I know that I would also gain some critical intelligence for Jim.

A Marine lieutenant working for Jim picked me up at Bagram Air Field in a pickup truck and drove me straight through an internal gate to the Special Forces area known as Camp Montrond. Jim was there waiting. As soon as I stepped out of the truck, he gave me a big hug and picked me up off my feet. He was in uniform and we didn't want to draw too much attention, so we walked side by side to his room.

"This is kind of awkward, isn't it?" he said, flashing me a smile as we tried to appear casual while soldiers walked by saluting him.

Jim was in Bagram putting together a detailed tribal strategy for eastern Afghanistan. In coming days, I helped him pore over documents to map out the key tribal areas. We also spent hours crafting official memos laying out the argument for Khas Kunar versus Marawara.

Jim briefed the Khas Kunar plan and won support from his Special Operations commanders—including Bolduc, head of the CJSOTF-A, and Lt. Col. Donald Lovelace, commander of the Special Operations Task Force–East (SOTF-E), which oversaw Special Operations Forces in eastern Afghanistan.

But opposition persisted from the battlespace owner, Col. Andrew Poppas, who had to approve the plan. Poppas was commander of the 101st Airborne brigade responsible for Konar and other eastern provinces. Like many other conventional commanders, Poppas tended to be fixated on areas of enemy concentration. A large influx of fighters had flowed into Marawara and other parts of Konar Province from Pakistan as insurgents stepped up attacks in the east. Poppas believed the enemy grip on Marawara was threatening Konar's capital, Asadabad.

I wanted to see the area for myself. On a cloudless morning in early October, I flew by helicopter to a small U.S. base in Konar where the battalion that fought in Marawara was stationed. Surrounded by steep mountains, Forward Operating Base Joyce was regularly hit by insurgent fire from recoilless rifles and other weapons. From there, I joined the battalion commander, Lt. Col. Joel Vowell, as he and his men patrolled on foot through the Taliban-held Dewagal valley as part of an operation to rescue a kidnapped British aid worker, Linda Norgrove.

"We didn't expect the volume and intensity of the fighting," Vowell told me as we stepped along a rocky streambed. There had been a 200 to 300 percent increase in attacks in his area compared with the previous year, Vowell said. He often felt his battalion was stretched too thin. "I'm out of Schlitz everywhere on combat power," Vowell said. "God help us if we hit an IED or are attacked." Both Vowell and Poppas were determined to keep up the pressure in what they considered a decisive area. So when they learned they would be gaining a Special Forces team, they wanted to send it to Marawara as a combat force.

Jim was more focused on identifying friends than foes. His strategy was based on local people inviting U.S. teams into an area, where they would build relationships, establish security, and expand outward from there. Given the tremendous weight of personal relationships in Afghan culture, Jim's history with the Mohmand tribe was a huge advantage. Yet Poppas didn't seem to grasp that, and repeatedly asked Jim why he wanted to work with the tribe in the same area where he had operated in 2003.

As the tug-of-war intensified over where Jim and the team would go, I unexpectedly got a front-row seat as I interviewed more American officers. One day Brig. Gen. Stephen Townsend, the deputy U.S. commander for eastern Afghanistan, took me on a daylong trip to see military operations around the region. On one stop, our helicopter landed at the U.S. base in Jalalabad, where Poppas had his headquarters. Townsend brought me into a briefing in Poppas's office. Right off the bat, Poppas, who knew nothing about our relationship, started talking with Townsend about Jim.

"Jim Gant wants us to put a Special Forces team here, in Khas Kunar," Poppas told Townsend, pointing out on a large map the district where Mangwel was located.

I suddenly felt as if I were a listening device planted in the room. I kept quiet and zeroed in on every word Poppas said.

"He's passionate about what he does. And I respect that," Poppas said. "But Khas Kunar has less enemy activity. It's not where I need that extra force right now."

Townsend nodded.

"Where do you want to put the team?" Townsend asked.

"I need them here, in Marawara, at the District Center," he said. "This is one of our biggest trouble spots, where we slugged it out in June. We have to keep up the pressure, and get some Afghan forces in place," he said.

"Done," Townsend said. "Marawara it will be."

*This is bad*, I thought.

But then Poppas spoke again. "It's not quite that simple," he said. "The Special Operations folks are saying that if they can't choose where the team goes, they won't send it to Konar at all. They may be bluffing, but if it comes down to that, I don't want to lose the team altogether," Poppas said.

"Understood," Townsend replied. "I'll work on it for you at my level."

*Thank God*, I thought. Poppas had unwittingly just revealed his bottom line.

We left the meeting, and as soon as I could I called Jim in Bagram to pass on the intelligence.

"You won't believe this," I said, relaying the conversation I'd just overheard. "You need to tell Lovelace and Bolduc to stick to their guns and tell Poppas they will withhold the team unless it goes to Mangwel in Khas Kunar. If they do that, Poppas is going to give in. He wants a team no matter where it goes."

"Damn, Ann—that's great news," Jim said. "I'll go brief Bolduc now." I finished my interviews and headed back to Washington.

Soon afterward, the conflict between the two chains of command was finally resolved in a meeting between two senior U.S. generals, Brig. Gen. Miller and Maj. Gen. Campbell, commander of the 101st Airborne Division. Energetic and affable, Campbell was the senior battlespace owner for all of eastern Afghanistan. As Jim's former battalion commander he valued Jim's expertise. The team would go to Khas Kunar.

By late October, Jim was back in Mangwel, laying the groundwork for the local security force. He was working with a Special Forces team from National Guard's 20th Special Forces Group, led by Capt. Geoff Burns. Burns based his team, Operational Detachment Alpha 2312, four miles down the road from Mangwel at Combat Outpost (COP) Penich. Penich was located at the foot of a mountain that allowed insurgents to target it easily with mortars and rockets. Insurgents had infiltrated the

Afghan Border Police unit that manned a checkpoint near the top of the mountain overlooking Penich, and the police helped direct attacks on the outpost. It was a glaring but not unusual example of the corruption of Afghanistan's national army and police forces, and underscored why Jim and many other American officers did not trust them. Indeed, some of the heaviest resistance to the plan to recruit local tribesmen to protect their villages came from national police officials. They argued that the weapons, funding, and training should go to their own force instead. Burns helped lobby provincial and local Afghan officials to support the plan, and negotiated how many men would be hired from Mangwel and surrounding villages.

Then one night in November, a Chinook helicopter flew into the base at Penich carrying the first shipment of guns for the tribal force, formally called Afghan Local Police: three hundred AK-47s, twelve hundred magazines, and about fifty thousand rounds of ammunition. Jim was ecstatic. He knew one could never underestimate the importance of guns and money to the tribesmen. Not only was the modern rifle the ultimate symbol of prestige to them, it was akin to life itself. Since the proliferation of arms factories in Pashtun tribal regions in the late 1800s, crude rifles emerged as a decisive factor in the tribes' prosecution of raids and blood feuds that were vital to their self-preservation. Jim invited Noor Afzhal to view the weaponry, leading him by the hand to a shipping container and opening a wooden crate full of Kalashnikovs. If Jim's loyalty and friendship had paved the way for his return to Mangwel, the delivery of the guns sealed it for the tribe.

Later that month, Jim and Burns arranged a meeting with Afghan district and provincial officials and a large group of tribal elders to formalize the plan. Miller also attended the meeting, called a "validation *shura*." Understanding the tribal elders' inherent mistrust of the government, Jim gathered them first outside in the courtyard. Squatting with them in a circle, he told jokes to set them at ease and then reassured them that they would choose and oversee the local men armed under the program. Just then, Miller walked through the courtyard. At first he didn't recognize Jim, mistaking him for a tribesman. He walked through again and did a double take.

"Jim?" he asked, lifting his eyebrows in surprise.

"Hey, sir, how are you?" Jim responded, standing up.

Miller smiled and walked back into the building.

A few days later, Jim was in Jalalabad when an email popped into his inbox, subject line: "Kudos."

"Great job," Miller wrote. Jim had earned a degree of trust and respect from Miller, who better appreciated the talents that came along with Jim's outspokenness. To Jim, who operated so often in the gray, outside conventional military bounds, support from his chain of command was critical.

In December, Burns and his men started training the first groups of Afghan Local Police for Khas Kunar. Soon afterward, Jim was given the go-ahead to move into Mangwel with his own small team. But far from the handpicked group of Green Berets that Adm. Olson had asked Jim to muster two years before, Jim was assigned a hodgepodge group of soldiers from a Kansas-based infantry battalion who were completely unprepared for the mission. It would be the biggest leadership challenge he'd ever faced.

# CHAPTER 14

IN TEN YEARS OF covering the Pentagon beat, I'd met hundreds of military officers. Many were talented, brave, and smart, but none seemed as single-mindedly driven by the love of his men as Jim. In the winter of 2011, as he prepared to lead a new team, I wondered what gave Jim that quality. Bits and pieces he'd shared with me from his past came to mind, like the sepia family photos and scrapbook clippings about his high school basketball career.

One story was especially heartrending. As a young boy of about four, he liked to play in the dirt lot and irrigation ditch across the street from the duplex where he lived in Las Cruces, New Mexico. But his father, James Karl Gant, told him not to, saying he could be hit by a car crossing the road. Gant, a school teacher and principal, strictly disciplined his only child, whom he called by his middle name, Kirk. One day Gant arrived home and caught his son playing with a toy car in the empty lot. He walked over, picked him up, and carried him back across the street to the side of the duplex. Then Gant went back to the dirt lot. "Kirk, come here," he said. But as soon as the boy went to him, Gant grabbed him and beat him. "I told you never to cross the street again!" he yelled, and took him back to the other side. Then Gant returned to the lot. "Hey, buddy, I am over here now. It is okay, you can come over," he coaxed. Jim went to his father. His father beat him again, even worse, dragging him through the dirt. Then he carried him back across the street. "Okay, Kirk," he said. "It's all right. I'm your dad. I wanted to teach you a lesson. You can come back. You can trust me," he said. By then the boy was sobbing, but he did what his father asked. Gant, frighteningly calm, struck him again, and took him back. At last, his father stood on the other side and for about ten minutes begged the boy to come to him. The boy stared at him, tears streaming down his face, but didn't move.

Jim told me that as a child he never felt his father loved or even liked him. His father wanted him above all to be tough, psychologically and emotionally, and made him earn everything he ever got. He was also ultracompetitive. He never once allowed Jim to beat him in any games, not basketball, not marbles, not even checkers. "He never said 'Good job.' He never said 'You are getting better.' He never really said 'I love you.'"

Years later, after Jim left home and went to war, he and his father mended their relationship and became extremely close. At the most critical points of his adult life, he said, his father was there for him. Once his father even traveled all the way to Afghanistan for two weeks to spend time with Jim and meet Noor Afzhal. His father admitted he'd made mistakes, and Jim gave his father more credit than anyone for helping make him the man he is. I had to agree.

When Jim was in elementary school, his father goaded him to fight other children. "Don't come home if you lose a fight," he'd say. One night his father and some friends got drunk and forced their sons to scrap with one another. After a few rounds, Jim had pinned an older boy on the ground, clearly defeating him. But his father urged him to keep hitting the bloodied youth. When Jim refused, his father grabbed him and mocked him.

Then one summer day when Jim was twelve, his father approached him as he shot baskets in the driveway of their one-story, blond-brick house in Hobbs, New Mexico. "Son, take care of your mother," he said. Then he hugged him, said goodbye, and drove away. Jim sat outside on the sidewalk by his front door for three or four hours, holding his basketball and crying.

For the next year after his parents split, he watched his mother, Judy, barely able to get out of bed. She broke down sobbing every day. For five years she didn't celebrate Christmas or other holidays. She blamed herself for not standing up to her husband. Then, finally, she got better. Seeing his mother slowly pull herself out of deep depression was Jim's first real lesson in courage. "I saw my mom push through the kill zone, wounded, and continue to fight," he told me. Jim and his mother were baptized together. He grew fascinated with the Bible, read it several times through, and tried to live by its teachings. For more than a decade, he did. "I didn't smoke, I didn't drink, I didn't use bad language. I tried very hard to be a good Christian."

His father's abrupt departure triggered something else inside Jim: it motivated him to go to extremes to excel in basketball, in what would become a lifelong quest to try to make his father proud.

The day after Jim watched his father disappear down the dusty street in Hobbs, he started shooting five hundred baskets a day. He practiced in the wind, rain, and sleet, and when it was so cold he could barely feel the ball in his hands, let alone bounce it on the snow-covered driveway. He practiced when he was tired or sick. And he never let his last shot be a miss. If he missed on the five hundredth shot, he would take it over until he made it. He went to school an hour early each morning to train at the gym. After school, he attended practice, went home to finish shooting baskets in the driveway, ate dinner, and did his homework. At night he lay in bed and played one or two entire games in his mind before he fell asleep. By middle school, he was consistently scoring thirty to forty points a game, chalking up school records, and went on to play on the Hobbs High School powerhouse team under legendary coach Ralph Tasker. Still, he felt his father's love eluded him.

In May 1986 Jim graduated from high school and accepted a basketball scholarship for a junior college in Hobbs. But, standing on the court one midsummer day at a gym in Las Cruces, he had an epiphany. It dawned on him he would never make a living as a basketball player and would have to coach. *That's not what I want with my life*, he thought. A week later, he was standing before a military recruiter. His sudden decision to enlist was not entirely out of the blue. About the same time he had his change of heart on the basketball court, he had found himself enthralled by Robin Moore's fictionalized account of Green Beret exploits in Vietnam. Aside from the combat and daring depicted in the book, there was something exotic and intriguing to Jim about the idea of fighting with an indigenous force. He was also under considerable pressure to make his own way in the world. His father, then on his third divorce, had made it clear that Jim was to be out of the apartment they were sharing in Las Cruces by the end of the summer. At nineteen, Jim was about to get married and had his first child on the way. He decided to make the rounds of the military recruiting offices, which were lined up in a row of storefronts in downtown Las Cruces. First he walked into the Navy office and said he'd like to be a Navy SEAL. The recruiter gave a chuckle, so Jim immediately turned and left. He then went into the Army recruiting office next door. "I want to be a Green Beret," he said. "Sit down," the recruiter replied.

Three months later, in mid-November 1986, Jim left for Army basic training at Fort Jackson, South Carolina. He took the same obsessive drive he'd shown on the basketball court to the training fields, seeking to prove himself and master the skills he needed to lead men in combat. Underneath it all, he was filling the void of love he had felt as a child by forging bonds of camaraderie and brotherhood with his men.

Jim spent the next two years struggling to pass the grueling trials required to earn his Green Beret. He volunteered for the Special Forces Survival Evasion Resistance and Escape (SERE) school, set up in 1981 to resemble a Viet Cong prison camp and train the soldiers to resist physical and mental torture and interrogation. Soldiers were stripped, beaten, blindfolded, deprived of sleep, water-boarded (tied down with water poured over their face to make them feel as if they were drowning), and confined in small cages for hours. Jim passed, and soon went on to formal Special Forces training. The first phase involved more arduous physical trials. In one test of his skill in using a compass to find his way through the woods, he slipped off a log, broke a rib, and had to start over. During a parachute jump he suffered a concussion, and had to start over again.

At the end of 1988 he began the last phase of the Special Forces qual-ification course, or "Q course." Known as "Robin Sage," the final trial is a four-week unconventional warfare exercise carried out over thousands of square miles in the backwoods of North Carolina. Soldiers are organized into twelve-man teams, just as they would be during real operations. The teams infiltrate a fictitious country known as Pineland where they link up with a guerrilla force and its leader, or "G-chief." The goal is to train and advise the guerrillas as they liberate Pineland from an oppressive regime. The exercise had a rocky start for Jim. Soon after arriving at the guerrilla base camp, the G-chief and some of his men summoned Jim and, at gun-point, took all his communications gear, stripped him naked, made him hug a tree, and tied his arms and legs around it. His team had to spend half their operational fund to get him released.

A formative lesson Jim learned at Robin Sage—one that would stay with him in coming years—was that it didn't really matter how he accomplished a mission, as long as he got it done. Sometimes instructors assigned missions that would be impossible if done by the book.

One afternoon, for example, Jim was told to carry out a reconnaissance of an extraction point twelve miles distant and return by midnight, just a few hours away. Realizing he'd never cover the distance in time, he led his teammate to a farmhouse and asked a man for a ride. The man readily agreed, gave the two soldiers some blueberry pie, and had them toss their rifles in the back of his pickup truck and climb in. Jim accomplished the mission with hours to spare.

"How the fuck did you do it?" an instructor asked.

"Hey, Sergeant, mission accomplished," Jim replied. He emerged with a provocatively unconventional mind-set that would be his trademark from then on.

Weeks later, the Robin Sage instructors gathered Jim's team together under the pine trees as they cleaned weapons at Camp Mackall, and one by one called out the names of the soldiers who passed.

"Gant!" his instructor said. "You made it."

He had earned his Green Beret and crest with the motto "De Oppresso Liber"—to free the oppressed. "It was the proudest day of my life," Jim told me.

Jim was assigned to the 5th Special Forces Group. Green Berets are organized into five Special Forces groups, which focus on different regions of the world, and the 5th Group concentrated on the Middle East, so the next stop for Jim was Arabic language school. After Iraq invaded Kuwait in August 1990, the 5th Group took part in the first Gulf War, the U.S. military campaign Desert Storm. Jim was a sergeant and was involuntarily extended beyond his four-year contract to go to war. His unit worked with an Egyptian mechanized battalion that was part of the central thrust from Saudi Arabia into Kuwait. His unit called in air strikes, breached obstacles, and targeted an Iraqi military base before moving into Kuwait City to help secure the Kuwait International Airport. By then a staff sergeant, Jim left the war disappointed that it was a completely uneven contest and he had not experienced real combat.

In 1992, Jim finished his enlisted time and decided to enroll in

Reserve Officers' Training Corps (ROTC) at New Mexico State University to become an Army officer. He was working as a security guard and martial arts instructor to support his wife and five-year-old son, Chant. But he was unhappy in civilian life, and the pressure strained his marriage. He divorced and later remarried. By early 1997, he was a lieutenant assigned to the 25th Infantry Division in Hawaii and had just been given his first platoon.

As a twenty-nine-year-old former Special Forces staff sergeant who'd already been to war, Jim felt intensely the imperative of training the thirty men in his platoon for combat. "What struck me is how badly those men needed leadership, and how little they knew," he told me.

Giving it his all, Jim plunged into the study of small-unit tactics. He began poring over Army infantry manuals. He scrutinized maneuvers such as the ambush and how to walk point in a patrol. He also sought to become a technical expert in the weapons systems his men used. Always one to question the written word and test it against "ground truth," he looked for lessons in firsthand accounts from recent wars—not the grand set-piece battles of World War II but the close-in jungle warfare of Southeast Asia. "What I wanted to know is how these platoons fought in Vietnam," he said.

Over time, Jim wrote all of his own standard operating procedures, or SOPs, and developed his own training regime for close-quarters battle and light infantry operations. His goal was to train his men as he wanted them to fight. Sometimes that meant flouting conventional Army wisdom, much to the consternation of his senior commanders.

Grenades were a good example. Army rules required a safety officer to stand next to a soldier and instruct him point by point in how to use a grenade. Jim didn't want his men to have that crutch—there were no safety officers in combat—so his men learned to toss live grenades by themselves.

Once after he led his platoon through a live-fire drill taking down a building as they practiced urban combat, a training officer, Maj. Brian D. Prosser, offered a critique. He advised Jim and his men that when stacking up before rushing in to clear a room, they should hold their rifles with either their left or right hand, depending on which direction they would have to shoot. Jim thought that was crazy, and immediately spoke up in front of all the other soldiers.

"Hey, sir, what you are telling my men will get them killed. Before you come out here and give AARs [after-action reviews], you need to know what you are talking about," he fumed.

Prosser, stunned, told the outspoken lieutenant they would talk "offline."

That night, during another live-fire drill, Jim was inside a darkened building when Prosser walked by at the end of the hallway talking with another officer. "Goddamn that Lt. Gant," he said. "If he didn't know his stuff so well, I'd fuck him up." Soon afterward, Prosser became the executive officer of Jim's battalion, 2nd Battalion, 5th Infantry Regiment.

What unnerved his superiors even more was that Jim's platoon was by far the scraggliest in the battalion. Jim himself let his hair grow too long, never polished his boots, and allowed his men to take candy and cigarettes with them to the field. Jim called his soldiers by their first names, hugged them, and shunned formalities such as salutes. But Jim trained his men hard and was a stern taskmaster if they messed up.

The commander of 2-5 Infantry was then Lt. Col. Campbell, a smart, hardworking officer well liked by his men. Campbell, like Jim, was Special Forces qualified and had served in the 5th Special Forces Group, where he commanded an ODA, or A-team. They were from the same Army "tribe." "Jim stuck out because he was prior enlisted, he was a little more cocky, and had definite ideas on how things would go," Campbell told me in his Pentagon office years later, when he was the Army's Deputy Chief of Staff for Operations. Still, at the time Jim's maverick leadership proved almost too much for Campbell.

One day Campbell was watching all the platoons in his battalion go through a rehearsal for a night live-fire training exercise. It was a dangerous drill in which the platoon tried to capture a hill that was fortified by mock enemy guns and about three hundred yards of trenches. Each platoon had to navigate a minefield, cut through three layers of barbed wire, and then clear the enemy out of the trench system that surrounded the hill. Meanwhile, the platoon's own machine gunners fired real bullets to suppress the enemy firing positions. But there was a problem: the machine gunners could not see the soldiers as they moved through the trench to clear it. The answer, according to Army Field Manual (FM) 7-8 on infantry platoon tactics, was for one of the soldiers to run along in front of the rest carrying a chemical light and a fifteen-foot-tall pole with a reflective neon orange panel at the top. That way, the machine gunners could fire well in front of the flag as a safety precaution to make sure none of their own soldiers were shot. But in Jim's view the flag had a serious drawback—in a real combat situation it would alert the enemy to the platoon's position, allowing for an easy counterattack. Jim again wanted his men to train as they would fight.

During the exercise rehearsal before live guns were fired, all the platoons followed the Army manual instructions and carried a flag through the trenches—except for Jim's, which was the last to go through.

"I have a better idea," Jim told his men. "We aren't using the flags. Keep your head below the trench line. Anyone outside the trench is getting shot!"

His men followed orders, but as soon as Campbell and other officers observing realized what was happening, all hell broke loose.

"Jim! What the fuck are you doing?" Campbell shouted through a bullhorn as he peered down at the soldiers through his black-rimmed Ranger glasses.

"It's not my fault if you have eight fucked-up platoons!" Jim shot back. Outraged, Campbell ordered him to lead his platoon through the exercise again, following FM 7-8.

"I want you in my office Monday morning," Campbell said tersely.

At the appointed time, Jim appeared before the stern-faced Campbell. "This is a letter of reprimand," Campbell said, handing Jim the document, an administrative chewing-out that can hurt an officer's career. Jim signed the letter, and Campbell told him he was dismissed. Jim turned, saluted, and walked toward the door.

"Jim," Campbell added before he could leave, "I thought about it all weekend. You were right." Later, on Campbell's recommendation, Jim trained his entire battalion in close-quarters combat and other drills using his unconventional tactics and techniques.

Over time, Jim's men proved among the most combat ready in the battalion, beating out other platoons in live-fire and other competitions. In turn, Jim began winning approval from superiors. Campbell gave him stellar reviews and considered him the most outstanding of the more than thirty lieutenants in his battalion.

In 1998, Jim went on to lead a second platoon in the 2-5 Infantry, a scout platoon that conducted reconnaissance. He was disappointed at the paucity of information in the Army's infantry field manual on patrolling. So, in his ongoing study of Vietnam tactics, he wrote a detailed pamphlet called *The Point Man*. It was designed to train the soldier walking point at the head of a patrol as well as the men behind him how to best detect and respond to enemy contact. *The Point Man* so impressed his superiors that toward the end of Jim's time in command, he was entrusted with training all the platoons in his battalion in this more sophisticated way of patrolling. In ten days of intensive live fires and maneuvers in a lush, jungle-like training area, none of the battalion officers came to observe. Then late on Friday afternoon, two hours before the training ended, one officer arrived at the range: Maj. Prosser, who had clashed with Jim early on.

It was June 1999, and Prosser was about to leave Hawaii. He came to give Jim a gift, one that would change Jim's life. He handed Jim a hardback copy of *Gates of Fire*, the historical novel by Steven Pressfield about the three hundred Spartans and their legendary stand against the Persians at Thermopylae. Jim returned exhausted from his training, and later that night opened the book. In it, he found this inscription:

Jim,
We didn't start off well. . . . I thought you were a young punk platoon leader. Things change—you're older now. . . . Of all the officers in 2-5 Infantry, I will miss you the most. Your energy, personality and drive are refreshing to be around. No one can say that Jim Gant's heart is not in the right place. Belief in yourself and your duties are more than half the battle. . . . and along with persistence and drive make all the difference. You have all these qualities and more. It has been my honor and privilege to serve with you.

Prosser went on: "This is a book about warriors . . . about men of old. Although fictional, I found it fascinating and full of lessons on character and life. . . . Remember, circumstances do not make a man, they reveal him."

Jim began reading the novel, which is narrated by the lone survivor of the battle, a squire of the Spartan warrior Dienekes. He was astonished by what he found. Every page of Pressfield's intimate account of the making of Spartan warriors and their heroics at Thermopylae seemed to sharpen and illuminate Jim's own emerging beliefs about combat. "It spoke to me like the Bible," he said. It was an intellectual and spiritual affirmation of truths that had come to him in the trenches and drill fields, about leading men in war. "I immediately knew those people [the Spartans]. I knew that time. I was meant to be a Spartan, perhaps I was. Every single part of that touched me. It was as though I had a very focused black-and-white picture, and *Gates of Fire* gave it color." Jim identified powerfully with the Spartan culture, and with the practice of young boys leaving their birth families to become part of a brotherhood of men-at-arms.

The key to motivating his soldiers, he had learned as a lieutenant, was simple: he loved them. "I try to treat people under my command the way I would want someone to treat my son. When you put it in those terms, it is easy," Jim explained to me. In coming years, he would act in many ways as a surrogate father to his men. He prepared them for combat, held them and cried with them, and risked his life to protect them and get them home. He told them again and again that when they were wounded in combat, the faces they would see and the voices they would hear comforting them—and the hands responsible for keeping them alive—would be those not of family members but of their fellow soldiers. He also expected his men to love him in return, as much as they loved their own fathers. In giving and expecting so much, he bonded with his soldiers in a way that went well beyond the norm for commanding officers.

Years at war followed, in which he honed his leadership skills and principles. Then one chilly night in 2011 in Afghanistan, he found himself face-to-face with his most inexperienced group of soldiers yet—with only two days to prepare them for combat.

# CHAPTER 15

SGT. EUGENE "SONNY" BOLES, a thirty-two-year-old Army cook, stepped out of the darkness into a tent at Combat Outpost Penich. He waited uneasily to learn why he had been sent to Konar Province, Afghanistan. It was February 2011. Only a few months earlier, Sonny and other members of his platoon from the Army's 1st Battalion, 16th Infantry Regiment, based in Fort Riley, Kansas, believed they were deploying to Iraq to escort convoys. Then rumors began flying about a security mission in Africa. But around Thanksgiving, their company commander announced they were going to Afghanistan. They were slated for an unusual mission with the Special Forces. It was all Sonny knew that night standing out in the cold.

In fact, the assignment of a conventional Army battalion to conduct a Special Forces mission had never happened before. Gen. Petraeus made the decision after realizing that the U.S. military did not have enough twelve-man teams of Green Berets to expand the Afghan local defense initiative, at least not as quickly as he believed necessary. "It was completely unprecedented," Petraeus told me later.

Special Forces had been in heavy demand in Iraq as well as in Afghanistan since the U.S. military invaded in 2001, and although the Army was working to increase the overall number of teams, they remained in short supply.

In October 2010, still waiting to get back to his tribe, Jim had worked on plans for expanding his strategy over the long term. Asked by his command to brainstorm ways to use regular infantry to supplement the

Special Forces effort, he and Lt. Col. John Pelleriti, who oversaw Special Forces operations and plans in eastern Afghanistan for SOTF-E, proposed three different models for an "infantry augmentation plan." One model involved dividing Special Forces teams and adding regular Army soldiers to each split team, in effect doubling the number of Green Beret–led teams available for the mission.

In a mid-November video teleconference with Special Operations commanders, Petraeus asked Bolduc what manpower he needed to expand. Bolduc described proposals including Jim and Pelleriti's plan to integrate conventional forces with his Green Berets. While Petraeus offered Bolduc an infantry battalion to "thicken" his teams, it was not without overcoming resistance from the conventional-force leadership about having their unit dispersed across Afghanistan. But within a week, the Army had approved the deployment of the 1st Battalion, 16th Infantry Regiment from the 1st Infantry Division at Fort Riley.

Back at the neatly mowed garrison in northeastern Kansas, news of the deployment elicited a mixture of reactions from Sonny and his comrades of 1-16 Infantry. It was an eclectic group of soldiers. Because its infantry line companies had been severely undermanned, the battalion had just undergone a chaotic reorganization. Senior leadership decided to pull soldiers from the support company—made up of medics, cooks, mechanics, drivers, and an array of other noncombat troops—to fill out the infantry ranks. The men had not trained together, and some were complete strangers. Many were not happy to be there.

Sgt. Jeremiah "Doc" Harvey had learned his wife was pregnant just two days before the word of the deployment came down; he canceled a deal to buy a house.

"A lot of the guys were out-and-out scared," said Sgt. Justin Thomas, a thirty-eight-year-old mechanic from Kansas City, Missouri. "They said, 'I didn't sign up for this crap and to go around carrying a rifle.'"

Justin, a former cavalry scout, was more enthusiastic about the idea of working for Special Forces, as were the handful of infantrymen in the platoon. For Sgt. Michael Taylor, a twenty-five-year-old veteran of combat in Iraq, the assignment was a blessing in disguise. "I was really glad to link up with ODAs [Special Forces teams]," he said. "Doing a regular coalition force mission, we would have got slaughtered. We never really went to the range."

For Mike, Justin, Sonny, and everyone else in the tent at COP Penich that night, though, the mission was a leap into the unknown. And none of the twelve platoon members was prepared for what happened next.

Wearing a baseball cap, jeans, and a jacket, with his beard now fully grown, Jim walked into the tent and introduced himself. He briefly told the men they were going to live in an Afghan qalat in the village of Mangwel. There they would eat, sleep, and fight alongside a tribe, and expand the local defense network as far as possible in Konar and beyond.

"We are trying to win this motherfucker," he said. "I need guys around me who are willing to do that." He went on: "What you are going to do is not okay. It's not fun, it's not safe, it's highly stressful. I do not expect all of you to make it home."

The room fell silent. The men stared at Jim, their faces ashen.

"Now I am going to give you my initial counseling. In the morning, the guys who are standing in front of me on the range at 0700 are the guys who are going to go with me. If, after receiving this counseling, you decide this is not what you want to do, man up and tell me now."

What Jim referred to as his "initial counseling" was for him a manifesto, one he had shared with four prior teams of men he'd led in combat. This team, he already sensed, would be the most challenging. Over the next two hours, he explained his philosophy of leadership and command. He told the men what he expected of them and what they should expect from him, and he asked them to sign a statement saying they understood. Last, he read a short essay about his own creed as a warrior, entitled "Who I Am."

Jim opened with a quotation from a Pressfield novel about the fifth-century BCE Athenian-turned-Spartan general Alcibiades, one who had never been defeated, and whom Jim most admired: "A commander's role is to model *arête*, excellence, before his men. One need not thrash them to greatness; only hold it out before them. They will be compelled by their own nature to emulate it."

Turning to his own beliefs, he told them: "All that we do must focus on combat. Our preparation for battle and our execution on the field of battle must be flawless."

However, he continued, combat expertise will not be enough. "This unit must show advanced skills in cross-cultural understanding. We must care about what happens to Afghanistan and its people.

"We will protect our village, our valley, and our tribe—at all costs. We are here to win our part of the war . . . we will be extremely brutal and incredibly empathetic in the blink of an eye. How will we accomplish our mission? Here is how. One year from now, if you can return to the States and look in the mirror and say, 'I did everything I could, every day while I was in Afghanistan,' then I promise you, we will succeed. . . . If you will give me, the other team members, and those Afghans we are fighting for all you have, I know for certain that when you are an old man and you start losing your memories, you will beg God to take the memories of the time you spend in the Konar last."

Jim said he expected his men to be loyal to the team, to provide him with honest feedback, to work hard, and to take the initiative. "Don't be a private," he said. In return, he promised to always lead from the front, to make good tactical decisions, to listen, and to treat them as grown men without regard for their rank.

Jim paused and glanced around the room. "It is imperative that you know who I am and what is important to me," he said. "We will live and die based on how well we learn to work with one another."

He looked each man in the eye, then continued.

"Who am I? I am a warrior. My physical, emotional, and spiritual self revolves around being a warrior. I believe war is a gift from God. . . . I am not a patriot or mercenary. I fight to fight. . . . I believe if you want to kill, you must be willing to die. I am willing to do both, whichever the situation calls for. I am a student of war and warriors. There will be no blood on my hands because I or my men were not prepared for battle. I will prepare for battle every single day. I will love my men as I love my own children. I will take my men places and show them things that they never believed possible. . . . I will give my life for them as readily as I kiss my children at night and put them to bed. I will be their protector and their avenger if necessary. I will always expect the impossible from them. . . . I believe in God but I do not ask for his protection in battle. I ask that I will be given the courage to die like a warrior. I pray for the safety of my men. And I pray for my enemies. I pray for a worthy enemy. . . . I believe in the wrathful god of combat. I believe in Hecate. The gods of war have received their sacrifices from me. . . . I have a huge ego. It feeds my daimon," he said, referring to a tutelary spirit. "It is me and I am it. But I know it is there. Passion is power. Passion feeds my soul. I will seek passion out in others. . . . I am my children, my parents, my friends, my tribal family, the men I have gone into battle with, and my enemies. They reside in me. It is for them that I do battle. I want, need, and long for their acceptance. I want them to be proud of me. I will be loyal to being a warrior for all time. I will prepare a place in Valhalla for the warriors whose paths I have been blessed to cross. I will be with them in this life and the next. I am a warrior."

Jim surveyed the room and saw the fear in the soldiers' eyes. *Good*, he thought.

"That's all I have, men. Strength and honor," he said, and walked out of the room, leaving a stunned silence.

*Holy shit, who the fuck is this guy?* thought infantry squad leader Staff Sgt. Robert Chase, who had qualified as an Army Ranger. Chase considered himself one of the more experienced members of the team. But Jim was a wild card that caught him completely off guard. *He's crazy. He's off the charts*, he thought. *Here's all my training out the window.*

Sonny, too, was in disbelief. The cook from Harker Heights, Texas, had never encountered an Army officer remotely like Jim. "It scared me shitless," he said later. "I was in complete, utter shock." He searched for words to describe the encounter. "It was like *Dances with Wolves* meets Charles Bronson," he said.

Justin felt nervous but excited. Like many conventional soldiers, he had always thought of Green Berets as capable of anything—but also nutty. Still, he found Jim articulate and inspiring.

In the end, only two of the twelve soldiers in the tent that night quit. Jim was more than Staff Sgt. Scott Fitzpatrick could stomach. Fitzpatrick had just been promoted and planned to leave the Army after the deployment. The last thing he wanted was a dangerous mission led by an officer as zealous as Jim was about combat. He decided then and there to quit the team. Fitzpatrick asked Spec. Francis Mitchell, a twenty-two-year-old mechanic from Peabody, Massachusetts, to take his place. Mitch liked Jim's irreverence, his shunning of rank structure, and his "big-boy rules." He also liked one line in particular in Jim's counseling statement: "Don't be a pussy."

"Sure, I'll go," Mitch said.

Another young private, Pfc. Russell Kiggins, was on the fence. Kiggins had just gotten married and was always sneaking off to the phones to call his wife. He decided to drop out. Taking his place was Pfc. Andrew Gray, a twenty-two-year-old machine gunner from Stokesdale, North Carolina. Drew had nearly been charged with a felony after high school for destroying a string of mailboxes with his fist. Drew believed all Afghans—all Muslims, for that matter—were bad. Never in a million years had he thought he would be living on an Afghan qalat. But he wanted to go to Mangwel because he believed Jim would build a close-knit military team.

The next morning at 0700, Drew took his place with eleven other soldiers lined up in front of Jim at the training range on COP Penich. Each one of them handed him his signed counseling statement. As soon as they launched into individual weapons training, Jim realized the magnitude of the task before him. Many of the soldiers didn't know how to use their night vision goggles or rifle lasers. They couldn't load, clear, or fix a malfunction on the machine gun. He was stunned and angry at how ill-prepared they were. It meant Jim would jam three months' worth of training into a day and a half straight.

Then on February 12 they headed for Mangwel.

# CHAPTER 16

IT WAS BELOW FREEZING and sleeting on the afternoon of February 12, 2011, as Spec. Chris Greenwalt nervously got behind the wheel of Jim's Humvee for the first time. Chris was a communications soldier and, by his own admission, drove like a grandmother. Now he would steer the lead Humvee of the combat patrol as the American team rolled out of COP Penich in five vehicles and headed southwest through insurgent-influenced territory toward the village of Mangwel. There, the handful of Americans would begin a bold and unprecedented experiment in the war, living in one of the most contested areas of the country under the protection of an Afghan tribe.

"Guns up, Sonny!" Jim called.

"Roger," the cook replied, loading the weapon for his first stint as a machine gunner in Afghanistan.

A broad, sandy stretch of the Konar River valley gave way to terraced fields and farming villages. They passed a cemetery, with the shale gravestones aligned one way for men and perpendicularly for women. Then the road twisted back and forth as it hugged a steep, rocky hillside in a natural ambush location, where two weeks earlier four U.S. soldiers had been wounded by a roadside bomb. It was a prime location for insurgents to place IEDs. Jim named that portion of the road "Zombieland," in honor of Burns's 20th Group Special Forces team, which had the call sign "Zombie."

"Center road!" Jim told Chris, who squeezed the steering wheel tightly. Jim always rode in the lead vehicle—or sometimes on the hood—to look for signs of bombs and to let the enemy know he did not fear them. Being his driver was one of the most demanding jobs on the team. Standing at the gun, Sonny scanned the ridgeline, his jaw taut.

The patrol continued through the tribally mixed village cluster of Kawer, where Taliban insurgents moved freely. Afghans came out of shops and qalats and watched the U.S. vehicles curiously. They knew the Americans were moving into Mangwel, but they were uncertain what it would bring. More killing? More broken promises?

Next the vehicles rolled past a boys' school and entered Mohmand tribal territory, where Noor Afzhal had sway. Through the window of the Humvee, Chris saw the first layered walls of Mangwel village rising to the east and giving way to a long mountain valley. Dense gray clouds obscured the peaks. Fifteen miles beyond them lay the Mohmand tribal areas of Pakistan.

At the far end of the village, Jim told Chris to turn off on a dirt road. The road led a quarter mile past flat land covered with scrub brush and stopped at what appeared to be the largest qalat in Mangwel. The outer mud walls of the qalat were about twenty feet tall and formed a rectangle some 120 feet wide by 200 feet long. The qalat was missing a front gate, but the walls were topped with four guard towers. Looking more closely, Chris got his first glimpse of a tribal fighter, wrapped in a woolen blanket and armed with an AK-47, peering down from a tower at the American patrol.

Umara Khan, his piercing green eyes missing nothing, immediately spotted his old friend Jim as the first Humvee rolled into the qalat. A wiry man no more than five feet tall, he slung his rifle over his back, scurried down a wooden ladder, and went to meet Jim, his angular face bright with excitement.

"*Salaam aleikum, tsenge yay, wror?* Peace be with you. How are you, brother?" Jim said, placing his hand over his heart in the Afghan way and hugging Umara Khan.

"*Khe yum*, I am well. It is good to see you again, brother," Umara Khan replied. He had moved into the compound along with a dozen or so other brand-new Afghan Local Police to secure it for Jim's arrival.

Noor Afzhal, too, came to welcome Jim, as did his sons Asif and Azmat, each embracing him in turn.

But Jim's eyes kept returning to Umara Khan.

A farmer, Umara Khan had known Jim since the American first came to Mangwel in 2003. He was dirt poor, even by the standards of Mangwel. He had seven children ages four to sixteen, and only a tiny plot of half a *jreb* (a quarter acre) of land. Illiterate like most of the villagers, Umara Khan wasn't sure how old he was—he guessed forty-five—and when he smiled ripples of deep wrinkles creased his cheeks and forehead. But his eyes flashed bright. He spoke with energy and intensity. He was a fighter, and he loved Jim.

Jim laughed. The tension of the move, the weight of responsibility for his men, the urgency of protecting the qalat—all for a moment felt less daunting. He looked up at the four guard towers and saw tribesmen in each one, their AK-47s pointing outward. He knew all they had to do was pivot and they could gun the Americans down without a contest. But something about Umara Khan's look, quizzical and intent and warm, assured him they would not. *This just might work*, he thought. He was betting his life and those of his men that it would.

For Chris and the other Americans, the Afghan qalat was a rude awakening. Apart from the sheer danger of being in the middle of a remote village in Konar, the compound had no electricity, running water, or latrines. COP Penich was just five miles down the road. But in terms of living conditions, the isolated military compound with its chow hall, hot showers, gym, wide-screen televisions, and Internet, all surrounded by reinforced walls and concertina wire, might as well have been a thousand miles away.

Inside the qalat walls, there were not yet any interior structures or buildings, just an expanse of ankle-deep mud. The soldiers barely greeted the Afghans as they struggled to unload the vehicles, sloshing through the muck in the freezing rain. The sun was going down, so they hurried in the last light to erect a tent for themselves to sleep in. Once darkness fell, it would be too dangerous to use flashlights or lanterns—that would signal their position to insurgents who could easily see into the qalat from an adjacent hillside the Americans nicknamed "Bad Guy Hill." They managed to raise the tent and connect its heater to a generator. Soon they bedded down for the night, taking turns on guard duty, scanning Bad Guy Hill and the surrounding high ground using a .50-caliber gun mounted on an armored vehicle.

Jim observed his team, then quietly went to lie down in the unheated, leaking tent where Umara Khan and the other tribesmen were staying. Beside Jim was his close Afghan comrade and interpreter, Ismail Khan, nicknamed "Ish." Jim had interviewed and turned down dozens of possible Afghan interpreters over several months, and had almost decided to go to Mangwel without one. Then, in February at the base in Jalalabad, he met Ish. Tall and handsome, with a reserved but focused manner, Ish was a schoolteacher. He was twenty-five years old but seemed wise beyond his age. There was an unmistakable sincerity in his large brown eyes that made an impression on Jim. "There was something in his eyes. He was very strong but humble, intelligent. I could see he had a very good heart," Jim recalled. "He was not what I was looking for, but exactly what I needed." Thirty minutes later, Ish was hired. During the most critical and delicate period of the mission in Mangwel, Ish never left Jim's side. That first night set the tone. It was miserably cold, and Jim, chilled to the bone, barely slept. Ish curled up on the ground next to him. Together they shivered through the night. The next morning, he gathered his men outside in a cold drizzle and taught them their first lesson in tribal engagement.

"We need security—that is our number one priority now," he told them. "So what is your most important task today?" he asked. They began rattling off a list: better fighting positions, sectors of fire, sandbagging enclosures in case of mortar attacks, observation posts. "All good," Jim said. "But you're missing what is most important." The men were silent; they didn't get it. "Your top priority is to set up a heated tent for the *arbakai* so they can rest as comfortably as you did last night," he said, using the Pashto term for traditional tribal police. (The force was officially called "Afghan Local Police," but they called themselves *arbakai*, or sometimes *mahali police*, which means "local police" in Pashto.) "We must live like they do."

The message, one Jim would drill into his men over and over in coming days and weeks, was that their lives depended on the relationships they would build with the arbakai in particular, and beyond that with the people of Mangwel. Many of Jim's men arrived in Afghanistan with the mind-set, reinforced by military training videos, that no Afghan was to be trusted. To some of them, Afghans were barely human. Afghans, they believed, were out to kill them, to stab them in the back. So the idea that relationships with Afghans would keep them alive was counterintuitive. Moreover, it was a deceptively simple concept, but dif-

ficult and complicated to carry out. To forge true friendships with Pashtun tribesmen meant sitting patiently with them for hours, studying
their language, culture, mannerisms, humor, and proverbs. It meant
grasping on a visceral level what was important to the tribesmen. It also
meant being genuine and knowing themselves. "Relationship building is
the weapon, time is the bullet," Jim would often say. Paradoxically, in
order to be understood as Americans, they had to become to a degree
Afghan. Jim, who had spent years honing his ability to connect with
foreign fighters, saw the bonds deepening in stages, ultimately leading
to cultural assimilation—signaled when the Afghans began adopting
some American ways.

The Americans set up the tent for the arbakai that day. But wariness
and mistrust lingered in the days after their arrival—on both sides.

Some tribesmen were suspicions about the Americans and their
intentions. One villager named Ehsan Ullah, a relative of Noor Afzhal,
had lobbied to keep the Americans out of Mangwel, believing their presence would draw more Taliban attacks. Once he learned they were coming anyway, he maneuvered to become the commander of the three
hundred arbakai in the district of Khas Kunar. He lied and separately
told Noor Afzhal, Jim, and Burns that the others had each picked him. It
was, it turned out, a typical Afghan method of deceit. Jim knew that the
average Afghan would lie to him, often very skillfully, and that he was
operating in a world of gray. A part of him admired their ability to deceive,
and he enjoyed the challenge of unraveling fact from fiction. But Ullah
went too far. With a beak-like nose and high forehead, he often wore his
shirts unbuttoned, revealing a mat of chest hair, and spoke and walked
with a certain swagger. He had an abrasive personality and rubbed Jim
and others the wrong way. In a culture that revolved around hospitality,
he managed to be rude. Worse, his son, who then was an interpreter at
COP Penich, was reporting to the insurgents on Jim's movements.
Before long, Jim and everyone else, including Noor Afzhal, were calling
Ullah "Shitty Man."

Within two days of the Americans' arrival in Mangwel, Ullah told
Jim that the arbakai were not going to guard the qalat—that it was not
their job. In the Pashtun context, it was a direct affront to Jim's honor
and power. Jim knew that if he did not respond forcefully, that one incident had the potential to undermine his entire mission. If the arbakai

followed Ullah or sensed he was a stronger leader than Jim, it would be over. Jim immediately kicked Ullah and all the arbakai off the camp. Shaming him in the Pashtun way, Jim refused to meet with or speak to Ullah. But by the next morning, the twenty arbakai assigned to guard the qalat returned on their own and apologized to Jim. Ullah was effectively marginalized.

Not long after that, Ullah was driving from Mangwel to the Khas Kunar district center down the road from Mangwel when he came under attack from Taliban insurgents firing machine guns, rifles, and rocket-propelled grenades. He took cover and survived the ambush. After that, he rarely returned to the qalat.

In the wake of the showdown with Ullah, Noor Afzhal, acting at Jim's request, named his third son, thirty-two-year-old Azmat, to command the roughly eighty arbakai in Mangwel. It was the one and only time Jim requested that a specific person be put in charge of one of the tribal forces. Jim was keenly aware of the need to let the tribe make its own decisions, but the leadership of the arbakai in Mangwel was too important to be left to chance. He trusted Azmat completely.

Azmat was good-natured and loyal. His thick eyebrows framed a round face that always bore a ready smile. A wheat farmer, at the age of twenty-seven he picked up his own weapon to police Mangwel, taking charge of a checkpoint on the edge of the village. The Taliban attacked the checkpoint repeatedly using heavy machine guns, killing or wounding some police, but each time Azmat and his men managed to fight them off. Once a member of the police tied to the Taliban let the insurgents into the checkpoint tower at night. They wounded three policemen in a brutal attack that lasted several hours. Azmat and his men managed to hold the checkpoint and gather the wounded. But the Taliban took control of the winding stretch of road opposite Kawer, effectively cutting off the route to the closest hospital in the main town of Asadabad, which was across the river. So they loaded the three wounded men onto a *jala*, an Afghan raft, and crossed the Konar River to reach the road to Asadabad. One of the men perished on the way.

Azmat knew that taking charge of the arbakai put him at risk of retaliation from the Taliban, but he said he could fight as well as they could. "I'm not scared of the threats," he said. "If I die one day, I die. I don't care what they say. They can't keep me from doing my job."

With the command of the arbakai more solid, Jim and the team set to work with the tribe establishing security in and around Mangwel. They built a firing range using earthen barriers and trained with the arbakai on basic marksmanship and qalat defense drills. Together they reinforced the guard towers with sand bags, and put in a metal gate on the qalat. Jim was careful, however, to place all the fortifications inside the walls so that—seen from the outside—their camp looked like any other Afghan qalat. They practiced first aid, putting bandages and tourniquets on one another. To extend the security perimeter around the qalat, they went on long foot patrols in the mountain valleys surrounding Mangwel. One moonless night after heavy rains, about two dozen arbakai and Americans walked high into a valley until they reached a stream so swollen they could not cross it. Seeing how easily the Afghans in sandals outpaced them moving up and down the mountainsides, the U.S. soldiers realized that if they wanted to be able to pursue Taliban insurgents over the hilly terrain, they had to shed their body armor. Jim never wore his, and many of them chose to follow suit.

Jim constantly practiced the kind of close and loyal relationship he expected his team to build with Afghans, talking and joking with the arbakai in Pashto, treating them when they fell ill, and rolling out the gate to aid them at the first sign of trouble. He often wore Afghan clothes and hats and carried an AK-47 on missions, and went out of his way to mirror Afghan customs and show respect for their Islamic faith. He made it clear to everyone that the team's three Afghan interpreters were trusted advisors and comrades and would be treated as such. The interpreters, Ish, Ibrahim Khan, and Imran Khan, were brothers from a large and influential family from Konar Province, and all had combat experience with the U.S. military. Ish was with Jim from the start in Mangwel. A couple of weeks later, Ish's older brother, twenty-nine-year-old Ibrahim, nicknamed "Abe," arrived. Ish had told Jim over and over that Abe would be his right-hand man. Immediately Jim could see from the way that Abe talked, walked, and carried a weapon that he was a fighter. He could operate all the weapons systems, the radios, and the vehicles as skillfully as any Special Forces soldier Jim knew. Over time Jim grew to feel he had never met anyone who was as similar to himself as Abe. Only later did Jim learn that Abe was with Army Special Forces Staff Sgt. Robert Miller and his team when they were surrounded by enemy fighters in a treacherous January 2008 battle near the border of Konar and Nuristan provinces. Abe saw Miller gunned down and killed leading a charge against insurgents that saved the lives of his teammates, and later helped recover his body. Miller was awarded the Medal of Honor. Abe received no recognition. From then on either Abe or Ish was with Jim on almost every mission. Imran, the youngest brother, arrived last. A scholar and a romantic, he was the most earnest of them all. Jim considered Ish, Abe, and Imran his brothers and he would not tolerate them being slighted in any way.

Soon every day, the U.S. soldiers, now also wearing Afghan clothing, began walking through Mangwel in twos or threes together with an equal number of arbakai to get to know the villagers and become a more steady presence. Like the Afghans, the Americans patrolling Mangwel wore no helmets or body armor, which would have signaled mistrust. Before long, everywhere they went, they were invited into Afghan homes.

Day after day, the Americans and Afghan tribal fighters ate the same food, rice and beans prepared in a cement-floored hovel by a slight, cross-eyed Afghan cook named Salim, who seemed to put the same Pakistani spices in every dish. Afghan workers arrived and at one end of the qalat built a row of Afghan-style latrines, crude outhouses with a key-shaped hole designed for squatting over. They built a simple shower and well, and an earthen pit where they cooked thick rounds of flatbread each day.

Many nights, at Jim's urging, the U.S. soldiers gathered to sit cross-legged around a long plastic mat and eat and drink tea together with the tribesmen. Afterward, they held informal Pashto-English classes. They taught each other greetings, numbers, and how to tell time. "This," Jim said during one of the sessions, "is where the war is won."

For the American soldiers, some of whom had never before set foot outside of the United States, the experience was life-altering. It was like living on another planet. Jim himself described feeling as though he were "walking on the moon."

Seemingly small and insignificant moments, multiplied dozens of times, created unbreakable bonds between the Americans and Afghans. One afternoon, Justin climbed into a guard tower to speak Pashto and English with one of the arbakai named Ghani Gul, nicknamed "Honey" by the Americans. It was cold and rainy, and they spent two or three hours just sitting and talking. There was something about Ghani's willingness to learn and to teach that gave Justin comfort. Looking out over the wintry landscape of snow-capped mountains and lowlands shrouded in gray, Justin knew he would never forget the Pashto word for rain—*baran*—and also from that moment, he began to believe that Jim's strategy would work.

Within weeks, a security regime was well established at the tiny camp, occupied by the roughly twenty arbakai, three Afghan interpreters, fifteen Americans, and a few dogs. In the ultimate sign of trust, Jim gave the arbakai in the towers M240B machine guns. No Special Forces team in Afghanistan had so quickly integrated itself into a village, or relied so exclusively on the local people to protect them and warn them of insurgent activity or attacks. At times in coming months, there were days when only three or four Americans stayed on the qalat, their lives completely in the hands of local Afghans.

Jim made sure his men stayed on their toes.

*Boom!* A smoke grenade tossed by Jim exploded in the center of the qalat and began spewing red vapor to simulate an attack.

"We've been hit!" Jim yelled.

Several arbakai and U.S. soldiers grabbed their rifles, threw chest racks of ammunition over their T-shirts, and raced to reinforce the guard towers.

"Vehicles! Come on!" Jim screamed. Azmat and several other arbakai jumped into two white Afghan Local Police trucks. Staff Sgt. Ryan Porter, a twenty-seven-year-old former paratrooper with multiple tours in Iraq, scrambled into the back, and they rolled out the gate. Other U.S. soldiers mounted the team's two Humvees and followed.

Jim designed training in Mangwel to be as close to actual combat as possible, which meant taking risks and firing live ammunition in close proximity to one another. It was a radical departure from the training his soldiers had received in the conventional Army. The challenge was magnified because Jim had given his men wide leeway in deciding which jobs they wanted to fill, and constantly pushed them to take the initiative. "Don't be a private," he reminded them over and over. Chris, a communications soldier, had no experience with the Special Forces radios and signals gear he was expected to operate, and learned how to use them on the fly. Harvey, a medic, had deployed to Iraq and Afghanistan but had never worked without the supervision of a physician's assistant. And every drill was more complicated because the men could never lose sight of their role working with and for the Afghans. Alongside scores of tribesmen in native clothing carrying AK-47s, the threat of friendly fire was increased exponentially.

Jim held meticulous critiques of training and missions, or "after-action reviews," in which he solicited views from the team on what went right and wrong, and invited questions about his decisions. That night, he held the session together with the arbakai and made sure to hear from Azmat and his deputy commander, Shamsul Rahman.

After a blow-by-blow assessment of the drill, Jim appealed to the Afghans in language they understood best.

"We have the best tactics in the world, the best equipment in the world, the best plan in the world, but when this happens it's going to come down to ghairat, your bravery and your courage to fight," Jim said. Ghairat—one of the words tattooed on Jim's wrist—was a core tenet of Pashtunwali, the code of behavior by which Pashtun tribespeople lived. It meant personal honor and valor, and was perhaps the most important measure of the character and manhood of a Pashtun tribesman. The arbakai nodded.

Finally, Jim asked Noor Afzhal for his thoughts. Wearing a white tunic and a 9 mm pistol in a shoulder holster with shiny bandoliers of ammunition framing his chest, Noor Afzhal stood tall and spoke sternly.

"You are here for us. We will protect you and this qalat, whatever the cost. We will fight to the last man," Noor Afzhal said, raising a hand and stabbing the air with his finger.

"The qalat is yours," Jim said.

"And if we have more guns and ammunition, the entire Pakistani military can cross the border and we will fight them off!" Noor Afzhal said, flashing a smile of sheer bravado.

The entire room burst into applause. Jim looked at his soldiers sitting cross-legged among the Afghans and knew they had earned their call sign: Tribe 33.

# CHAPTER 17

IN THE EARLY DAYS after arriving in Mangwel, Jim made it a habit to venture out of the qalat alone. Wearing an Afghan tunic and pants and carrying an AK-47, he headed into the village and surrounding hills either on foot or on horseback. The horse was intended to transport caches of water and ammunition into the mountains for operations. Another appeal of the horse for Jim was Noor Afzhal's description of him as nangyalee, the warrior on the white steed. The horse turned out to be old and scrawny, but he rode it anyway.

Jim wanted to explore, to take in the sights and smells and sounds of the Pashtun countryside, and to blend in among the people as free as possible from the trappings of a U.S. military officer. He also sought to project an image that would resonate and enhance his influence with the local Afghans—one of a man who was brave and would walk in their shoes. If he appeared eccentric, a bit wild, that was good, too. He wanted people to believe he was capable of both unusual compassion and violence. Indeed, to the casual observer Jim's interactions with Afghans might seem spontaneous and unscripted, but the opposite was true. He meticulously planned his words, gestures, and clothing based upon whom he was meeting and what he wanted to achieve. He weighed, for example, where to sit among a roomful of tribal elders, whose hand he should kiss, and whose he should refuse to touch. He decided when to pull out a string of prayer beads, conveying respect for Islam, or to clean his teeth with a wooden stick that he kept in his pocket in the Afghan style. He reflected on whether or not to drink *chai* or take a dip of opium-laced *naswar*, popular among Afghan men. Clothing also sent messages; he donned his finest Afghan *jami* suit in a show of respect, or wore a dirty, ragged one as an insult. Sometimes he dressed all in white, the preferred color of the Taliban, to subtly let people know that he understood his enemy.

When on foot, Jim wrapped himself in a woolen blanket, or *tsadar*. With his beard and round *pakol* hat, from a distance he was indistinguishable from the tribesmen. Sometimes he stopped to drink tea, but often he just walked for miles, to clear his mind. He passed through villages that smelled of smoke from ironwood fires and got to know the barks of dogs at different qalats. He crossed fields on raised dirt paths or along irrigation canals, watching men farming and children playing and fighting. He imagined how it would feel to be Afghan—to endure the bitter winter cold without a heated tent to return to, or to swelter in the summer sun with no chilled bottled water to quench his thirst. He returned to the qalat and told his men: "Don't judge these people because they steal wood or trash. You are going home to the United States. They are not."

Word spread quickly up and down the Konar River valley that Jim was back, that he had moved into Mangwel bringing guns and money and was improving life for the people there. Security in the vicinity of Mangwel and beyond improved almost overnight, as the arbakai, wearing black vests bearing the Afghan flag, became a visible presence safeguarding their homes and villages. Mangwel started out with fifty arbakai, and Jim and his men soon trained some thirty more, giving Mangwel the largest force out of the total of three hundred arbakai in Khas Kunar District in early 2011. Jim drilled into both the arbakai and his U.S. soldiers a clear and simple mission statement: "Protect the qalat, protect the village, protect the valley, and expand village stability operations." What made all the difference was that the Americans were not commuting to Mangwel from a U.S. military base but were living there. This was a basic counterinsurgency tactic, but only a small fraction of the teams engaged in the local defense initiative were practicing it. One beauty of the strategy was that once Tribe 33 was inside the village, the Taliban would have to attack Mangwel—to fight their own people—to get to the Americans. Moreover, residing in the village, the small American team was able to hear and immediately respond to gunfire or any other signs of trouble. Tribe 33 came with a lot of firepower: an 81 mm mortar system, .50-caliber machine guns, armored vehicles, and access to aerial reconnaissance, attack aircraft, and medical evacuation. The team reacted to not only insurgent threats such as suicide bombers but also to family feuds, floods, and fires.

Insurgent activity in the area fell, including the emplacement of deadly roadside bombs, as the armed tribesmen, their families, and residents as a whole grew more confident in reporting Taliban activity. Information traveled rapidly by word of mouth through the tribe and its web of connections, creating a swift and robust early warning network against potential attacks. Jim often said he had not one checkpoint but four thousand—they were the four thousand pairs of eyes and ears of the population in Mangwel and surrounding areas. Tips from local people and sources began flowing in by cell phone. Jim gave some villagers phones and small payments to report on insurgent activity, helping him to stay one step ahead of the enemy. Meanwhile, gunfire rang out from the range next to the qalat and the explosions from mortars fired by Jim's team thundered down the valley, in a warning to insurgents who might challenge the tribe and their American comrades. The Mohmands, already relatively unified, were emerging as the strongest tribe in the area under the alliance of Noor Afzhal and Jim.

The Taliban, led by Abu Hamam, was caught off guard by the rapid spread of arbakai in the district and immediately tried to disrupt it. The insurgents had influence outside the Mohmand area such as in the village cluster of Kawer, along the stretch of road known as Zombieland where they had frequently placed bombs. In February, Jim received a report that insurgents had placed another bomb there. He found the bomb booby-trapped with a grenade, which he detonated by throwing a rock at it, the blast perforating his eardrum. Unlike Mangwel, Kawer was not unified by one strong tribe, giving the Taliban more opportunities. A U.S. military patrol from COP Penich killed four of Hamam's fighters in Kawer in February, angering the insurgent leader. The arbakai commander in Kawer, a taxi driver named Gujar, was said to be playing both sides. The Taliban thought he tipped off the Americans who killed its fighters, and Jim believed he was involved in placing the bomb that injured him. One day, when Gujar was returning from the city of Jalalabad, the Taliban attacked, shot him dead, and took off with his body. Gujar's force of eighteen men quickly disbanded. Jim drew an early lesson from that failure: go in hard, fast, and big with the arbakai.

But apart from the setback in Kawer, the arbakai in Khas Kunar District stood their ground against threats from the Taliban.

The security gains in Mangwel translated into economic and development opportunities. The arrival of Jim and his team brought the community badly needed jobs, not only for the arbakai but for workers on the qalat and laborers performing development projects in the village. Umara Khan had been working for the past six years serving tea at a girls' school, paid between 1,800 and 3,500 rupees a month. Others had worked as security guards or police far from Mangwel. Most were unemployed and survived on farming and day labor. The team launched projects to refurbish and dig wells, improve irrigation and water systems, build retaining walls, increase the supply of electricity, put in streetlights, and provide school supplies and better corn seed.

Hearing life in Mangwel was changing for the better, dozens of families that had fled from the village to Pakistan during the 1980s Soviet occupation of Afghanistan began moving back to their homes and farms. Attracted by the improvements in security and the creation of jobs, they started to invest in repairing some of the scores of qalats damaged and gutted during decades of war.

Meanwhile, elders from other clans of the Mohmand tribe began descending on the qalat in Mangwel from villages and valleys nearby and across the region, asking Jim to bring the local defense program to their areas. Seventy-five elders arrived from the neighboring Sarkan valley, a highland area just behind Mangwel. The Sarkan clan was the one that had a blood feud with Mangwel, with three killed from each village since the 1930s. Sarkan was locked in a land dispute with Mangwel when Jim first arrived there in 2003. The Sarkan elders remembered Jim, who had thrown his weight and a threat of violence behind Noor Afzhal and Mangwel in the conflict. They wanted first to make amends to Jim, and then to work with him to create tribal police in their area.

Other elders came calling all the way from Khewa, Kama, and Goshta districts, as much as a two-and-a-half-hour drive south in northern Nangarhar Province. The Khewa elders appealed to Jim to bring arbakai to their area, complaining there were no Afghan security forces or government presence in the district on their side of the river. They invited him to visit and discuss the program, and he agreed to go, the next day.

The enthusiasm in rural communities for tribal and village security forces was tangible, indicating that the program had potential well beyond Mangwel. What was less clear to Jim was the ability of the U.S. military and Afghan government to support a rapid expansion of the program. The view emerging from the ground, living with Afghans in a qalat, was increasingly at odds with that of the military and government establishments—sometimes dangerously so.

The next morning, Jim, Noor Afzhal, and several arbakai and U.S. soldiers dressed in Afghan garb headed down a rocky dirt road in two Hilux pickup trucks toward Khewa. Ish was driving. As they traveled, two U.S. military attack helicopters flew by overhead, then banked and turned back. Apparently one of the pilots wanted to check out what looked like a group of armed Afghans. "They are circling," Ish said. Suddenly, one of the helicopters fired a 30 mm cannon into an adjacent hillside. "Hold!" Jim said. The trucks screeched to a halt. "Everyone get out and put your guns down!" Jim said. He pulled out a reflective panel to signal to the pilots that they were friendly forces. Pfc. Miah Hicks jumped out one back door and Noor Afzhal the other. Umara Khan bailed off the back of the truck. The choppers circled once more and left.

The team traveled on to Khewa, where the tribal elders were gathered in a traditional council, or jirga. "We are very interested in arbakai, we have many men willing to join," said one of the leading elders, Ustad Ghafoori, a vice chancellor of Nangarhar University. Other elders agreed, as they spoke back and forth, drinking tea and eating sweets. They told Jim they would move forward. "We are at your service," he said as the team prepared to leave.

Jim turned to Miah as they walked to the trucks. "It's a hundred times more dangerous now than it was on the way here," he said. He was mainly concerned about protecting Noor Afzhal. After they traveled a little ways, he stopped both trucks in a dip in the road. "Get in the other vehicle," Jim told Noor Afzhal, having him switch from the white truck to the black one. But when the threat came, once again it was not from insurgents.

A U.S. Kiowa helicopter came into view and flew close to the trucks. A second later, it fired a rocket into a nearby ridgeline. It was unclear whether the rocket was a test fire. Jim stopped the trucks and threw out a green smoke grenade to signal they were friendly forces. "The Americans are the most dangerous people on the battlefield," Jim said as he returned to the truck, "by far."

The men of Tribe 33 frequently traveled disguised as Afghans, and the ability to slip in and out of towns and cities in low-visibility operations was critical to skirting insurgent attacks. Even on the street, the Americans were regularly mistaken for Afghans. Justin went on a mission with Jim, Ish, and some arbakai and stopped in a town outside Jalalabad. An Afghan man approached to shake hands with the arbakai and then greeted Justin exactly the same way in Pashto, not realizing he was foreign. Yet ironically, the more they blended in, the more vulnerable they became to strikes by their own U.S. comrades. Looking like Afghans caused other problems. Jim and his team sometimes had difficulty gaining access to U.S. bases. And once Jim was angrily confronted as he washed his hands in the men's room of COP Penich by a U.S. Army private.

"What the hell are you doing in here by yourself?" the private yelled at Jim.

Jim looked at him.

"If I see you in here again by yourself," the private continued, "I'm gonna kick your ass!"

In one move, Jim stepped to about six inches from the private's face, and replied in perfectly good English.

"Hey, buddy, first thing: I will stomp the living fuck out of you," he said slowly.

Then Jim pulled out his military ID card, which he happened to have in his front pocket because Penich guards had lately started to hassle him at the gate.

"Number two, you cock-sucking motherfucker, get at attention!" Jim said.

The private, by then blanching, snapped to.

"Third. I should beat the shit out of you anyway, because the only reason you are treating me this way is because you thought I was an Afghan," Jim said. He turned and walked out.

He told the private's first sergeant about the incident, and the private was relegated to guard duty—at least temporarily.

Such incidents increased the sense of alienation Jim and his men felt from their conventionally minded military peers. The more they dispensed with their Army uniforms, flak vests, and armored vehicles and integrated themselves into Afghan life, the more they grasped what their true mission was, how uniquely they were executing it, and how divorced their operations were from the enemy-obsessed, highly kinetic, door-kicking approach of other military units in Afghanistan.

"We are doing something 180 degrees different than other people," said Mitch after a walk around Mangwel one spring day. "I like the fact that we can walk outside and immediately when we get to town five or six people greet us and invite us to have chai and go into their qalat. We are not walking around with weapons and a crapload of armor. We are showing them we are not afraid of what happens, or of them," he said. "We try to be as close to them as we can."

Drew, the machine gunner, felt the same way. "If they were doing this all over Afghanistan, the war would be over," he said. "This works. It's something you have to see to believe. It's a different kind of warfare. Sometimes you use bombs and bullets, and sometimes you need another method—relationships."

All sorts of people began arriving at the qalat seeking aid and protection. An old, nearly blind man who had lived in Mangwel his entire life and was struggling to support his family came and found Jim, who gave him a job handling trash. "You are the only person who cares about me," the man said, his cloudy eyes welling with tears. Chevy, an eleven-year-old boy whose father had been killed by the Taliban and whose mother was ill, rode a bicycle several miles to the qalat and was taken in by Jim to wash clothes and do other chores. A man from the contested village of Kawer, Fazil Rahman, was transported to the qalat deathly ill with an infection from a gunshot wound. A security guard who had helped defend the Special Forces camp inside the U.S. military base in Jalalabad, Rahman had been shot in the back by the Taliban when responding to an ambush near his village. He fled to Pakistan, and the Taliban burned down his house. Jim asked the U.S. military to send a helicopter to evacuate Rahman and provide medical care—but the answer came back no. "I might as well order a goddamn aircraft carrier to sail up the Konar River," Jim fumed. Eventually Jim prevailed and Rahman was taken to a U.S. military hospital, but it was too late and he died a few days later. Jim supported Rahman's widow and children for months in Mangwel until other male relatives arrived to care for them.

Such actions spoke far more loudly to the local people, and did far more to counter the insurgency, than killing any number of Taliban.

One day Jim was stopped on the main road in Mangwel by an old man.

"I am praying for you," the man said simply. "May God bless you for what you are doing."

Jim, a non-Muslim and so by definition an infidel to Muslims, had never been told such a thing by an Afghan, and was moved almost to tears.

Not long after that, Jim and Ish were on a walk when a girl of about twelve approached them on a footpath beside a field. The girls in the village had been largely invisible since the Americans moved in. But this girl did not turn away or hide. As she passed by she looked at Jim and addressed him in Pashto.

"My uncle, how are you? It is good to see you," she said.

# CHAPTER 18

IN MARCH 2011, AS Jim and his team forged ever closer ties with the tribe, I landed in Jalalabad, capital of the eastern Afghanistan province of Nangarhar, on another trip to cover Jim's story. There were just four months left before President Obama's Afghanistan troop surge clock ran out. The small propeller plane descended past fields and mud-brick houses and skimmed over the teeming streets of the city of nearly a million people. An ancient crossroads some ninety miles east of Kabul on the main trade route between Afghanistan and Pakistan, Jalalabad was exotic, much like a beautiful woman with a scarred face. No stranger to war, the city had fallen prey in recent years to suicide bombings and other sensational attacks by Taliban insurgents. Violence had escalated in Jalalabad, as it had in Kabul and other major cities. Ten years into the conflict, it was more dangerous than it had been on day one. Militants were staging complex attacks with bombs, rockets, and rifles. Many of the biggest strikes were aimed at killing Americans on the city's massive U.S. military base—the one at which I had just landed.

The plane taxied down the landing strip on Jalalabad Air Field, located inside Forward Operating Base Fenty. Once a Russian base, the sprawling facility spanned an area the size of twenty football fields and was surrounded by high, reinforced concrete walls and dirt-filled barriers topped with barbed wire. I was able to fly directly onto JAF because of a gaping security loophole that Jim and I had figured out how to exploit. To be honest, the loophole wasn't that big—we just had the brazenness to use it. But we both knew it could close at any time. Would it happen this trip? If military authorities found out I was slipping under their noses, Jim's entire mission could come to a crashing halt.

I unzipped the top pocket on my backpack, checked my passport and official DOD "letter of authorization," and mentally rehearsed my cover story. I was a U.S. Defense Department contractor, providing educational aid as well as outreach to rural Afghan women. It was all technically true—but just a glossy veneer, enough to appease a casual inquiry by a base security officer. If anyone dug deeper into what I was doing in Afghanistan, we'd have a problem. To buy my air ticket on DFS Middle East, the expensive charter service that flew contractors onto JAF, I had to submit a required form with passenger details such as my name and passport number. The form also had a required section on "badge details" with boxes labeled Option 1 and Option 2. Option 1 asked for specifics of my military, DOD, or contractor badges, of which I had none. Option 2, however, simply asked the purpose of my visit and my "contact person" on base. Bingo. I listed Jim's rank, name, and phone number, and got my ticket. Once I arrived at JAF, though, I had to get to Jim before anyone else got to me.

The flight attendant opened the aircraft door. I put on my sunglasses, readjusted the flowing blue Afghan scarf around my face, and smoothed my embroidered tunic and pants. Complicating the operation, I had to step off the plane dressed as an Afghan woman. On a military base full of Americans, it would have been better to wear a T-shirt and khakis. But once Jim met me, we were heading directly into Jalalabad, where I had to blend in as an Afghan. I walked down the steps of the plane to the runway, picked up my duffel bag from a pile of luggage unloaded from the plane's belly, and began hoofing it across the runway to the air terminal with the other passengers. Heat waves rose from the tarmac, and sweat dripped down my neck under the scarf as I walked. Just then, I saw a middle-aged soldier in a crisp uniform and a cap rapidly approaching. I gripped my bag and looked ahead. The soldier seemed to be headed straight for me. I scanned the area for Jim and spotted him in the distance, waiting for me at the flagpole in front of the terminal as planned.

"Ma'am," the soldier said, blocking my way.

"Yes?" I said, my heart beating faster.

"Need help with that bag?" he asked with a southern twang, breaking into a smile.

"Sure," I said, trying to stay calm.

We walked side by side, chatting, until we were almost to the flag-pole. Jim, wearing a white Afghan tunic and pants with a plaid scarf around his neck, was watching us.

I stopped, and as if on cue, Jim stepped forward.

"Thanks, bro," he said, addressing the soldier. "I'll take it from here." He took my bag from the soldier, who looked surprised but said nothing.

I followed Jim quickly down a dusty gravel road to his white Toyota Hilux pickup truck—the same model favored by the Taliban. He swung my duffel into the truck bed and opened the back door for me.

"Hop in," he said.

I slid in and spotted on the seat a leather holster with the 9 mm pistol Jim had taught me to shoot. He'd also taught me to fire both his rifles and use grenades. I put the pistol in my lap, where it was ready but out of sight.

Jim got in the front passenger seat, his AK-47 rifle propped against the dashboard.

Jim was one of a limited number of U.S. officers with standing approval from his command, under what was known as a Level One CONOP, to navigate the country on low-visibility operations. He took full advantage, driving Afghan-style vehicles, wearing Afghan clothing, and carrying an AK-47. Since he never wore body armor, neither did I. Although traveling as Afghans had its risks, at times it was a huge advantage; in that guise we bypassed suspected bombs and once pushed straight through a Taliban ambush on an Afghan army convoy.

"Let's go!" Jim said, turning to a blond, bearded American in a tan Afghan suit who was hunched over the wheel. A friend Jim nicknamed "Mullah," the driver was a former Special Forces sergeant who was on Petraeus's counterinsurgency advisory team and was writing about local security efforts.

"Roger that, kimosabe," the driver replied. "And good to meet ya, Ann," he said, grinning at me in the rearview mirror as he put the truck in gear and stepped on the gas.

The truck rolled slowly down the road through the small military town that was JAF, past brigade headquarters bristling with antennas, chow halls, and gyms fortified with concrete T-barriers, yards stacked high with shipping containers, and convoys of heavily armored vehicles. Then we turned down a narrow side road that ended at a gate made of chain-link fence. It was locked.

"Shit!" Jim said. "Where's my guy?"

We were trying to get out—just as Jim had come in—through the low-vis gate at JAF, a back door to the high-security base intended to allow Afghan sources to come and go quietly and quickly. The gate had far less traffic and fewer guards than the base's main gate and was used both by the CIA and Special Forces.

Jim climbed out, paced for a minute, and lit a cigarette. Five minutes later—it seemed like forever—a Special Forces intelligence soldier in civilian clothes came hustling toward the gate.

"Sorry, man," he said, pulling a key out of his pocket.

"No worries, bro," Jim said, tossing down his cigarette butt and crushing it with his sandal. He climbed back in the truck and off we went.

Mullah swerved past a serpentine series of concrete barriers to the outer wall of JAF and drove through a final checkpoint manned by Afghans. I let out a sigh of relief. We'd made it—at least over that first hurdle. Now, what awaited us in Jalalabad?

In an instant, we escaped the drab, monochrome base and were swallowed up by the teeming, kaleidoscopic streets of Jalalabad. We passed fruit and vegetable markets and small shops laden with tea, spices, and big sacks of rice. Clothing stalls fluttered with brightly colored dresses and woven garlands of paper flowers that Afghans wore at weddings and other special occasions. Groups of schoolgirls walked by dressed in flowing black uniforms topped by ghostlike white head coverings that extended to their arms.

As I soaked in the city, Jim looked out his window, scanning the crowded streets for danger. Had we triggered an insurgent early warning network when we left the base? He knew the enemy watched all the gates, even the low-vis one. All it would take was for a spotter to flag our vehicle and target it with a VBIED, or vehicle-borne improvised explosive device—his biggest concern. He ran through a mental list of other threats and contingencies for each: a U.S. military or Afghan checkpoint, a vehicle breakdown, friendly Afghan intelligence picking us up and passing it on to an American unit that would mistakenly target us . . . or simply a random attack.

Yet as we passed down roads in central Jalalabad swarming with rickshaws, bicycles, sedans, and minivans, we were largely unnoticed by the Afghans passing us just feet away. Jim had learned—and taught his men—the tactical fine points of avoiding looking like an American: no wristwatches or Oakley sunglasses, the right Afghan headgear, M4 carbines kept out of sight. Jim had also consciously toned down the heightened sense of danger he projected, deliberately looking through the Afghans rather than at them, not locking their eyes with his killer gaze as he once had.

The landscape opened up as we approached the outskirts of the city, two-story buildings and high-walled compounds falling away and turning into open dirt lots with views of mountain ranges in the distance on either side. Mullah turned down a long gravel drive toward our destination: a large, secluded guesthouse known as the Taj.

The Taj was run by Dr. Dave Warner, a brilliant and irreverent hippie from California who was rumored to have ties to the intelligence community. Dave was the only person I'd met who had received an honorable discharge from the Army for using hallucinogens and failing a rehabilitation program. I checked in with Dave regularly, as I was technically working for him, and he and his organization were providing cover for my presence with Jim in Afghanistan. Dave had an MD and a PhD and ran an offbeat organization called the Synergy Strike Force, which carried out Internet-centered development projects in Afghanistan. Dave considered himself the supreme ninja warrior of the government-contracting world—and had the deals to prove it. For critical periods, Dave provided me with official Defense Department letters of authorization and funds for travel. In exchange, I would assist with some education development projects in the village of Mangwel. Jim was taking me to Mangwel to live with him the next day. That night, in a brief respite, we were going to enjoy being back together.

A single Afghan guard stationed in front of the Taj waved us past, and the main gate swung open. They were expecting us. We pulled up to the side of the building, its grounds replete with well-tended flowering plants, giving it an exotic feel. We stepped down a walkway and into a grassy courtyard. On one side was a bamboo bar with a patio and lounge chairs. Beyond that was a small but deep swimming pool. The Taj at the time had one of the only bars in eastern Afghanistan. It served as an enclave where like-minded foreigners and Afghans working for various nongovernmental organizations could relax and share information outside the constraints of the conservative Muslim society. If Graham Greene or Sean Flynn were alive, they'd be at the Taj. We settled into our room, with high ceilings, a double bed, and a private shower. I pulled an orange, lemon, and lime out of my bag along with a bottle of Jim's favorite tequila, Patrón, that I'd picked up at the duty-free store in Dubai. He mixed us a couple of margaritas and we sipped them by the pool, and then spent the first part of the night dancing on the roof to Jay-Z. Mullah put on a favorite slow dance song for us. As I swayed in Jim's arms while the sun dipped behind the distant mountains, I felt happy and excited about our trip to Mangwel. After a couple more drinks, we went to bed.

The Taj was only lightly guarded, and that night, as usual, Jim slept with his M4 carbine and AK-47 rifle next to him. But a few hours later, a loud explosion rattled the windows of our room, waking us up. In the darkness, we heard a U.S. helicopter swoop overhead, followed by more explosions. Rockets. Jim jumped up, shirtless, and grabbed his rifle.

"Stay here!" he told me, and handed me his 9 mm pistol. "Shoot anyone who comes through that door!" He went out of the room and tried to get up to the roof to see what was going on, but the door to the upstairs was locked. He listened and heard no activity at the Taj. So he came back and barricaded our room with chairs, a table, and every other piece of furniture he could push against the door. He pulled a couple of grenades out of his ammo pouch. With no way of knowing whether the Taj was safe, he stayed awake and on edge the rest of the night.

In the early morning darkness, Jim suddenly got violently ill as a result of some combination of liquor and pharmaceuticals he'd taken the night before. Sitting on the floor under a steaming shower, recovering, he wiped his face with a wet cloth. I showered and sat next to him. He looked at me with distant eyes.

"Ann, you need to know something before we get back to Mangwel," he said slowly. "I am so deep into this now. Deeper than when you were here before. The tribal elders are coming to me. They trust me," he said.

Since arriving in Afghanistan just under a year earlier, Jim had worked as both an advisor and a commander to help expand the local security program to more than a thousand Pashtun tribesmen in eastern Afghanistan, including Mangwel.

"I'm afraid this won't end well," Jim confided, his voice low and anguished. "When the U.S. bails out of here, these people who have sided with me, and believed in me, are going to be slaughtered."

I looked at him intently. Jim trusted his gut because it almost never let him down—many times it had kept him alive. The problem was, even if Jim was right, we both knew there was no turning back now.

The next morning, Jim pressed ahead with his mission. We drove from the Taj to the heavily guarded Jalalabad palace of Gul Agha Sherzai, the governor of Nangarhar Province. Located down a long drive lined with carefully manicured hedges and gardens, the palace was a symbol of Sherzai's opulence and power. Sherzai fought alongside U.S. Special Forces in the southern province of Kandahar during the 2001 campaign to overthrow the Taliban regime, earning his name, which means "son of a lion." He was one of several influential strongmen Jim was meeting to push his tribal strategy forward. Sherzai had been accused of corruption as had many senior Afghan officials. What mattered was getting the power into the hands of the local tribes, and Jim believed Sherzai could help with that. Jim had gained approval to set up Afghan Local Police in three districts of northern Nangarhar, but government officials there were interfering with the ability of local tribal elders to select men for the job. Jim had not gained authorization from his U.S. military command to travel to Jalalabad and meet with Sherzai, but he knew that unless Sherzai intervened, the local police program would fail in northern Nangarhar. So he reached out to Sherzai directly and was invited to the palace.

We were ushered into a large reception room lined with plush chairs and tables laden with cups of tea and platters of almonds, dried chickpeas, and raisins. A moment later a heavyset man with a thick black mustache walked in and sat at the head of the room, next to Jim. It was Sherzai. As often happened in such meetings, the two men established their credentials as warriors.

"I had fifty Special Forces guys fighting with me in Kandahar," said Sherzai, referring to the 2001 battle to take the Taliban stronghold in southern Afghanistan. Jim's boss, Bolduc, later worked with Sherzai in Kandahar. "Aircraft were dropping bombs everywhere," he said.

Jim wanted to know where Sherzai stood on the local police. He explained the government interference in northern Nangarhar, and asked Sherzai for his advice.

"More than a hundred years ago when Afghanistan had a king, there was no national army or police force. It was just a tribal army, like you are building now," Sherzai said, taking a puff of his cigarette. President Karzai, he explained, had been opposed to the American plan to raise tribal police. "At first, Karzai was concerned that tribal forces would just get the weapons and break the law, but I took the tribal elders to see him and we convinced him to support it," Sherzai said. "It is important for the tribal elders to guarantee who becomes arbakai. I will invite all the district leaders and tribal elders from northern Nangarhar here to discuss it."

"That is a good idea," Jim said. "I will be here when they meet."

Another step forward.

We moved into a vast dining room with a long wooden table, where dozens of Sherzai followers sat down to a lavish meal of soup, rice, chicken, and spinach with Jim and me. Jim and Sherzai traded jokes about growing up and being beaten by their fathers. I discussed tribal history with Sherzai and one of his advisors. Meetings with Afghans always involved long periods of social conversation and brief spurts of business. Jim knew that and practiced it religiously, never wearing a watch to such meetings. It was one of his twelve principles for working with Pashtuns.

We said goodbye and got back into the Hilux to head for Mangwel. Jim handed me the 9 mm pistol in a holster to wear under my gray-blue burkha, the flowing cloak that Afghan women wear to cover themselves when outside of the home. With only a small, netted opening for my eyes, it was sweltering but necessary. Jim was concerned about not triggering the insurgent early warning networks as we left Jalalabad. Having a woman in a pickup truck without a burkha would be noticed. We crossed the Behsud Bridge over the Kabul River and headed northeast along the Konar River. The biggest threats on the route from Jalalabad to Mangwel were roadside bombs and direct-fire ambushes. If we were spotted at the bridge, an insurgent there could phone ahead and identify the truck, saying we were the only white Hilux headed their way. Down the road, the Taliban could lay out an IED watched by two or three attackers with a machine gun—and we would be done.

We passed through farming villages and ripening cornfields and crossed the Konar River at the *Spin Jumat*, or White Mosque Bridge.

Ten minutes from there, the road twisted as it curved around the ambush spot nicknamed Zombieland. After that, we entered the territory of the Mohmand tribe and came to the first rectangular, mud-walled homes of Mangwel. At the far end of the village, we turned down a dirt road to the qalat where Jim and his team lived with the arbakai—and now I too would stay. I was excited but apprenhensive. I knew I had to win acceptance from them all.

I got out of the truck in my Afghan dress and stood there, unsure what to do next. Just then, a small Afghan man with green eyes approached, smiled at me, and greeted me respectfully. I recognized him as Jim's old friend from 2003, Umara Khan.

"*Tsenge yay, khor?* How are you, sister?" he said, placing his right hand over his heart.

"*Ze khaiem, manana.* I am fine, thank you," I answered in Pashto, imitating him and putting my hand on my heart. Other tribesmen looked on but kept a distance.

My relationship with the Pashtun men in Mangwel was an extremely delicate matter. Mangwel was light-years away from the cosmopolitan city of Jalalabad—it was a traditional, rural Pashtun village that had changed little in recent centuries. It had no electricity, running water, or computers. Its people still lived and breathed the strict tribal code of Pashtunwali. Under that code, a Pashtun man had to fight for his honor at all costs by protecting what was his—his land, his women, his guns. Jim had gathered all the arbakai in a tent on the qalat before my arrival and told them about me. I was his wife, his *khuza*, and he expected them to respect me and protect me as such, he said. In bringing me to Mangwel, Jim was taking an incredible risk. If any of the tribesmen disrespected me in the slightest, he would be honor-bound to fight them, a conflict that could endanger his hard-won relationship with the Mohmand tribe. I could also inadvertently provoke such a conflict if I did not take care to live according to Pashtunwali—to cover myself in the baggy clothing of rural Pashtun women, to always walk behind Jim in Mangwel, to act demurely around Pashtun men, and follow many other rules that were foreign to me. The sun dipped behind the mountains, and I settled in for my first night.

Early the next morning, Jim came to me excitedly. "Sitting Bull is here to see you!" he said.

I pulled on a black embroidered dress, billowing maroon pants, and a scarf that I had bought at an Afghan market in Virginia, and followed Jim into a tent on the qalat. Noor Afzhal stood from a cot to greet me, touching his chest with his hand. He smiled and offered me a rich, sugary flatbread with homemade butter, plus milk tea—all carefully transported from his home.

"Welcome," he said in a deep voice.

"I am very happy to be here," I replied in Pashto. I gave him a gift—a warm vest—and some sweets for his wife.

"Have some tea," he said, filling my glass cup from a silver kettle.

As we talked, we heard the patter of rain on the tent roof and then the rumbling of thunder in the distance.

"You are good luck, you bring the rain," Noor Afzhal told me.

"Do you like it here?" he asked.

"It is beautiful," I said. "The mountains and river remind me of my home."

Noor Afzhal smiled again and took a sip of tea. Then his face grew serious.

"Jim is my fifth son, and you are my daughter-in-law," he told me. "Here you must follow Pashtunwali and live the Pashtun way."

"I understand, and I will," I promised. I was the first American woman Noor Afzhal had ever met and allowed into his village this way.

Jim's decision to bring me to Mangwel marked an ultimate sign of his trust in the tribe, and Noor Afzhal's willingness to accept me also signaled Jim's status with the tribe.

But I had to play my role as a strange hybrid of an American and an Afghan woman. And apart from morphing as best I could into a Pashtun wife, I had to continue to hide my presence from military authorities and other prominent visitors to Mangwel, and also to accept Jim's total control over where I went and when. Jim decided whether I went on missions or not—he bore the overwhelming responsibility of having the woman he loved in his care in a war zone, a burden that, try as I might, I could never completely understand. It was as it should be. Still, for a war correspondent used to pushing the limits—wearing combat boots and baseball caps and rolling out the gate on every mission—it was sometimes easier intellectually than emotionally to stay in my place.

The first day in Mangwel, I moved into our tiny room in the qalat.

Jim and I lived inside an eight-by-sixteen-foot room built inside a shipping container. The container was sequestered at the end of a narrow corridor in one corner of the qalat. The floor of the room was tilted, and inside there was just enough room for a bed and a few shelves. A sign on the door read: "TOP SECRET—Stay the fuck out if you do not have a TS SCI clearance—TOP SECRET." Jim and his team and a handful of close confidants were the only elements of the U.S. military who knew I was living there. No senior officer in Jim's chain of command was aware of it. Whenever generals, senators, Obama administration officials, or other VIPs came to visit Mangwel, I stayed in the room. It was a particularly strange feeling because I knew most of the visitors and had interviewed them repeatedly—and they knew me by name as well. I was itching to grill them on everything I was learning about the strategy in Mangwel, but I had to stay out of sight and bite my tongue.

The same was true for missions. When planning missions, Jim referred to me as "X." I would check his mission diagrams, drawn with colored markers on butcher paper on an easel in the operations center—to see if there was an X in one of the vehicles or not. In the beginning, I would ask him if I could go. Later, I learned not to.

The dangers were real, as I was reminded the day after we arrived. Despite our efforts to stay under the radar as we traveled from Jalalabad, insurgents had spotted our faux Afghan caravan. A report made its way to the military that night that Abu Hamam had talked about killing an American woman. Hamam was the Taliban commander for the nearby Shalay valley and surrounding areas. Thirty-five years old, with long black hair and a black beard, Hamam was known for kidnapping and ruthlessly killing Afghans he suspected of supporting the Afghan government or U.S. forces. He made videos of the executions to intimidate other villagers. He had also ordered attacks that had killed several U.S. soldiers. Jim knew from sources inside the Taliban that Hamam had a crude sense of humor and was also an uneducated but effective commander. He made the most of his scraggly band of sandaled insurgents who lived in the mountains and survived on whatever food and shelter they could get from the people. They waged a primitive form of warfare, using smoke signals to warn of approaching U.S. troops and then ambushing them with old Russian weapons. The fighters were agile, unencumbered by armor, and knew every footpath and village in the valley. Hundreds of dry streambeds snaked through the Shalay, allowing them to hide and stage lethal attacks virtually unseen. Ghostlike, they vanished as quickly as they appeared.

A military friend—a smart young officer in Jim's chain of command who respected Jim's expertise and knew about me—alerted him about Hamam in an urgent email: "Jim, heads up . . . where is Ann?? We have some [information] from Khas Kunar/Abu Hamam where they are discussing plans to kill an American woman. They are also mentioning you by name."

Jim, disturbed and angered by the intelligence, spoke with me harshly that night in our room.

"Ann, you must never leave the qalat, even to step outside the gate for a minute. Do you understand?" he said, almost glaring at me. "If I ever come back and I do not know where you are, this is over," he said.

"Yes," I said with a nod, looking straight at him and stiffening. I knew I could say nothing more. I felt what it was like to take orders from him—a twinge of what his men felt every day. I also knew it was because he loved me, as he loved them.

"Always know where your AK-47 is. If I ever ask you and you do not know where that rifle is, I will be furious. Do you understand?" he continued, his voice rising and his face as stern as I had ever seen it.

"Yes," I answered again.

"They are watching us. Every minute. They know we are here, they know when we leave, they know about you . . . I can hear them out there, right now," he said. Then he walked out, letting the wooden door to our room shut loudly behind him.

I gathered my composure, put my head scarf back on, and climbed a wooden ladder into one of the two unmanned guard towers in the corner of the qalat above our room, listening, and looking at the moon.

# CHAPTER 19

A FEW DAYS AFTER arriving in Mangwel, I went on one of my first long walks around the village. It seemed a little surreal as I followed Jim and Abe—just the three of us—down a dirt path that ran between rustling fields of green cornstalks on the outskirts of the village. I felt a strange mixture of sensations as I took in the tranquil beauty of the Konar River valley.

We wore Afghan clothing and no body armor. Our only overt protection was the rifles Jim and Abe carried. I felt freer but also vulnerable without the heavy plates I'd lugged around for years on patrols in Afghanistan and Iraq. I had on my one black embroidered Afghan dress and sandals, and tried to keep my head scarf firmly on while watching my footsteps to make sure I didn't trip on the uneven path.

Ahead of us the fields stretched for about a quarter mile to the Konar River, which fed the narrow strip of crops with water channeled through irrigation ditches. Across the river a string of mountains rose under a clear blue sky. This was the heart of rural Konar. I knew from my research that about four hundred thousand people lived in the province, 98 percent of them Pashtun and 96 percent rural. Konar covers some sixteen hundred square miles, an area about the size of Rhode Island, and is shaped like a sideways bow tie resting on the border with Pakistan. The Konar River runs through the middle of the bow tie, roughly northeast to southwest and parallel to the rugged Pakistani frontier. The hills and mountains that make up 90 percent of the province rise from the flat plain of the valley. The idyllic landscape is deceiving, though. Konar is one of the deadliest places in Afghanistan. As we moved down the path, I was uneasy knowing that cornfields often provided cover for insurgent attacks.

The path entered a shady grove of trees, and Jim stopped to talk with an elderly Afghan farmer and his son. I noticed through the trees in the distance a few Afghan women watching us in front of a mud-brick home. Then, unexpectedly, Jim turned to me.

"Go talk with them," Jim said, tilting his head in the women's direction.

I gave him a questioning glance.

"It's okay. Go," he said.

Jim had to keep his distance. It would have been a cultural misstep and violation of Pashtunwali for him—or any man outside their family—to approach the women and speak with them. As a woman, I could do so. Still, Jim wasn't sure how the Pashtun tribeswomen would react to me, likely the first American they had ever met. I didn't realize it at the time, but Jim was testing both them and me.

I walked toward the women slowly but deliberately, smiling as I went.

"*Salaam aleikum.* How are you?" I greeted them as I got closer. To my surprise, they didn't vanish inside the walls of the house.

"*Aleikum salaam,*" replied an older woman, dressed head to toe in black. She smiled and looked at me curiously. "Are you Commander Jim's wife?"

"Yes," I said.

Her face lit up. News traveled fast. Using my basic Pashto, I began to ask her about her family. How many people were in the household? How many sons did she have, and daughters? It turned out the woman was a cousin of Fazil Rahman, the man whose life Jim had tried to save after Rahman was fatally wounded by the Taliban. She was clearly grateful.

A few moments into the conversation, the woman invited me to sit next to her on a typical Afghan bed, made of woven rope on a wooden frame that was about the size of an Army cot. She poured me a cup of green tea from a kettle that was steaming on a small fire in front of us.

Then she reached out and touched the wide, beaded sleeve of my dress and pointed to the fitted ankle bands of my pants.

"You are wearing gypsy clothes!" she said, and burst out laughing.

Afghan gypsies own no land and the men stay home while the women work, begging or telling fortunes.

Realizing the costume I bought in the Virginia market looked a bit comical, I laughed, too, and replied: "That is because I am a gypsy!"

I had passed the first test and gained a degree of acceptance from local women. In a small but powerful way, my first foray into a world that was closed to Jim and his men deepened our connection to the tribe and also our safety.

Word traveled quickly. In the days that followed, visitors from the village and surrounding area began arriving at the qalat bringing gifts of clothing for me. One of the villagers who had relatives in Jalalabad gave me a glittering pink dress, pants, and scarf in the more fitted fashion worn by women in the city. The local police chief brought another "gypsy" outfit, a turquoise gown with hundreds of shiny pieces of mirror sewn onto it. But my favorites were the hand-stitched dresses made by the wives of the arbakai in the rural Pashtun style—simple tunics with straight sleeves worn over matching baggy pants.

Then one afternoon Noor Afzhal arrived at the qalat and asked to see me. I met him in the arbakai tent. As we sat together, he took out two cardboard boxes and handed them to me.

"These are for you," he said.

Inside one was a beautiful handmade red tunic, scarf, and pants, ornamented with multicolored embroidery and sequins. I opened the other and found wrapped in tissue paper a shiny pair of red plastic sandals. All were my size.

"Thank you very much, it wasn't necessary," I said, using a Pashto phrase I had learned was polite.

"You and Jim can come to my home tonight, we will all have dinner together," he said, a statement as much as an invitation.

"I would be delighted to," I said.

I had shown Noor Afzhal I would respect the Pashtun ways, and so, it seemed, I was passing the test with him, too.

Late that afternoon, Noor Afzhal swung open the heavy wooden door at the entrance of his house and invited Jim and me into his *hujira*, or guest room. His sons Asif, Azmat, and Raza Gul were waiting there. We sat on a crimson patterned carpet and leaned against large pillows, talking. The hujira was the formal greeting place in Afghan homes where men and male visitors gathered to discuss weighty issues. I never saw an Afghan woman in the hujira, but as an American, I was allowed not only to be present but to participate, as I did that night. After a short time, though, Noor Afzhal got to his feet.

"Come with me," he said with a smile.

Walking with his cane, he left the hujira and ducked through another doorway into the inner courtyard of the home. I followed, leaving Jim and the others behind. Suddenly, I was surrounded by a group of women— Noor Afzhal's wife, Hakima, two of his five daughters, and three of his sons' wives—as well as several children. Dressed in black, Hakima took my hand and motioned me to sit on a pillow placed on one of the woven rope beds. Looking around as we talked, I immediately felt how lively and relaxed that space was—how it was the heart of the Afghan home.

The inner courtyard was open to the sky and had a packed-dirt floor. In one corner, a wood fire burned in a small pit lined with stones. One of Noor Afzhal's daughters stirred a pot of stewed chicken for our dinner. Another baked wheat flatbread in a pan on the fire. Adjacent to the fire pit was a small manger enclosed by a low mud wall and covered with a thatched roof. In the manger, a goat, two cows, and a donkey chewed on hay. Several hens strutted around the courtyard flapping their wings, as the barefoot children chased them and played on a swing hanging from a roof beam. Along one wall of the open courtyard were four bedrooms, one for Noor Afzhal and his wife, and the rest for his three married sons and their families. The rooms were to me like art pieces of simplicity and functionality. Babies and toddlers slept in small cloth hammocks hung by the sides of the wooden beds. Brightly colored pillows and blankets lay neatly folded on the beds. Shiny aluminum pots and kettles and painted enamel bowls on wooden shelves decorated otherwise bare walls.

In coming weeks and months, I would spend many hours in the interior of Noor Afzhal's home, speaking in my broken Pashto with his family members about their daily lives. His wife, Hakima, and his daughters Farmeen and Naida showed me how they made bread, yogurt, and other Afghan foods. They tried to teach me how to milk their cow and laughed at the dribble I squeezed from the udder into the pail. They joked about how hard they worked both at home and in the fields. But even then, they delighted in showing me how they beautified themselves. They sewed clothing for me on a hand-powered machine, and braided my hair with colorful tassels made of thread. Naida painted my palms, nails, and fingers with ornate designs in henna. She mixed a paste made with dried, ground henna leaves and squeezed it onto my hands like frosting from the tip of a tube. After letting the paste dry for about an hour, I washed it off, revealing an intricate reddish brown design. Then she squeezed glittery plastic bangles onto my wrists.

I played with the children and prompted the school-aged ones to practice English with me and recite the alphabet. I told them about my own children and how I missed them, and they nodded with understanding. One of my happiest moments was when Hakima allowed me to cradle Noor Afzhal's newborn grandson, swaddled in a tiny bundle.

I visited the separate schools for girls and boys, met with the teachers, and learned about their needs. Later through donations I was able to provide all the children with backpacks, notebooks, and pens and pencils, and outfit the schools in Mangwel and other villages with basic supplies from chalk to carpets. I also offered to train local female teachers to expand the grades at the girls' school.

From then on, whenever I went on walks with Jim, women from families in the village would peek out of the big wooden doors of their homes and beckon to me (in the Afghan way with their hand downturned), inviting me in to chat with them. I would disappear inside for as long as I thought Jim could comfortably wait without getting too worried.

Still, I was surprised during one visit when a young woman I had been speaking with excused herself and then returned with a box. She opened it and took out a heavy book wrapped in cloth. Removing the cloth, she showed me a beautiful golden Koran. Then she began reading a prayer to me. Most Afghan women are illiterate, including 90 percent of those in Konar, but she had learned some prayers in Arabic.

"Now you say it!" she prompted.

"*Ashadu Allah Ilaha Illallah, Wa Ashadu Anna Muhammed Rasulullah.* There is no god but God, and Muhammed is the messenger of God," I repeated the Muslim creed, which I recognized, in Arabic. I had partly memorized the prayer a few years earlier at the urging of my Jordanian Arabic teacher. She wanted me to be prepared in case I was ever kidnapped in Iraq, and believed that if I recited the creed to my captors, it would save me from beheading or some other gruesome death. I carried an English transliteration of the prayer on a worn green note card in my wallet.

"*Ho*, good!" the young woman said.

She was delighted, and kept adding more lines in Arabic that I didn't know. The harder I tried to repeat it, the more excitedly she taught me.

Paying no heed that I was an American woman and an infidel, the tribeswomen were teaching me how to live as a Muslim and a Pashtun.

I WANTED TO BETTER understand my adopted Pashtun family and its origins, and turned to Noor Afzhal. At the age of nearly eighty, he had lived twice as long as most Afghan men. I asked him to share with me the story of his life. He agreed. We spent hours sitting together, and I filled page after page of composition notebooks with his reflections on his past. I treasured the time with him, and found that he, too, enjoyed our conversations and looked forward to them. He started arriving at the qalat every morning to talk with me, often bringing small bouquets of fragrant pink flowers he had picked and bound with straw to a stick.

Then one warm spring day, Noor Afzhal asked Jim and me to go with him to visit what he called the "special place."

The next morning, we followed Noor Afzhal as he slowly walked up the valley behind Mangwel, his worn wooden cane in hand. I was glad to see him use the cane, made of twisted hickory, that Jim and I had given him several months earlier. The cane, engraved with his name and ours, rarely left Noor Afzhal's side. It steadied his gait, but he wielded it just as readily to break up a fight or strike anyone who got out of line. With his shoulders thrust back, he leaned lightly on the cane as he navigated a rocky footpath familiar from his boyhood toward the stone ruins of his ancestral village, nestled between two ridges.

It was midmorning, and the sun beat down on Noor Afzhal, dressed in white jami and kandari hat. He pulled a cotton scarf from his chest pocket and mopped his brow. As he climbed the path, he passed on the northwest side of the valley a large face of stratified rock etched with primitive drawings of herds. Farther along, he reached an outcrop of boulders at the head of the valley just above the remnants of the village. A spring flowed from the rocks.

Noor Afzhal stopped next to one of the boulders, joined by Jim and me and a small group of Afghans and Americans who had made the pilgrimage with us to the home of his tribal ancestors. Noor Afzhal turned to his youngest son, Raza Gul, who was by his side. Lifting his cane, he pointed toward the far hillside.

"Fire!" he said.

Raza Gul squatted on his haunches and shot off several celebratory rounds, sending echoes crashing down the valley like so many breaking waves.

Noor Afzhal stood silently, his jaw set and lips pursed in thought. His large hands rested one upon the other atop his cane. His eyes traveled over the skeletal foundations of a lone qalat and down half a mile to the mouth of the valley that opened onto the corn and wheat fields of Mangwel. He told me that as a child of six or seven years old he had herded his family's sheep in the same rock-strewn hills. I could see that something was weighing on him. I wondered how the tribe he had led for decades would remember him. Would his people donate money to feed the poor, or plant poles festooned with colorful flags around his grave? Would the children know his name?

Later back in the qalat, he told me proudly what he remembered of his forefathers. Like many rural Afghans who are illiterate and rely mainly on oral history, Noor Afzhal was not certain when he or his ancestors were born and had few papers documenting events in his long life.

"My great-great-grandfather settled here from Manzarichina in the Mohmand tribal territory," he said slowly, referring to a town in what is now the Mohmand Agency, a Pashtun tribal region in Pakistan.

"Our way is Pashtunwali," he said, speaking of the code that Pashtuns have lived by long before and after they adopted the Muslim faith beginning in the seventh century.

The Pashtuns, I learned, are a people that since the first millennium BCE have lived in a region stretching from the Hindu Kush mountains in Afghanistan to areas west of the Indus River in Pakistan. The Pashtun tribes relished their independence and were never fully subjugated by the Afghan government. Afghan rulers depended on the support of tribal networks, and at times on powerful tribal confederations that installed them, but a distance always remained between the central government and the tribes. Tribal leaders and traditional laws were the only effective authority in large parts of the country where the government had no reach. While the Pashtuns are one people, the Pashtun tribes have long fought over land and other resources, but have come together when foreigners arrived.

His voice grew stronger as he spoke.

"The Mohmands are one of the largest and most powerful Pashtun tribes," he said proudly. "Our territory reaches south from Afghanistan's Konar Province into Nangarhar and across the border into Pakistan.

"My grandfather was a malik in the Mohmand tribe, and he served in a *lashkar* that fought against British forces," Noor Afzhal went on. When faced with an external threat, Afghan rulers called on each major Pashtun tribe to raise a militia, a tribal army called a lashkar. The grandfather, Fatih Mohammed, was part of the Mohmand tribal militia that fought in decisive battles of the Second Anglo-Afghan War.

In 1919, Noor Afzhal's father, Sher Mohammed Khan, joined tribal forces that rallied against British troops in the Third Anglo-Afghan War. Noor Afzhal remembered his father's tales of brave Pashtuns battling the British with swords, slingshots, and homemade explosives.

"The British were treating the people badly in Peshawar," he recalled. "So the big tribal leaders and maliks got together, chose a commander, and put a turban on him so he could lead all the tribes to fight. My father was one of them," he said. "He wanted to fight to the last breath."

Like many of his fellow tribesmen, Sher Mohammed was fighting not for the Afghan nation but for the Pashtun people. Born in Mangwel around the turn of the century, he deeply resented the British and their splitting of the Pashtun lands, Noor Afzhal said.

In a draconian measure, the British launched aerial bombing strikes on Kabul and Jalalabad during the war, and after about a month of fighting, the war ended in an agreement in which Pashtuns in Afghanistan were forced to accept a division of the Pashtun land. (In 1893, a demarcation called the Durand Line had been drawn down the middle of the Pashtun world, and half was appropriated for the British Empire's Raj, which included modern India and Pakistan.) Britain in turn recognized Afghanistan's independence in internal and external affairs. Afghans from then on considered 1919 as the year their modern nation was born, but Pashtuns also see it as the year when their territory was torn apart in the service of Western politics.

"When the fighting was over my father returned to this village to farm. He married and thanks to God started a family." Mangwel then had a few hundred homes, and the villagers—all from the Mohmand tribe—were mostly subsistence farmers who grew rice, corn, wheat, and opium poppy in fields irrigated by the Konar River.

Noor Afzhal's father, Sher Mohammed, was a kind but strict husband and father to three wives and five sons. Noor Afzhal was born around 1935, the youngest son of the brood.

"When I was a boy, I remember him saying, 'You are forbidden to go to village wedding parties,'" Noor Afzhal said with a smile. "Father believed all the music and singing and smoking of hashish was against Islam.

"But Ann, I will tell you, I loved the songs of those who came from Jalalabad with sitars, rababs, and drums. When I was this high"—he raised his hand to the height of a boy of eight or nine years old—"I began to be too tempted by the music and happiness of those weddings, and I and some others would sneak off to the feasts. I never ever smoked, never ever. But the food, how I remember the delicious food, spiced rice and fragrant meat slaughtered for the occasion.

"You know my father could not read, but he wanted all of us, his sons, to learn reading and writing," he said. "But my brothers and I had to work in the fields, too, and so after a few years, I decided to quit. Our land needed tending, and I was drawn to that kind of life. For my older brothers, things were different; they studied and one became a police officer and the other a doctor. For me, my education came sitting at my father's side. He had been chosen senior malik in Mangwel and leader of the shura for the entire district. He was an honorable and strong man and I wished to follow his example.

"I know in America things are different than they are here. Here we have the shura, where we sit with all the tribal and religious elders and government men. I would sit and listen, and each and every day men arrived at our house and sat in the guest room to ask my wise father to solve their problems. This is how I learned how to mediate conflicts between my people, by watching and listening."

As the head malik, Sher Mohammed helped resolve disputes and struggles with neighboring tribes. His reputation for wisdom and fairness grew.

"It is hard to give the people justice," Noor Afzhal told me, recalling the councils. "But even if your own family is guilty, you have to be able to make that decision," he said.

As he learned about the internal workings of the tribe, Noor Afzhal also absorbed his father's beliefs as a Pashtun nationalist. In 1947, when he was twelve years old, Pakistan, a Muslim-dominated territory that was part of greater India under the British, gained independence. On behalf of the Pashtun tribes, Afghanistan's leadership in Kabul tried to negotiate the dissolution of the Durand Line. The effort fueled the movement for the creation of Pashtunistan, a sovereign state that would unite all Pashtuns in Afghanistan and Pakistan. To no avail.

Noor Afzhal's apprenticeship in the Pashtun ways of leadership sealed his future when his father passed away during the reign of King Zahir Shah in 1955. As the family grieved, dozens and then hundreds of people arrived at the qalat from far and wide to pay their respects. The governor of Nangarhar Province came all the way from Jalalabad with an entourage of more than two hundred to honor the malik, bringing a wooden coffin. He also brought money for a feast for the poor, a tradition at Afghan funerals. The guests overflowed from the qalat into a cemetery in front of the home, where Noor Afzhal's grandfather was also buried.

After the burial, all the Mohmand tribal elders gathered and sat to discuss who would become the next malik. Noor Afzhal sat on the outskirts of the circle, listening. A few names came up, often that of his brother, Sher Afzhal, the policeman. Then it came time to vote.

The oldest of the elders, Malik Amin, tallied the results. The governor whispered to Malik Amin. He stood, unfurled a green striped cloth, and wrapped it into a turban. Then he walked over to Noor Afzhal and placed it on his head.

At the age of twenty, these voters believed, Noor Afzhal had the wisdom and judgment to lead the tribe. His heart was both heavy and proud. He had lost his beloved father but had won the trust of his people.

Two years later, the young malik was conscripted into the king's Afghan security forces and ventured out of his tribal area for the first time. He served as a member of a local guard in Jalalabad for a year and was then transferred to a tribal border police force called the *jandarma* and assigned to the mountainous border districts of Kamdesh in Nuristan and Sirkanay in Konar.

Tensions over the border rose as the Pashtunistan movement intensified. Pakistan and Afghanistan pressed to gain territory on either side of the disputed frontier, in part by encouraging resettlement. Fighting broke out along the border in 1960, when Afghan prime minister Daoud Khan, a member of the royal family, ordered troops and tribal forces into the Pakistani region of Bajur, including Noor Afzhal's jandarma unit.

Noor Afzhal was put in charge of ten men from the Mohmand tribe. Other fighters came from the powerful Safi and Shinwari tribes whose territory neighbored the Mohmand's in Konar and Nangarhar provinces and elsewhere in eastern Afghanistan. Like the rest of the jandarma, Noor Afzhal had no uniform and wore jami and a pakol. Other tribesmen who lived in the area joined the clashes, armed with knives, axes, and other tools.

The Afghan and Pakistani forces and civilians faced off for weeks along the border. Noor Afzhal and his men built fortifications and fighting positions. In a last-minute bid to diffuse the conflict, the Pakistani employed the Pashtun concept of *nanawati*. They sent over a small delegation of women carrying Korans and a promise not to fight.

But the cease-fire was short-lived. One morning not long afterward, a single shot rang out on the Pakistani side, followed by men yelling in Urdu for an attack.

Pakistani militia opened fire on the Afghan border positions. Noor Afzhal and other armed jandarma shot back. Civilians from Bajur joined in the attack, clashing with a group of Safi tribesmen, known for their ferocity, who battled with knives. The fighting raged for a day and a half, but the Pakistani militia, armed with machine guns on the high ground, proved overpowering.

As the Afghans began to fall back, many were gunned down, with the Safis taking heavy losses. While retreating with his army, Noor Afzhal's unit was ambushed by the Pakistanis.

"Drop your weapon!" a Pakistani fighter ordered Noor Afzhal.

He refused. A brief standoff ensued that ended when Noor Afzhal's older brother arrived with forty police from his checkpoint.

Despite the Afghan losses, Noor Afzhal gained honor as a fighter in the eyes of his tribe. An honorable man will never give up his weapon. The border clash solidified his support for Pashtunistan and his admiration for Daoud Khan, whom he considered one of Afghanistan's most talented leaders.

Pakistan and Afghanistan severed relations in 1961 as a result of the conflict, which ultimately contributed to the decision of King Zahir Shah to force Daoud Khan to resign. Years later, in 1973, however, Daoud Khan overthrew the monarchy and created the Afghan republic, making himself president. He ushered in a period that some Afghans, including Noor Afzhal, considered a golden era because of the country's relative stability and his strong support for the Pashtun tribes. Noor Afzhal returned to Mangwel.

"For a few years I lived with my oldest brother, named after my father, Sher Afzhal. I helped him on his farm, but it finally became very clear that my brother was jealous of my status as malik. I could see it gnawing on him, this jealousy. Then one day it just went too far and he exploded at me. I still remember his words and how badly they hurt me: 'You aren't doing anything!' I knew there was only one thing I could do: leave. I packed a small bag and all my money and just started walking down the dirt road toward Jalalabad.

"My second brother, the doctor, he tried to stop me, but it was no use. I just kept walking. I can still hear my kind brother pleading, 'Stop! Come back!' But I did not look back."

It was 1970 and Noor was young, strong, and tall, his jet-black hair wrapped in a traditional turban. He continued walking. After a few weeks he crossed into Pakistan and bought a one-way rail ticket to Karachi, the bustling port city on the Arabian Sea. For the first time in his life, he boarded a train, and rode for two days and two nights on a wooden bench, eating a bit of bread from Mangwel. As the train rolled through the countryside, taking him ever farther from home, the reality of what he had done overwhelmed him. He was angry and downcast. He had never wanted to leave his village. Now he was on his way to a strange city. As the train neared Karachi, he was bombarded by unfamiliar sights and sensations, and gawked at the paved roads, traffic, and tall buildings. What shocked him most of all, though, was seeing the ocean.

He felt isolated and alone, torn away from the tribe that gave him dignity and purpose. Little did he know that he was a vanguard for his people, who would soon follow him in an exodus from Mangwel.

A DECADE AFTER NOOR Afzhal left Mangwel, catastrophe struck for the people of his home village.

At dusk one day in October 1981, Soviet tanks crested the western bank of the Konar River, heading toward Mangwel. Several tribesmen witnessed the Russian attack on their village.

Umara Khan, the arbakai with big green eyes, was just a teenager with peach fuzz that year. He saw the tanks and ran from the village up to a nearby hillside. He knew the Russians were trying to press young tribesmen into the communist Afghan government's army, and hid behind some rocks to evade capture. But that night the Soviet forces would take no prisoners. Russian bombs killed some three dozen people in Mangwel and destroyed more than half the homes in the village. It had been nearly two years since the invasion of Afghanistan on Christmas Eve 1979 ended a period of relative stability and plunged the country back into war.

A Pashtun nationalist coup in 1973 had brought Daoud Khan to power. Daoud Khan ruled with an iron hand, which made him unpopular, although Pashtun tribesmen, including Noor Afzhal, valued his staunch nationalism and considered him a great leader.

When Daoud Khan was executed by a Soviet-backed communist regime in 1978, the new government soon announced reforms such as land redistribution and women's rights that struck at the heart of traditional Pashtun society. Pashtun tribesmen revolted in eastern Afghanistan, and the Konar valley became a stronghold of resistance. Desertion spread throughout the Afghan army and the Soviet Union dispatched helicopters and other weaponry and a growing number of military advisors to put down the rebels. In one early atrocity in August 1979, Russian helicopter gunships destroyed villages and killed dozens of civilians in Konar. The incident galvanized the resistance. Soon a new fighting force coalesced in the tribal areas. They were called *mujahideen*, "soldiers of God."

By the spring of 1980, mujahideen were ambushing Russian armored units, and the war raged in the Konar River valley. Mangwel residents recall a mujahideen attack on a Russian convoy that destroyed fourteen large military trucks carrying fuel, food, guns, and clothes.

While the fighting continued, the refugees streamed out of the villages. Thousands joined a mass exodus from Mangwel. Fearing more Russian attacks, families just wanted to escape with their children. They packed a few of their belongings, loaded them onto donkeys, and traveled east—to Rawalpindi, Islamabad, and other Pakistani cities.

*All photographs courtesy of the author unless otherwise noted.*

Maj. Jim Gant patrols the rural village of Chamaray, Konar Province, April 2011. Less than two months earlier, Jim and his small team of U.S. soldiers moved into the area as part of a strategy to recruit Afghanistan's ethnic Pashtun tribes to defend their territory against Taliban insurgents.

Malik Noor Afzhal (left), a leader of the Mohmand, a Pashtun tribe, talks with Jim in the entrance of Noor Afzhal's *qalat* (walled family compound) in the village of Mangwel, Konar Province. About four thousand members of the Mohmand tribe live in Mangwel and nearby areas. The relationship Jim builds with Noor Afzhal is the cornerstone of his early success in tribal engagement. August 2011.

Noor Afzhal (front row, second from left) sitting with Jim at Noor Afzhal's home in Mangwel, together with Jim's second-in-command, Capt. Dan McKone (front row, left), his interpreter and advisor, Ismail Khan, nicknamed "Ish" (front row, right), and Noor Afzhal's two sons Azmat (back row, left) and Raza Gul (back row, right). August 2011.

Jim speaking with Mohmand tribesmen who are members of the *arbakai*, or tribal defense force, officially known as Afghan Local Police (ALP) in Mangwel. The tribesmen, wearing black vests as their uniform and carrying AK-47 rifles, are paid to protect their home, village, and valley, reinforcing the role traditionally played by Pashtun tribesmen. Jim and his U.S. military team trusted the tribal fighters, who were their primary security, with their lives. July 2011.

Umara Khan, a Mohmand tribesman and member of the Afghan Local Police, manning an M240 machine gun in a guard tower of the qalat in Mangwel where Jim and his small team of U.S. soldiers lived. The team's canvas tents are visible on the left, and the training range on the right. April 2011.

A group of Afghan Local Police from Mangwel set off on a "show of force" patrol into the nearby Taliban-influenced village of Kawer in September 2011, as part of an effort to prepare for the establishment of ALP in Kawer. The qalat where Jim and his team lived with the arbakai in Mangwel is visible behind the patrol.

Jim and Ann standing on a hillside above the village of Mangwel just after he asked her to marry him on January 14, 2012.

Ann holding the newborn son of Mohmand tribesman Raza Gul, the fourth son of Noor Afzhal, in the interior quarters of the family's qalat in Mangwel. Each of Noor Afzhal's sons and their wives had rooms inside the qalat, and as a woman, Ann was allowed to spend time in the inner portions of the qalat that were off-limits to males outside the family. August 2011.

Children at the school in the village of Chinaray, Chowkay District, Konar Province, gathering around Ann as she practices basic English vocabulary with them. Ann helped provide supplies and other support to schools in the tribal areas where Jim worked. Jim and his team had moved into a qalat near the school as part of his strategy to engage the Safi tribe that dominated the area. February 2012.

Jim and his team of U.S. soldiers and Afghan advisors, with the call-sign "Tribe 33," dressed in Afghan clothing in the qalat where they lived in Mangwel. From upper left: Pfc. Jeremiah "Miah" Hicks, Air Force Staff Sgt. Andy Deahn, Staff Sgt. Robert Chase, Ismail Khan "Ish," Jim, Sgt. 1st Class Tony Franks, Staff Sgt. Justin Thomas, and Pfc. Andrew Gray (next person's name is withheld for privacy). Front row, from left: Pfc. Kyle Redden, Capt. Dan McKone, Spec. Chris Greenwalt, Sgt. Mike Taylor, Sgt. Jeremiah Harvey, interpreter Imran Khan. All the U.S. soldiers apart from Jim, Capt. McKone, Staff Sgt. Deahn, and the unnamed soldier were members of the 1st Battalion, 16th Infantry Regiment. August 2011.

Jim and his Iraqi comrade and interpreter, Mohammed Alsheikh, riding on the hood of a Humvee near Haifa Street in Baghdad, Iraq, in 2007.

*Photograph courtesy of Jim Gant*

Gen. David Petraeus (second from left front with cap), the top U.S. commander in Afghanistan and a key supporter of Jim and his tribal engagement strategy, visits Mangwel with other officers in Jim's chain of command, May 8, 2011. Col. Don Bolduc, commander of the Combined Joint Special Operations Task Force-Afghanistan (CJSOTF-A) (on right next to Jim) and Lieut. Col. Robert Wilson, commander of the Special Operations Task Force-East (SOTF-E) (upper right, directly behind Bolduc). Col. Mark Schwartz is in cap behind Petraeus on the left.

*Photograph courtesy of U.S. Navy Chief Petty Officer Joshua Treadwell*

Jim meets with Senator John McCain and Senator Lindsey Graham when they visited Mangwel in July 2011. The senators were among a string of VIPs who traveled to the village in 2011 to better understand the local security initiative.

*Photograph courtesy of Jim Gant*

The original six—then-Capt. Jim Gant and members of his Special Forces team, Operational Detachment Alpha 316—at their camp in Asadabad, Konar Province, Afghanistan, in 2003. From far left, back row: Sgt. 1st Class Chuck Burroughs, Staff Sgt. Dan McKone, Staff Sgt. Tony Siriwardene, Staff Sgt. Scott Gross. Seated from left: Sgt. 1st Class Mark Read, Interpreter Khalid. Seated on floor front: Capt. Jim Gant.

*Photograph courtesy of Jim Gant*

Noor Afzhal, Jim, and Ann
at the Mangwel qalat where
Jim's team, Tribe 33, was
based. April 2011.

Jim has an emotional reunion with Noor Afzhal in Jalalabad, Afghanistan, in August
2010. It is the first time the two men have seen each other since Jim's first deployment to
Afghanistan in 2003. The meeting also marked the beginning of a powerful alliance that set
the stage for Jim's tribal engagement in Konar.

*Photograph courtesy of Jim Gant*

Jim giving Noor Afzhal the Joint Service Commendation Medal, which Petraeus awarded Jim during Petraeus's visit to Mangwel in May 2011. Jim believed Noor Afzhal deserved the medal as the man most responsible for the success of the tribal strategy, and the act also symbolized Jim's loyalty to the Pashtun tribe.

*Photograph courtesy of Jim Gant*

Noor Afzhal greets Jim's father, James Karl Gant, who visited the village of Mangwel in July 2011. The visit was another demonstration of Jim's trust in the tribe and his respect for Noor Afzhal.

Mohammed Jalil (left), a Taliban sympathizer and member of the Mohmand tribe, meets with Noor Afzhal and Jim in Noor Afzhal's home in May 2011. Jim's outreach to the Taliban, an effort to persuade insurgents to quit fighting and return to the embrace of their tribe, was a key facet of his strategy to bring lasting security to the area.

From left: Ismail Khan "Ish," Afghan Local Police member Safe, Haji Jan Shah (son of Safi tribal leader Haji Jan Dahd), and Jim meet to share intelligence at the Chowkay qalat, March 2012. Jim's work with the Safi tribe, the most powerful in Konar Province, marked a major expansion of his tribal engagement plan.

Taliban insurgent-turned-Afghan Local Police commander Niq Mohammed (center, with white cap) looks over maps with Jim (back to camera) as they meet in Mangwel for their first serious discussion on establishing an Afghan Local Police force in the insurgent-influenced nearby village cluster of Kawer. Noor Afzhal's son Asif, second-in-command of the Mangwel ALP, sits on right next to Niq. September 2011.

Jim meets with Mushwani tribal leader Haji Ayub (center) in the city of Jalalabad in May 2011 at the home of Ismail Khan "Ish" (right), Jim's comrade, advisor, and interpreter. The visit marked the beginning of an important opportunity to work with the Mushwani tribe, powerful in northern Konar Province and across the border in Pakistan.

Jim and Ibrahim Khan, nicknamed "Abe," on a patrol deep in the Mangwel Valley, behind the village of Mangwel. Spring 2011.

*Photograph courtesy of Jim Gant*

Azmat, the third son of Noor Afzhal and leader of the eighty Afghan Local Police in Mangwel, standing in a guard tower in the qalat where Jim lived with his team, Tribe 33. Mangwel is surrounded by foothills of the Hindu Kush mountain range, visible to the right, and its farmland borders the Konar River. July 2011.

Shafiq, a member of the Safi tribe, worked as a mechanic and driver for Jim. Shafiq stands in the qalat in the village of Chinaray, Chowkay District, wearing traditional Afghan leg warmers used when climbing mountains. Shafiq fought to defend the qalat alongside Jim and his new U.S. infantry team, call-sign "Tribe 34," and the Afghan Local Police they recruited from the area's dominant Safi tribe. March 2012.

Capt. Dan McKone tries to spot insurgent positions while returning fire and directing his American teammates and Afghan Local Police during a Taliban attack on the qalat in Chowkay District, Konar Province. March 2012.

Mohmand tribesmen recruited as Afghan Local Police in Mangwel are photographed together with Noor Afzhal, Jim, and Bolduc (center front, in uniform). February 2011.

*Photograph courtesy of Jim Gant*

Lieut. Gen. John Mulholland (left), commander, United States Army Special Operations Command (USASOC), meets Noor Afzhal in Mangwel in the spring of 2011. Mulholland was impressed by the security gains Jim and his team made by working with the tribe.

*Photograph courtesy of Jim Gant*

Jim hosts a lunch for Maj. Kent Solheim (right), commander of Special Forces company 3430, and his senior enlisted advisor, Sgt. Maj. Brian McCafferty (center), at the qalat in Chinaray village, Chowkay District, March 12, 2012. Solheim was in charge of all Special Forces teams operating in Konar and Nangahar Provinces. Two days later, Solheim led a mission to Chowkay to relieve Jim of his command and relieve Dan from duty.

*Photograph courtesy of Sgt. 1st Class Fernando Gonzales*

1st Lieut. Thomas Roberts playing with a monkey at COP Penich, February 2012. Roberts, a platoon leader with 2nd Battalion, 3rd Infantry Regiment, along with several of his men, was placed under Jim's direct command in Konar in December 2011. Jim's new team, Tribe 34, moved from Mangwel to COP Penich in January 2012. Jim then took several infantrymen from the team, not including Roberts, and moved with them to Chowkay to work with the powerful Safi tribe. In March, Roberts submitted a statement to his superiors accusing Jim and Dan of misconduct, prompting a formal investigation.

*Photograph courtesy of Sgt. 1st Class Fernando Gonzales*

Safi tribal leader Haji Jan Dahd speaks with a group of tribesmen from his home in Chowkay District, Konar Province, September 2011. Jan Dahd held such meetings almost daily and played a key role in resolving tribal and local disputes. Jan Dahd, a former mujahideen commander, was wounded repeatedly while fighting Russian occupation forces in the 1980s, battled the Taliban regime in the 1990s, and served as the governor of Konar Province immediately after the U.S. military invasion toppled the Taliban in 2001. Jim, who first met Jan Dahd as a captain in 2003, reestablished ties with the tribal leader starting in 2011 to vet recruits for the Afghan Local Police as part of a security plan for the area.

Jim, Dan, and other members of Tribe 34 at the qalat in Chowkay on the day Jim and Dan were relieved of duty and taken away by Solheim and his men. From front left: Dan, Jim, Ish, Sgt. 1st Class Fernando Gonzales, Sgt. Danny Bird, Pfc. Chad Armstrong, Pfc. Richard Lerma. Front seated: Spec. Fernando Ruiz. Back row from left: Sgt. 1st Class Tony Carter, Staff Sgt. Ed Martin, Pfc. Jonathan Bartlett, Air Force Senior Master Sgt. Wesley Brooks, Pfc. Jonathan Salyer. March 14, 2012.

*Photograph courtesy of Sgt. 1st Class Fernando Gonzales*

Afghan Local Police commander Abdul Wali fires his AK-47 rifle at Taliban insurgents attacking the qalat in Chowkay District in March 2012. Abdul Wali is fighting alongside U.S. infantryman Pfc. Richard Lerma (rear of vehicle) and Pfc. Jonathan Bartlett (in gun turret). Abdul Wali was one of eighteen tribal fighters who guarded the qalat in nine-man shifts and went on missions with Jim's team.

Abe helps Jim put on his ammunition rack in Chowkay, March 2012, before a mission to the nearby insurgent-held Dewagal Valley. They are wearing red, green, and black shoulder patches bearing the Pashto word *ghairat* or "honor."

The qalat in Chinaray village, Chowkay District, in the heart of the territory of the powerful Safi tribe, on the day Jim and Tribe 34 moved into the compound in February 2012. Insurgents staged attacks from positions in the mountains behind the qalat until Jim sent Safi tribesmen recruited as Afghan Local Police into the high ground to set up observation posts.

Abe fires an M240 machine gun during an attack by Taliban insurgents on the qalat in Chowkay in March 2012.

Jim holds a team meeting in the rudimentary operations center inside the qalat in Chowkay in March 2012. Haji Jan Shah (seated, wearing an Afghan pakol hat), son of Safi tribal leader Haji Jan Dahd, attends the meeting. From left to right: Senior Master Sgt. Wesley Brooks, Pfc. Richard Lerma, Pfc. Jonathan Salyer, Haji Jan Shah, Ish, Abe, Shafiq, Sgt. 1st Class Fernando Gonzales (standing), Capt. Dan McKone.

Afghan Local Police member Mahmud Dwaher carries his commander, Abdul Wali, into a room in the qalat after Wali was shot in the leg by Taliban insurgents during an attack on the qalat in March 2012.

Abe plucks a white Taliban flag from a compound deep in the Dewagal Valley, Chowkay District, Konar Province, during a mission in August 2011. Jim stands nearby as Ish takes a picture. The move was meant as a signal to Taliban insurgents operating in the valley that Jim and his team would challenge them head-on.

Jim congratulating a new group of Afghan Local Police recruited from the Safi tribe in Chowkay in March 2012. Jim and his team quickly recruited and trained scores of tribesmen in a few days after moving into Chowkay in February 2012.

Former Taliban commander Obeidullah, who decided to switch sides with Jim's backing, in a meeting at the qalat in Chowkay in March 2012.

Ann, dressed in a traditional Afghan *burkha* to blend in, travels through Jalalabad in March 2012 after evading the U.S. military in Konar.

Tribal leaders and Afghan Local Police commanders from Konar Province gather at the compound of Governor Fazllulah Wahidi to protest Jim's removal from his command and ask for his return. March 2012.

Jim and Ann during a walk through the village of Mangwel. August 2011.

Noor Afzhal, established in a shanty near the sea, brought his family to Karachi. And now that the tribe was abandoning the village wholesale, many other families followed, gravitating once again around his leadership. As the Soviet war continued in their homeland, nearly two hundred families from Mangwel made the fifteen-hundred-mile trek to Karachi, and Noor Afzhal was their malik.

Every day Noor Afzhal went to work at the port, unloading huge bags of wheat, tea, sugar, fertilizer, and other goods from ships. The labor was backbreaking and the pay uncertain, but it fed his family. In time, Noor Afzhal gained a Pakistani ID card and became a foreman at the port, in charge of a crew of ten laborers. He and Hakima had six more children—in all, four sons and four daughters—and sent them to government-run schools in Pakistan. He knew that the greater opportunities for his family lay in Pakistan. He had tasted life in the westernized world. But as the years passed, Noor Afzhal still longed to return to Mangwel with his tribe.

One chilly day in 1995, Noor Afzhal walked down the main road into Mangwel. His oldest brother, the police officer Sher Afzhal, had passed away, and when Noor Afzhal attended the burial in Mangwel, the other exiled tribespeople appealed to him as their malik to return to the village and bring them home. He agreed.

At the time, the country was in the grip of civil war, as mujahideen factions had divided the country into warlord fiefdoms. Noor Afzhal passed one qalat after another with crumbled stone walls and courtyards overgrown with weeds. He turned down the familiar path to his qalat, his birthplace, and found it gutted. Not a shred of cloth, a single washbasin, or any other remnant of his past life there remained. His beloved village was little more than a ghost town.

Noor Afzhal moved back to Mangwel and for weeks afterward stayed up all night, guarding his home against bandits who roamed the area beating and robbing anyone who was unarmed. To feed his large family, he also bought land, and sold his house in Karachi to buy a tractor. In Afghanistan a family's land is divided among the offspring, with sons getting a full share and daughters a half share, causing the portions to shrink from one generation to another. A *jerib*, the Afghan unit of measure for land, is about half an acre. Noor Afzhal inherited three jeribs of land from his brother and bought two more, for a sum of about two and a half acres. That produced enough grain to feed his family with some left over to sell.

But as Noor Afzhal and other families struggled to rebuild their lives in Mangwel, they faced a continual threat that even their meager gains would be stolen from them. "Everyone was so tired of the mujahideen and wanted to get rid of them," Noor Afzhal told me.

The chaos created an opening for the Taliban. In September 1996, word spread through Mangwel that Taliban forces had launched an offensive in Jalalabad, the main city in Nangarhar Province that bordered Konar to the south. The Taliban was a radical Islamic movement that claimed it would restore law and order and purge the country of the impurity of the mujahideen, whose leadership had been widely discredited. The Taliban took its name from the students at Islamic madrassas, or *taliban*, who made up most of its ranks. It was an overwhelmingly Pashtun movement, led by the reclusive Mullah Omar.

Within days of taking Jalalabad in mid-September 1996, the Taliban captured Nangarhar and Konar provinces. Seeing the writing on the wall, the mujahideen leaders in Khas Kunar District, where Mangwel is located, put up no fight and disappeared overnight, fleeing across the border to Pakistan. Two or three Taliban officials arrived first and took charge at the abandoned district government, and not a shot was fired. Like many Pashtun tribal leaders, Noor Afzhal decided to throw his support behind the Taliban as the best alternative to the anarchic days of mujahideen rule.

"I will take responsibility for my village," Noor Afzhal told the Taliban officials. "There will be no problems here, nothing will happen to you."

Initially, the Taliban practiced consensus leadership under a shura system based on the tribal jirga, or council. Whereas the jirga was purely tribal, the shura also included religious and government officials. As a senior malik in the area, Noor Afzhal had become a member of the district shura under the mujahideen and continued during the Taliban period. The Taliban's military organization, too, borrowed from the tribal militias, or lashkars, which were localized and fluid in composition. But at its core, the Taliban was mistrustful of Afghanistan's traditional tribal authorities and sought to undermine their leadership.

Ultimately, Noor Afzhal rejected the Taliban regime, too, and in particular, its strict version of Islam. It was this transition that paved the way for his alliance with Jim in 2003.

One late-summer morning we sat on a cot outside the arbakai tent and drank milk tea.

"How do you balance decisions for the tribe when a conflict arises between Islam and Pashtunwali?" I asked Noor Afzhal.

"This is not an easy question," he replied as he looked up toward the mountains. "I should say that as a Pashtun I would always follow Islam, but honestly Pashtunwali comes first," he said. "Yes, that is true."

What impressed me most about Noor Afzhal was his ever-enduring Pashtun ways, which somehow allowed for great pragmatism and moderation at the same time. Upon his return to Mangwel after years in the metropolis of Karachi, he had become more worldly, and that likely explained in part how he was able to show such tolerance and openness toward an American woman in his midst.

As my time neared to make a trip to the United States, Noor Afzhal seemed to understand, but appeared sad and asked when I would come back again.

"I would like Jim and you to live here in Mangwel, and visit your country when you want to," he said.

I smiled at him and took a sip of tea. He had no idea how appealing his words sounded to me, but I could make no promises.

"Father," I replied simply, "I will see you soon."

# CHAPTER 20

ONE WARM SPRING EVENING under a sky brimming with stars, Noor Afzhal rose from a wooden cot in the courtyard of his Mangwel qalat. The village was quiet except for the soft chorus of neighbors' voices and a donkey braying in the distance. He stepped through a doorway and walked past the front gate to the hujira, or guest room. There he slipped off his plastic sandals and sat cross-legged on a carpet. Reclining slightly against a large cloth pillow, he was content to be surrounded by his sons and grandsons.

Jim sat in his usual spot, by Noor Afzhal's left side.

Raza Gul, Noor Afzhal's youngest and most cheerful son, rolled out a rectangular plastic mat on the floor. Then a ten-year-old grandson came through the door carrying a basin and a long-spouted water pitcher, or *loota*. He placed the basin in front of each man in turn and carefully poured water over their outstretched hands, then offered them a towel. Raza Gul returned with a stack of warm, homemade wheat bread and placed one of the flat rounds in front of each man. Boys followed with large metal platters of fragrant rice and smaller plates of stewed chicken. They hunched over and began eating, breaking off pieces of bread and dipping it in the chicken stew, and scooping up rice with their right hands and pushing it into their mouths with their thumbs in the Afghan manner.

Noor Afzhal ate heartily. His health had revived since Jim brought a heart surgeon to the village to examine him and prescribe medicine for his heart and high blood sugar. Jim watched the elder's intake of sweets, constantly checking to make sure he was drinking his tea without sugar. Noor Afzhal was visibly stronger and more robust than when Jim had met him in Jalalabad a year earlier, in the summer of 2010. He walked farther and sometimes sportily rode on the back of Azmat's motorcycle.

Jim tore off a piece of chicken meat and turned to Noor Afzhal.

"You are getting younger and I am getting older, *wailay*, why?" Jim asked him.

"It's because you have taken the responsibility," said Noor Afzhal with a smile.

After they finished eating, Raza Gul took away the food and folded up the plastic mat with the leftover bread. A short while later, another grandson arrived with a pot of green tea and a tray of glasses. Then Jim's favorite of the grandsons, an active boy with a mischievous grin whom he nicknamed "Little Malik," came in, and Jim called him over to play.

"God made a mistake," Jim told Noor Afzhal, taking a sip of tea. "I should have been born in Afghanistan."

"God does not make mistakes," Noor Afzhal replied, not missing a beat. "You were born in America so you could come here and help us."

The longer Jim stayed in Mangwel, the more he emerged as Noor Afzhal's de facto eldest son and his biggest protector and ally, even within his immediate family. Noor Afzhal set aside a room inside his qalat for Jim. He promised Jim the rights of a son, including the inheritance of a portion of his land. Jim, too, began to call the Mohmands his tribe and Mangwel his home. He was Mangweli—a son of Mangwel, everyone said. Jim was clearly empowering Noor Afzhal. But he also had to carefully calibrate how to use his growing influence, as he was drawn deeper into internal issues that simmered within the family and tribe.

Jim joined the Afghan family at a particularly delicate time, when the question of Noor Afzhal's succession was in the offing. In Pashtun tribes, the overarching structure is stable, but leaders wield power informally and there are few rules governing succession of maliks. Chiefs must continually prove themselves to their followers and adversaries, making tribal leadership inherently uncertain. Adding complication and intrigue are the tribal blood feuds, cycles of killing and revenge that rival the drama of an ancient Greek tragedy. In the case of Noor Afzhal, his younger brother, Dost Mohammed Khan, was vying to take his place. A school supervisor who had spent several years at a boarding school in Kabul, Dost Mohammed Khan had worked as an educator and shop-keeper in other towns for many years. In the fall of 2010 he had opposed Noor Afzhal by voicing reservations about Jim and his team moving into Mangwel, arguing it would make the village less secure. Jim realized the depth of the rivalry when Dost Mohammed Khan asked him for a rifle. Jim had given Noor Afzhal more than one handsome gun, a greatly empowering gesture for Pashtuns. But Noor Afzhal privately warned Jim not to give his younger brother a weapon.

"Do not give him the rifle," Noor Afzhal said. "He will kill me with it." Jim agreed to withhold the gift, and warned that Dost Mohammed Khan would regret it if he so much as touched Noor Afzhal.

Another candidate to succeed Noor Afzhal was his second son, Asif, who had worked as a contractor for development projects at COP Penich and Mangwel. Asif was always joking and came across as incapable of being serious, but one-on-one he was just the opposite. He had gone to school until eighth grade, longer than his two younger brothers. At the age of thirty-seven, he was an effective manager, working hard to build up his own network through his contracting jobs. Asif's personal life had been marred by tragedy. He had lost two sons, one who was still-born and another who had died in infancy. And he had suffered a hor-rific accident while harvesting wheat in 2005. He had caught his left hand in a threshing machine, which chopped off most of his fingers. He was very self-conscious about his mangled hand, and always wrapped the portion of it that remained in a bandage.

Noor Afzhal favored Asif to take his place, and Jim, as always, backed up his decision. A tribal jirga was convened to consider the matter. One factor favoring Asif in the decision was how important patrilineal rela-tionships were in Pashtun culture. The tribal leadership had been under Noor Afzhal's direct line for the last century. Moreover, the village elders realized that Jim, an immediate source of significant patronage, was allied with Noor Afzhal. The jirga convened and endorsed Asif as a malik and the successor to Noor Afzhal. After the succession conflict was resolved, Jim gave Dost Mohammed Khan an AK-47.

But with Noor Afzhal still in power, conflicts began to arise between father and son that exposed the generation gap between them.

Noor Afzhal was a traditional malik whose authority was rooted in his solid judgment and ability to settle disputes between his people and with other clans or tribes. His power was based in Mangwel and nearby rural communities of the Mohmand tribe. He lived by the code of Pashtunwali. He remembered the relatively peaceful Afghanistan of the 1960s and 1970s, and opposed the radical extremist dogma of the Taliban.

Asif, on the other hand, was forming connections well beyond Mangwel with urbanites in Jalalabad and elsewhere with whom he shared commercial interests through his contracting work. Unlike Noor Afzhal, he carried a cell phone, wore sneakers, and learned to speak some English. Asif was part of the generation of Afghans who remembered nothing but chaos and war—the same generation that gave rise to the Taliban. He had grown up with many young men who became Taliban insurgents, and he realized that after his father died, he would likely have to deal with them one way or the other.

For Pashtuns, a man's honor is derived largely from that of his father until he establishes his own reputation. Asif, still living in his father's shadow, pledged to carry on his traditions. "I will follow the way of my father. He is the expert," he said. But Asif took on the responsibilities reluctantly. "Maliks are bullshit!" he would say privately in English, only half jokingly.

After Asif was selected, Jim moved publicly to bolster Asif's power, while quietly mentoring him behind the scenes to encourage him to make good decisions.

"I am with you, brother," Jim told Asif one evening as they ate a meal of lamb kebabs Asif bought in one of his frequent trips to Jalalabad.

But the next day, as if testing his new status, Asif directly challenged Noor Afzhal as they, Jim, and other soldiers met at the qalat and discussed development projects for Mangwel. Asif wanted to build a well for a friend of his who did not live in the village. Noor Afzhal protested. Any wells should benefit Mangwel, he said, but the village already had enough. Other projects, such as irrigation ditches, would benefit the entire village more, he said. "You must solve problems for the people, they are depending on you for that," Noor Afzhal told Asif. The criticism stung.

"You are an old man," Asif lashed back. "Your brain is not working anymore."

That night, Noor Afzhal could not sleep. He asked his wife and other sons if his mind was failing him. The next day, he asked Jim his opinion. "Your mind is good," Jim told him. "If your mind becomes weak, I will tell you, and I will protect you even more then."

Soon afterward, Asif came to Jim, asking how to mend the rift.

It was the most serious crisis yet between father and son. Jim decided to address it by telling them each separately the same story, about a conflict he had with his own father. Jim sat down first with Asif.

A few years earlier, he said, he had gathered the men from his Special Forces team, those he loved and respected most, for one last party. He invited his father to join them. But in the middle of the party, his father, who was drunk, blurted out that he used to beat Jim up when he was a boy. Jim was furious. "Yes, you did," he told his father. "But if I wanted to, I could beat you now."

He told Asif, "It was the worst thing my father ever did to me. I didn't talk to him for two days, and then put him on an airplane. I love my father dearly. He loves me dearly. We make mistakes," Jim said. "He wishes for all he is worth that he hadn't said that to me, because it hurt me and embarrassed me. It shamed me and took my honor. But then, I turned around and embarrassed him and shamed him and took his honor as well," he said, speaking to Asif using the language of Pashtunwali. "Neither one of us was the better for it. We both lost. Sons and fathers love one another, but they both make mistakes."

Next Jim related the story to Noor Afzhal, and the rift with Asif seemed to dissipate.

Only time would tell what kind of malik Asif would become. He had a reckless side that one day caused a near disaster. Jim and Ish were walking down the main road through Mangwel, having received a report that a suicide bomber was on his way to the village. Just then, around a curve in the road, he heard gunfire and cars approaching. Jim took aim down the road with his AK-47. The first car came into view, with men shooting out the window with rifles. Jim tightened his finger on the trigger and was about to fire on the vehicle. He waited an instant to gain a better view. Just then he spotted Asif riding in the front seat. It was a wedding party, and Asif and his friend were reveling. Asif hadn't notified the Tribe 33 team about the celebratory fire, a customary part of Afghan weddings. Jim lowered his rifle, angry beyond words. He shuddered at how close he had just come to killing his Afghan brother and Noor Afzhal's chosen heir. Asif sheepishly apologized.

Not long afterward, Asif got into a fistfight with his brother Azmat in the qalat. The brothers clashed over an issue that was distressingly petty for Jim but one he nevertheless understood—trash. The people of Mangwel were so poor that almost any reusable commodity coming from the qalat was of value—scrap wood from crates of supplies, empty plastic water bottles, the metal casings of bullets fired on the range next to the qalat. A conflict escalated between one of the laborers on the qalat, who worked for Asif, and one of the arbakai, who worked for Azmat, over some of these resources. Heated words between the brothers turned to blows before some of the arbakai pulled them apart.

Later that day, Jim brought the two brothers together. "I have come from America to help you. I will fight anyone for you. I will die for you. But I am not here to see my brothers fight," he said. They lowered their heads; the shame implicit in his words was clear.

Then Jim pulled out two 9 mm pistols and racked a round in each of them. He handed one to Asif and one to Azmat.

"They work," he said. "Here are three magazines apiece. Next time you fight, don't fight with your fists."

As Jim played a greater role within the family, Noor Afzhal decided that Jim, too, should become a malik of Mangwel—and so made him one. "You are my malik," Noor Afzhal told Jim one day at his home. "I make decisions for the tribe. You make decisions for me."

Noor Afzhal gave Jim a gray turban and began involving him in tribal jirgas. Jim mainly attended the assemblies that involved concerns between Mangwel and neighboring areas, such as disputes over land, water, and grazing. He tried to leave problems within Mangwel to the village elders. Pashtun tribes are strongly egalitarian, and ordinary men have the right to speak their views at the jirgas. Afterward, the elders discuss the issue and together make a collective decision. Many U.S. military officers have attended jirgas and spoken at them, but what was different in Jim's case is that he was consulted as another elder of the tribe.

Jim's increasing sway within the family and Mohmand tribe put him at the center of a complex and shifting equation. He was now a son, brother, and malik of the tribe—but was also the only one who could reasonably act as an outside, honest broker. He wanted to wield his influence to strengthen the tribe and advance his mission, but if he overstepped he could easily go from empowering the tribe to ruling it.

He had to constantly calibrate where to assert himself, where to hold back. And he had to remind himself that what might appear to be petty, tribal politics was anything but trivial to the Afghans around him and could make or break the entire endeavor.

"I had used what I called my 'enemy eye' for many years to see my actions from the enemy's vantage point. But now I had to see myself through the eyes of friends. I found it vastly more difficult," he told me.

Guiding Jim's every move was his deepening understanding of Pashtunwali and, to a lesser degree, Islam. Every day, he scrutinized the people, their proverbs, and events for nuggets of insight into the Pashtun tribal code. What he learned guided how he operated moment to moment, how he made friends and dealt with enemies.

The fascination with Pashtunwali came naturally to Jim. He identified deeply with how a Pashtun man defended his honor by fighting for his land, the women in his home, the rifle on his shoulder—collectively known as namoos. The world of the tribe was one of bright colors, where the slightest insult could escalate into a life-or-death fight. It was a primitive world, one that was harsh and brutal but also starkly beautiful. Jim was living his dream, free from many of the constraints of modern civilization. He sometimes wondered what it would be like to convert to Islam, marry a Pashtun girl, and settle down in a qalat in Mangwel. With a knife on his hip, he was surviving in an austere land largely on his courage and his wits, surrounded by comrades, a family, and a tribe that he had grown to love. What more could he want?

At the center of this world was ghairat, or honor. In America, the word was almost quaint, an anachronism, Jim felt. But Pashtuns lived for it, and joining the tribe for him was like going back in time. A violation of honor was considered *peghor*, or "shame." That in turn required badal (revenge), often through killing or other physical violence or destruction, in order to regain lost prestige. Such feuds gave young Pashtun men a chance to prove their mettle. Tribal law and custom had its conciliatory aspects—for example, balancing the punishment for whatever crime was committed against honor. The dark side of Pashtunwali, for the tribesmen as for Jim, was the relentless competition for ghairat. To be in honor's thrall meant there could be no rest.

Once, a malik of the neighboring village of Chamaray, Mohammed Hanif, insulted one of Jim's men, Staff Sgt. Ryan Porter, when Port was visiting Hanif's qalat to check on economic development projects. Port and teammate Sgt. 1st Class Tony Franks had worked tirelessly to carry out projects in the area. Hanif complained to Port that the Americans had done nothing for Chamaray. In fact they had refurbished many wells, built six footbridges, hired thirty-one local men as arbakai, and brought security to the village. Jim knew that to disparage a guest in one's home is the height of disrespect, flying in the face of the Pashtun obligation to practice *melmastia*, or "hospitality." And since Port was representing him, it was an attack on Jim's honor. There had been other problems with Chamaray, such as some celebratory gunfire that included a few stray rounds fired at the Tribe 33 qalat. The people of Chamaray were ethnic Tajiks but followed Pashtun ways. Jim summoned Hanif to the qalat, letting him know to bring the AK-47 and knife that Jim had given him.

Under a cloudless sky the next day about 9:00 a.m., an hour before the meeting, Jim stepped out of the Tribe 33 qalat wearing a soiled set of Afghan clothes. It was already uncomfortably warm. Jim walked alongside the outer wall of the qalat toward a twenty-foot-wide, five-foot-deep pit where the team dumped and burned trash. He shooed away a dog and started rummaging through the garbage. He wanted to stink.

When Jim spotted Hanif and his son walking up the dirt road, he went back to the qalat gate and stood in front of it. Hanif approached and tried to shake Jim's hand, but Jim just stared at him, stone-faced, and refused.

"My men went to see if you needed help, and you disrespected them," Jim said. "You disrespected my men and me in your home. If I had been there, I would have killed you on the spot," he said. "I am more Pashtun than you are. I will not disrespect you inside my qalat."

Hanif started pleading, saying it was a misunderstanding.

Jim ignored him.

"Give me the knife," Jim said, narrowing his eyes. Hanif handed it over.

"I gave this to a friend who I thought had one face. I was wrong," Jim said.

Hanif looked at Jim as if he were twisting the knife in Hanif's gut.

"Listen, if one bullet is shot at my qalat from Chamaray, you are a dead man," Jim warned.

Jim turned to Hanif's son, who was holding the AK-47 Jim had given Hanif. He took the rifle.

"Port is seven thousand miles away from home. He is risking his life for you and your people. He is honorable," Jim said to Hanif. "I don't ever want to talk with you again, do you understand?"

Hanif looked shocked and said nothing.

Then Jim spat on the ground, turned around, and walked back into the qalat. He gave the rifle to Port, who thanked him.

Word of the incident spread quickly.

That day, Hanif was disgraced as a malik. He walked with his head down, mocked by his own people. Among the Pashtuns, no man should surrender his weapon as long as he is alive. By taking away Hanif's rifle—his namoos—Jim had destroyed his honor and set an example that strengthened the umbrella of protection around Jim's own men.

Jim had become more Pashtun than the Pashtuns.

# CHAPTER 21

TRIBE 33 HAD COME far since Jim first met his U.S. soldiers that cold February night at COP Penich, particularly in building relationships with the Afghan tribesmen. But the mixed team of infantry and noncombat soldiers still struggled with basic yet critical war-fighting tasks—maintaining weapons, communications, and carrying ample ammunition. Jim was riding them hard. They needed to stay alive.

At about noon on May 7, the team was put to the test. Jim got a call on his emergency Afghan cell phone, which he carried because military radios chronically malfunctioned. Another Special Forces team was caught in a complex ambush high in the Shalay valley. Taliban commander Abu Hamam and about a dozen of his fighters were dug in.

"Hey, bro, we need help in Shalay," the team chief told Jim. "We're in a big fight—how quickly can you get up here?"

"We'll be there in twenty minutes," Jim replied.

He turned to Sgt. Michael Taylor, who ran Tribe 33's makeshift operations center in a tent on the qalat, and launched a quick reaction force, that day composed mostly of Americans.

"QRF! Let's go!" Jim yelled, heading for his Humvee, called Vehicle One.

The team scrambled to get ready.

Chase, the Army Ranger and team sergeant of Tribe 33, climbed into the commander's seat of his Humvee, called Vehicle Two. "One, One, this is Two. Radio check," he called to Jim, his voice tense.

Several weeks earlier, a screwup by the team over radios had almost driven Chase and Jim to blows. It was a clear day in early March, and the morning sun had just started to take the chill off the air. The growl of Humvee engines starting up broke the quiet as soldiers loaded up gear for the mission. Jim had been up early reading over intelligence reports and was just finishing his coffee, anticipating a smooth joint patrol of Americans and the arbakai. Jim had briefed the mission in detail to the soldiers the night before. At nine o'clock sharp, he walked over to the staging area where the soldiers were readying the Humvees, expecting them to be good to go. Jim's driver Chris had just climbed into his seat next to Jim, when Jim tried to conduct a check of the radio on his vehicle. Right off the bat, Jim discovered the radios were not working—again. *This is going to get us killed*, he thought. He went ballistic.

"The Americans are fucked up! No Americans are going on this mission!" Jim yelled at the team. It was a huge put-down, but, in his mind, justified. The Afghans were by then handling the basics of running the camp, apart from a few technical tasks. He sometimes felt he could do without the Americans altogether. He needed them to step up, and was constantly frustrated when they underperformed. The American team was performing the mission of engaging the tribe incredibly well. But the soldiers had barely trained together for combat, and many of them were still learning on the job, including how to set up unfamiliar radios. They lacked a sense of urgency required to master core tasks, such as communications, that were essential for their survival. Jim had to push the men to higher standards.

Chase watched the men's faces fall. *That was a gut shot.* He walked into Jim's tiny room at the back of the operations center.

"Jim, I need to talk with you," he said tersely. "First, I don't kiss ass. It's hurt my career and got me in a lot of fights."

He looked Jim straight in the eye.

"These were my men before we left Kansas, and they will be my men when we get back," he said. "What you said was fucked up."

Jim flew into a rage, slamming his fist down.

"Have you ever seen anyone die in combat?" Jim yelled. "Have you had to speak with their families about it?" His face was crimson, and only inches from Chase's. "*You* are fucked up, Ranger!"

Chase stood his ground. He spoke as calmly as he could, calling Jim "sir."

"My men are tired, sir. Sonny was a cook. Chris worked in a little office in the communications shop. We're doing the best we can," he said.

Whether the men were tired or not, the mistakes were inexcusable, Jim said. But Chase refused to back down. "There has to be a different approach," he said.

Jim took a deep breath. Right or wrong, he had to respect Chase for speaking his mind. They stepped outside onto the qalat's gravel courtyard.

"I am not going to apologize," Jim said. "We are under a lot of stress. This mission is so important. I love you all. I don't want to live with one of you getting killed." His eyes welled up. Chase, too, held back tears. They embraced, and Jim went back inside the operations center.

Chase sat down and lit a cigarette.

The soldiers looked on, their eyes wide.

A few minutes later, Jim spoke to the team. "You have one of the best noncommissioned officers in the military," he told them. They had seen Chase bear the brunt of Jim's frustration over their poor preparation for combat. Jim deliberately let them see it as a catalyst to draw them together. In coming weeks, the team's track record on radios improved.

THE SUN WAS BEATING down on the qalat when the call came on May 7. Most of the team was just finishing a lunch of rice and flatbread, when they heard Jim shout, "QRF!"

Sonny, the thirty-two-year-old Army cook, jumped behind the M240 machine gun in Jim's Humvee. A large, broad-shouldered man, Sonny easily loaded the gun, pulling back the charging handle and thrusting it forward with a sharp clack. A sergeant, Sonny had served three deployments in Iraq, standing guard at Baghdad checkpoints and making supply runs. But as a support soldier, he had seen other commanders retreat from attacks. "They just ran," he recalled. Jim's ultra-aggressiveness came as a shock to him but also gave him confidence. The morning after he met Jim at COP Penich, Sonny volunteered to man the main gun on Jim's Humvee. "I just raised my hand," he said. Now, as he strapped on his helmet, he realized the next day was his birthday. Maybe this was his birthday present.

Pfc. Miah Hicks, a twenty-year-old fresh out of basic training, was helping ready the guns. The son of a Baptist missionary from Springfield, Missouri, Miah had wanted to join the military ever since he was a kid playing toy soldiers and hearing his grandpa's stories about the Korean War. He enlisted in the infantry in April 2010 after being kicked off the Missouri State University football team for fighting. A freshman middle linebacker, Miah had taken on a three-hundred-pound senior offensive lineman who was bullying one of his buddies. The Army was his element. He was hardworking, fit, and ambitious. He wanted to serve on an elite Special Operations unit.

When Doc Harvey ran out of the tent with the QRF list, Miah ignored the fact that his name was not on it.

"Sir, I'm ready to go. Do you need another gun?" Miah ventured.

"Get your shit!" Jim replied.

Miah flashed a smile and climbed into the back of Jim's Humvee.

Jim got into the Humvee and radioed the team caught in the ambush.

One of the arbakai unlocked the big metal gate of the qalat, and the two Humvees rolled out. They turned down a dirt path through fields of scrub brush to the main paved road that wound through Mangwel and other villages along the Konar River. After about four miles, the Humvees turned up a rocky trail carved into the red, rain-washed soil of the Shalay valley.

"Zombie 16, this is Tribe 33, we are turning up the Shalay road," Jim radioed to the other Special Forces team, ODA 3316.

"Tribe 33, this is Zombie 16," the ODA 3316 team leader, Capt. Matt Lommel, answered. "Hold at the clinic and wait for the ANA," he said, referring to a contingent of Afghan National Army soldiers.

"Good copy," Jim said.

Blue mountain ridges rose high on either side as the vehicles mounted the narrowing Shalay valley toward a roadside clinic. Lommel's team was about two miles farther up the road, separated from Jim's team by entrenched Taliban fighters. Dry streambeds crisscrossed the valley, creating a natural trench network for the Taliban that allowed them to maneuver almost invisibly. The scorching sun glinted off the vehicles as they passed a roadside clinic, nearing the ambush location. A half dozen Afghan soldiers arrived from their camp at the bottom of the valley and pulled up in a truck after Jim's team arrived.

"Sonny, are you ready? We are going to drive into these guys and let them have it," Jim called up behind him.

"Yes, sir!" Sonny replied.

The Humvees pulled into position about four hundred yards across the valley from a small village. The village contained a mosque and a cluster of four mud-brick qalats that the Taliban had been shooting from. Jim decided to move in and clear the qalats.

"Miah, grab those ANA soldiers and follow me!" he said.

Jim and Ish started maneuvering on foot toward a rocky slope that dropped steeply before rising again toward the village and insurgent position. Miah and the Afghan soldiers fell behind. Ahead was a village with a distinctive tall structure the soldiers had nicknamed the "smokestack," where Jim believed the Taliban insurgents were concentrated, reportedly protecting their leader, Abu Hamam. As Jim and Ish moved toward the village, bullets from a Russian machine gun and AK-47 rifles suddenly began tearing up the dirt just ten yards away.

"Oh, shit!" Miah yelled. Ahead he saw Jim and Ish get down. Alone with the Afghan soldiers, Miah led them sprinting back up the hill to take cover behind some rocks. At first the Afghans weren't shooting, but Miah shouted at them and they started firing their PKM machine gun toward the qalats until they ran out of ammunition.

The staccato of Taliban gunfire grew louder and closer to Ish and Jim, who lay prone but exposed behind a small rock. With bullets then kicking up dust a few feet away, Ish was spraying automatic rifle fire at the compounds. He reloaded his AK-47 and unleashed another burst at the Taliban positions. Jim scanned for targets with his M4 carbine. Still the gunfire intensified. Both men were sweating and breathing hard. They were pinned down.

"Cover me!" Jim told Ish.

Suddenly Jim spotted a Taliban fighter dart from a tree and move into a dry streambed, or *wadi*, a few hundred yards away. Jim's eyes blurred in the sweltering heat, and he lost the fighter from his sights. He knew he only had one shot; a stray bullet would alert the fighter to take cover. He steadied his breathing, took aim, and slowly squeezed the trigger. The Taliban fighter dropped.

Jim felt a rush of elation that came only from battle, or, more precisely, from killing. It was a high that many men in combat felt but few spoke of. To kill, Jim would often say, you have to expose your throat, to be willing to die yourself. To win in close combat, the Americans had to risk getting out of their armored vehicles in order to draw the Taliban out of their rock-covered positions. It was exactly the tactic he was using that day.

Jim and Ish needed to move. Just then, from about 150 yards away up a hill, Miah again opened up on the Taliban positions. He took a knee and fired with everything he had. At the same time, Sonny was hammering the compounds from a nearby terraced field with some of the three thousand rounds he pumped out that day. Jim always told his men to shoot when they needed to and promised he would back them up later by saying that he had given the order and taking any heat for it. When Sonny saw Ish and Jim pinned down, he got his chance.

"Die, you motherfuckers!" Sonny shouted as he squeezed the trigger on his M240 machine gun and held it down, the vibration of the gun running up his arm.

Together Sonny and Miah caused enough of a lull in the enemy fire to give Ish and Jim a chance to run to a little dip in the ground that offered better cover.

Miah maneuvered back toward the Humvees. Jim popped a white smoke grenade to help direct the Humvees carrying Sonny and Miah to a path between the terraced fields, giving them a better firing position.

Well-aimed insurgent bullets rained down on them again—this time from three directions. Tribe 33 again returned fire. One Taliban bullet slammed into the M240 machine gun ammunition can on one Humvee, just a few inches from the chest of Jim's medic. A rocket-propelled grenade burst in front of the vehicle.

Another bullet snapped by Chase's head, giving him goose bumps. "Son of a bitch!" he said. "This is awesome!"

Chase knew Jim was on the exposed side of his Humvee and thought he'd been hit. But then he saw Jim walk around the vehicle "calmer than shit."

Jim started laughing.

*This guy is nuts*, Chase thought. Then he broke out laughing, too.

Jim's confidence inspired his men as they rallied and counterattacked, giving Capt. Lommel's pinned-down team an opportunity to withdraw. Jim's team kept fighting until the insurgent gunfire dwindled into pop shots as the shadowy fighters slipped away. Mission accomplished.

As their two Humvees rolled out of the Shalay valley toward the qalat in Mangwel, the men of Tribe 33 were euphoric from the battle. To some of them it felt as good as or better than sex. To others, it was like a drug. They all wanted more. Tribe 33 was the fifth team of men Jim had gone to war with, and he was proud of them. *They fought bravely today,* Jim thought. *They had their first taste of blood.*

The team was helping Mangwel and other lowland villages push the Taliban out, creating a cocoon of security. So, the next day, Jim did not hesitate to bring Petraeus in.

# CHAPTER 22

GEN. DAVID PETRAEUS LOOKED out the window of his Black Hawk helicopter as it flew over the glistening Konar River, banked at the valley's edge, and dropped steeply—not, as usual, inside the fortified walls of a U.S. military outpost, but onto an open field in the village of Mangwel.

Petraeus pulled off his radio headset as the chopper whipped up a cloud of dust and touched down on a spot marked by a green smoke grenade. He was surprised by what he saw next. Not a single U.S. soldier guarded the austere landing zone. Instead, scores of tribal arbakai, with black vests as uniforms, stood watch on the adjacent ridgeline and hills and surrounded the field, their visages stern and proud. The general smiled.

A few months earlier, it would have been unthinkable for the American four-star commander to land in the middle of Konar protected only by Afghans. Even more telling, Petraeus arrived wearing no helmet or body armor—just a soft Army cap and field uniform. That was a testimony to the early success of the experiment in Mangwel, the only place in eastern Afghanistan at the time where U.S. forces were living in a qalat guarded exclusively by Afghan Local Police.

Since Jim's team moved to Mangwel in February, tribal forces there and in nearby villages had created a belt of safety in one of the most dangerous provinces in Afghanistan. In the May 7 firefight and other operations, Tribe 33 was moving deeper into insurgent-held territory such as the Shalay, in an effort to keep the enemy away. Taliban leader Abu Hamam had been in the Shalay during the fight, and afterward spread the word that Jim was trying to bring arbakai into the valley. Abu

Hamam had already made clear that he was determined to kill Jim—one of the clearest indications that the tribal strategy in Konar was working. Known as the key instigator of the initiative in Konar, Jim was specifically targeted by name by Abu Hamam and an array of assassins, suicide bombers, and insurgents emplacing deadly roadside devices. The Taliban propagandists reportedly called Jim the "father of evil." One report after another confirmed that the Taliban leadership saw the recruitment of local Afghans to defend their tribes and villages as a major threat, one the Taliban intended to attack.

Petraeus was eager to see how the strategy was unfolding from Mangwel. The lean general stepped briskly out of his helicopter and shook hands with a small entourage of U.S. military brass who had arrived in advance for his visit. Then Petraeus saw Jim, dressed in sandals and a flowing white cotton tunic and pants.

"Hey, Jim!" Petraeus shouted over the beating of the helicopter blades, throwing an arm over Jim's shoulder. "I guess we pulled that off!"

"Yes, sir, we did," Jim replied as they walked toward the qalat.

The feat Petraeus was talking about was getting Jim's deployment orders changed from Iraq to Afghanistan. But he could just as easily have been referring to the local security program as a whole. Petraeus had championed the initiative at the top levels of the U.S. and Afghan governments for the past year and a half. The program had taken off rapidly since Petraeus and his subordinate commanders, Brig. Gen. Miller and Col. Bolduc, launched it in the summer of 2010. With the U.S. military initially choosing the locations, distributing the weapons, and controlling the pay, U.S. Special Forces teams quickly recruited, armed, and trained thousands of local police around the country by early 2011. Bolduc cut through U.S. military bureaucracy to grant the Special Forces teams far greater leeway, streamlining and speeding up the convoluted approval process for operations by reducing the people involved from thirty-five to two. Bolduc also created "ops boxes"—geographic areas where the teams had authority to act without concern for interference from other military units. Petraeus was eager for the initiative to show results as he prepared to brief Congress on the progress in the Afghanistan war.

"Petraeus was getting ready to testify on whether the surge was working, and that was a huge political reality for all of us," said Lt. Col.

John Pelleriti, who was in charge of Special Forces operations in eastern Afghanistan at the time. "We felt a sense of urgency, that we had to make VSO [village stability operations] work, and we had to do it fast."

In March 2011, Petraeus returned to Washington to testify. Before the cluster of microphones and television cameras at the Senate and House armed services committees, he singled out the Afghan Local Police program, saying it had emerged as one of the most important factors creating positive momentum in the war. "So important," he said, "that I have put a conventional U.S. infantry battalion under the operational control of our Special Operations Command in Afghanistan to augment our Special Forces and increase our ability to support the program's expansion." Petraeus testified that seventy districts had been identified for establishing Afghan Local Police, and the police had been set up and "validated" by the government in twenty-seven of those.

Jim's superiors praised him for an unusual and significant contribution to that effort. Pelleriti credited Jim for allowing his command to expand from zero to more than thirteen hundred trained Afghan Local Police at fifteen locations across eastern Afghanistan. Bolduc upheld Mangwel as a shining success story in his official PowerPoint briefings on village stability operations. "Jim is a very gifted Special Forces officer and very effective in that particular area," Bolduc said. "That region of Konar has been changed from a non-permissive environment, to semi-permissive, to permissive, in a way no one thought would happen," he said. "I don't think anyone else could have done it."

Miller said he needed more Special Forces officers with Jim's skill and commitment. "I wish I had more Jim Gants, to be quite honest," Miller said. "He has given up a hell of a lot of his life to be doing that and has taken the risks. You are a small element there in the middle of nowhere. It's harder than sitting in Bagram and filling out permissions," he said. "The real question," Miller went on, "is can you get people to do this? Go live in Konar, that is your home. You don't have a PX, you live with the people and eat their food. It's kind of a romantic notion, but a lot of people don't want to do it."

Now Petraeus, the senior sponsor of the local security effort, wanted to hear from the man he considered an expert tactician carrying it out on the ground.

Petraeus and Jim walked into the Tribe 33 qalat, where Jim briefly oriented the general and then took him to meet his U.S. soldiers. Dressed in Afghan clothes with their beards trimmed to regulation and the required American flags Velcroed to their shoulders, they lined up at attention by rank.

"Sir, just so you know, these men have been a phenomenal team and I have been honored to serve with them," Jim said. "Just yesterday we were in a three-hour gun battle up in the Shalay. They performed flawlessly."

Petraeus shook the soldiers' hands, pressing one of his metal ISAF commander coins into each of their palms in the Army custom. A general with a common touch and genuine interest in his men, Petraeus put the soldiers at ease, speaking briefly to each one. He asked Chris what he thought of living in Mangwel.

"Sir, I love it," Chris said.

"How would you describe your mission?" Petraeus queried.

Chris answered straightaway. "Defend the qalat, defend the village, defend the valley, and expand VSO wherever we can, sir."

Petraeus smiled.

Down the line, he reached Miah, who wore light blue jami and a tan pakol.

"How are you doing, Private Hicks?"

"I'm doing fine, sir," Miah said.

Jim turned to Petraeus. "Sir, based on yesterday's actions, I am going to put Private Hicks in for an ARCOM with V [Army Commendation Medal for valor]."

Petraeus gave Miah a discerning look. "ARCOM with V, huh?"

"Yes, sir," Miah said, looking straight ahead.

"Done," Petraeus said, turning to a captain in his entourage and asking for the medal. "Congratulations!"

Miah grinned. It was the twenty-third award for valor received by men who had fought under Jim, and he had not lost a man yet.

Petraeus was encouraged. He saw that Jim had not only secured the area by relying almost entirely on the Afghan tribesmen but also effectively integrated the conventional infantry into the Special Forces mission. Moreover, Jim's team was highly unusual in Afghanistan because it was technically a Special Forces team but was composed almost completely of regular infantry soldiers.

As they left the qalat and headed down a small dirt path into

Mangwel, Petraeus was struck by the sense of normalcy and calm. What a difference, he remarked, from the last place he had seen Jim, Baghdad in 2007, where every few hours a massive car bomb would cause horrific carnage. "It was the most dangerous place on earth, just as you wrote in 'One Tribe at a Time,'" Petraeus said. "I will never forget those days," he said.

"I won't, either, sir," Jim responded.

Little Malik, Noor Afzhal's grandson, approached with a group of young Afghan boys. Jim shook his hand and introduced him to Petraeus. Jim explained that the schools and clinic were open, and families were moving back to Mangwel from Pakistan. They passed some fields and entered a narrow lane that went by a tailor's stall and the pharmacy to the heavy wooden door of Noor Afzhal's qalat. Noor Afzhal greeted them and they sat down for lunch in the guest room. As they broke bread, Noor Afzhal sat on a bench next to Petraeus and spoke of his affection for Jim, and how he considered him a son and a malik. Jim sat on the floor near Noor Afzhal's feet. The extent of Jim's connection to the people struck Petraeus as extraordinary. "He had a fantastic relationship with the locals in that particular area," he told me later. "I just kind of marveled at what he was doing." All in all, he said, "it was an uplifting day in a tough neighborhood."

Back at the qalat, Jim briefed Petraeus on his strategy for southern Konar and gave him a copy of a classified, long-term plan he had written for the area. The plan, informed by everything Jim had learned on the ground, was detailed and aggressive and pushed his unconventional style of warfare to a new level. The plan called for expanding tribal engagement beyond Konar into tribal areas of Pakistan. It involved creating surrogate forces that would operate across the Afghan-Pakistan border, conduct "red-on-red" operations that used Taliban to fight other Taliban, and promote reintegration of insurgent fighters. Jim's plan included specific timelines, locations, and numbers of forces as well as the cost.

As they discussed the way ahead, Petraeus agreed with Jim that leveraging tribal and village security forces was the only way to bring stability to the mountainous border province. "In places especially like those out in Konar—where you have very remote areas, very rugged terrain, limited numbers of coalition and Afghan forces—local security initiatives are the ultimate solution for those areas," he said. "That is how you keep the extremists, the Taliban, Al Qaeda, the TTP, whomever it may be, from establishing safe havens in those areas from which they attack into Afghanistan and go back into Pakistan." Petraeus fully endorsed Jim's vision of how to spread local security networks through the inhabited villages and valleys along the Konar River. "You go from Khas Kunar, and sort of work your way up the eastern side of the Konar River valley," Petraeus said. In conjunction with this plan, Petraeus authorized Jim to establish a foothold west of the river with the powerful Safi tribe and its key leader, Haji Jan Dahd. Jim was delighted—he had been laying the groundwork to engage the Safis for months.

But the views of Petraeus and Jim diverged when it came to the potential of the Afghan government and the Afghan national army and police to back up the local forces. Petraeus was adamant that the NATO-led coalition had to staunchly back the Afghan central government, saying it simply needed more time. Afghan national security forces were not a failure and would improve, he said. But on the ground, Jim was bombarded daily with examples of the pervasive corruption of the government and the ineptitude of its forces. He doubted their will and ability to support local security elements such as the one he had nurtured in Mangwel.

As the U.S. military shifted responsibility for security to the Afghan government, and as the government gained control over the distribution of guns and money, politics and corruption were already bogging down the Afghan Local Police program. The Kabul government began to play ethnic politics with the approvals for Afghan Local Police in different districts. Non-Pashtun politicians worried that the ALP could grow too strong in Pashtun areas, and so they wanted to strengthen the forces in Tajik, Uzbek, and other minority ethnic areas, particularly in the north of Afghanistan. They were preparing for the civil war expected after U.S. troops departed. By mid-2011 the minister of interior, Bismillah Khan, who is Tajik, had stopped approving *tashkils*, or authorizations for creating new police forces in Pashtun areas, demanding that the U.S. military first establish more of the forces in Tajik and other communities. Khan had insisted on creating ALP in the northern province of Kunduz. The program was a disaster. In September 2011, a Human Rights Watch report detailed how militias in Kunduz and other northern provinces morphed into ALP and committed serious abuses such as killings, beatings, and stealing.

A major, related problem was Afghan government interference with the distribution of guns—a form of currency in Afghanistan. Initially, U.S. military personnel drew the weapons—brand-new AK-47s—from the Ministry of Interior and took them directly to the Afghan district for distribution, sometimes dropping the guns in by air. That was how Jim and Burns's 20th Group team received the first batch of weapons for Mangwel in November 2010—on a Chinook transport helicopter. But then the Afghan government began requiring that the guns move instead from the Interior Ministry through the provincial and district police before reaching the Afghan Local Police—and the system came to a grinding halt. New AK-47s began disappearing in the bureaucracy, in some cases replaced by inferior, plastic Czech-made guns known as VZ-58s, which were notorious for malfunctioning. Jim refused to hand out VZ-58s to his tribal police, saying to do so would be immoral. Other issues emerged around pay for the local police and logistical support including fuel. Pay initially came from U.S. military funds known as Commanders' Emergency Response Program (CERP) funds. When the Afghan government began taking over, local police often went unpaid and their vehicles lacked fuel.

From his vantage point living in an Afghan village, Jim understood how little confidence the Pashtun tribes placed in their corrupt government and its often predatory national security forces. The longer he stayed, the more he questioned the entire premise of the U.S. counterinsurgency strategy—that under existing time, resource, and cultural constraints the United States and its allies could build a capable Afghan government and security force and somehow connect them to the people. To Jim, the premise was faulty. And the push to transfer responsibilities to the Afghan government, and the resulting problems with tashkils, weapons, pay, fuel, and other supplies, all bore this out.

"Sir, I need to tell you that there is no government of Afghanistan here. The district center is seven kilometers away, but it might as well be seven thousand," he said. "Out here there are no Afghan National Security Forces," he went on. "Any program that relies on the success of the Afghan government will fail. Any program that relies on the success of ANSF will fail.

"I don't have anything to tie the success to," Jim continued. "We are in a boat floating around and we have nowhere to anchor."

Still, Petraeus urged Jim to write "One Tribe at a Time II" and in it to discuss how to connect the tribes to the government. Jim thought about it, and a few days later, he did. The paper was a single page, on which he typed: "It cannot be done."

AT THE END OF a four-hour visit to Mangwel, Petraeus approached Jim in the tent that served as his operations center in the qalat. "Hey, Jim, I have something for you," Petraeus said. "In most cases, these awards do not accurately reflect what has been done. In this case, I think it does."

Petraeus took out a Joint Service Commendation Medal and asked an aide to read the citation. It said Jim "distinguished himself by exceptionally meritorious achievement while serving as a village stability platform team leader in Mangwel. Major Gant's unparalleled dedication to the development of the concept and strategy for village stability operations has made possible the unprecedented advancement of the campaign in Afghanistan. His tremendous knowledge of the Afghan culture, tribal and political dynamics at the tribal level has been very important to VSP Mangwel's mission success."

Petraeus pinned the medal, a gold star and ribbon, on Jim's Afghan tunic.

"Thank you, sir. It has been my honor to serve under you," Jim said. "And," he added, nodding at the U.S. soldiers and Afghans surrounding him, "it has been my privilege to lead these men."

Soon afterward, Jim walked Petraeus to his helicopter.

"Sir, how much time do we have?" Jim asked.

"Three years, for the coalition," Petraeus said. "For Special Forces, it's indefinite. And Jim," Petraeus said, grasping Jim's hand, "when you get out of here, don't get too comfortable."

He turned and strode to the chopper.

What Petraeus did not yet know was that Obama was preparing to announce within weeks an aggressive plan to withdraw a quarter of the coalition—thirty-three thousand U.S. troops, or the equivalent of the "surge" force that Obama ordered to Afghanistan in 2009—by the end of summer 2012. Most of the remaining coalition forces were to leave gradually and hand over security to Afghans by the end of 2014. In the June 22 announcement, televised from the East Room of the White House,

Obama told the war-weary American public it could "take comfort in knowing that the tide of war is receding." U.S. military commanders including Petraeus had argued vigorously for a more conservative time-line, advocating keeping more troops in Afghanistan until after the 2012 fighting season so that they could shift some forces from the south to the volatile eastern provinces bordering Pakistan. Obama disagreed. The clock, despite Petraeus's efforts to slow the pace of the drawdown, was rapidly running out.

LATER THAT DAY, JIM met with Noor Afzhal. Holding his hand and squat-ting at his knees, Jim read him the award citation, and then pinned the gold star and ribbon on his chest.

"Father, without you, there is no me," Jim told Noor Afzhal. "It is yours."

Noor Afzhal nodded.

"If you got the award, it means I got the award," Noor Afzhal said. "You and I are one."

"That's right," Jim said. "And there is no turning back now."

The message was clear. Jim was fighting not for his country but for his family, his men, and his tribe.

# CHAPTER 23

"DREW, GET DOWN!" JIM yelled to his gunner that day, Pfc. Andrew Gray, as our Humvee approached a blind curve in the steep road into the Dewagal valley in Afghanistan's Konar Province.

Drew crouched down from the turret of the M240 machine gun, and I grabbed his leg to steady him while the Humvee veered around the turn. As the vehicle straightened out, Drew, a heavyset blond with a ready smile, stood back up in the gun. I was in my usual seat, directly behind Jim and next to several green metal boxes of ammunition. My primary job on enemy contact was to hand Drew more bullets if he ran out. I had trained with the team on the shooting range and in live-fire exercises, and had practiced firing almost every weapon they had. As I usually did on combat operations, I was wearing a dark green U.S. military uniform and boots with my hair tucked up in a baseball cap to try to prevent insurgents from singling me out as a woman.

It was summer 2011, the heat of the Afghan fighting season, and Jim was expecting an ambush any minute. The small, two-vehicle patrol of a dozen Afghans and Americans was venturing deep into the insurgent-held valley, territory rarely covered by U.S. military units. The mission was dangerous, but it was critical to a much larger prize: Jim's recruitment of a new tribe—the Safis—the most powerful in Konar.

Taliban fighters and supplies moved freely through the Dewagal. Rising from the lowland villages along the Konar River, the Dewagal spilled into the notorious valleys of another river—the Pech. The Dewagal led into the Shuryak valley, where in 2005 Navy SEAL Marcus Luttrell survived a vicious Taliban ambush on his four-man team. A short distance beyond that lay the Korengal valley, where embattled U.S. troops manned several outposts until the military decided to abandon them starting in 2009. Hard-core insurgent commanders such as Maulawi Basir operated in the Dewagal. Basir was associated with the strict and violent Salafist strain of Islam. In September 2010, militants under Basir kidnapped British aid worker Linda Norgrove and hid her in a mud shack in the Dewagal until a U.S. military team accidentally killed her with a grenade in the chaos of a rescue attempt.

Jim was deliberately pushing high into the Dewagal with the smallest possible force and no aircraft overhead in an effort to draw out the enemy and expose their locations.

In the rocky ridges above, five Taliban fighters watched the patrol as it continued twisting up the road carved into the rugged mountainside, according to radio intercepts. Seven more fighters were dispersed in the fields of ripening corn below, preparing to strike. "We will attack when we are ready," one insurgent radioed.

Jim got word of the insurgent radio chatter. *Good. Let's see what they've got.*

Jim was in the Dewagal to stir things up, to make a statement to the Safi insurgents: *We have no fear, and we are moving into your valley, your tribe.*

Earlier that day, dawn cast a soft blue light on Jim and his team as they rolled out of the village of Mangwel and left Mohmand tribal territory. They crossed the Konar River as it coursed past the White Mosque Bridge. On the north side of the river, they entered Chowkay District and the domain of the Safis, the biggest and strongest tribe in Konar. Historically the neighboring Safi and Mohmand tribes fought over land and water, but came together during times of external threat. The Safis controlled a large swath of territory that extended from the western side of the Konar River far into the Pech River valley. One of the most dissident Pashtun tribes in Afghanistan, the Safis led a major revolt from 1947 to 1949, fighting a number of successful battles against government forces. Eventually, though, Afghan government troops suppressed the uprising, bombing Safi villages and arresting tribesmen. The government then forcibly relocated Safi leaders and their families around the country. But the Safis had once again proven fiercely independent.

For months, Jim had been courting the Safi tribe in Konar through its most influential leader, seventy-five-year-old Haji Jan Dahd. Feared and revered as a brutal former mujahideen commander, Jan Dahd had waged guerrilla warfare against Russian troops in the 1980s. He captured and killed scores of Russians and liked to roll up his sleeves and pants to show visitors his multiple gunshot wounds. He battled the Taliban regime in the 1990s and served as the first governor of Konar after the Taliban government fell and President Karzai took office in late 2001. As the reigning tribal chief in central Konar, Jan Dahd maintained access to hundreds of armed men, whom he could call on at a moment's notice. Jan Dahd had influence over a significant part of the province's illegal timber trade. His son, Haji Jan Shah, was the CIA-trained deputy chief in Konar of the National Directorate of Security, the Afghan government intelligence service. The U.S. military had at times placed Jan Dahd on its target lists, even while actively engaging him and seeking his help when critical events took place in Safi territory, such as the Norgrove kidnapping, or threatened protests when U.S. military air strikes killed civilians in the Dewagal in early 2012.

Holding court daily in his large qalat, set against a steep mountainside in Chowkay, Jan Dahd advised Safi elders who came to him in droves seeking his advice to resolve conflicts over land, water, and other disputes. A religious man and conservative Pashtun, he wore a long gray beard and plain, coarse clothing, and often went barefoot. But he displayed perhaps the ultimate power symbol in Pashtun culture—he wore no gun, protected instead by bodyguards. And his enemies dreaded him as a man who would kill without hesitation.

Jim believed that Jan Dahd, based on his stature and authority among the Safis, was the best person to raise a tribal force to secure central Konar. After Petraeus gave Jim the go-ahead in May to launch the local security initiative in Chowkay, it took little time for Jan Dahd to agree. He had heard of Jim's work with the Mohmand tribe and respected Jim as a fighter, remembering him from clashes they had in 2003 when ODA 316 blew up the weapons cache of Jan Dahd's border police.

"You shouldn't have destroyed our heavy weapons—we will need them now!" Jan Dahd joked.

In a gesture of respect, Jan Dahd traveled to Mangwel to see Jim and

Noor Afzhal, who put on his most handsome gray turban for the occasion. The two tribal elders had known each other for years but had not met since 2002. They embraced warmly and traded pleasantries, as is the Pashtun custom. Sitting together cross-legged in a tent on the qalat protected by Mohmand tribesmen, they drank green tea and shared stories on the Russian occupation and Taliban era. Then, after a meal of flatbread, chicken stew, and rice, Jan Dahd and Noor Afzhal turned to the real business—the relations between their two tribes.

"If we have problems with the Mohmands, we will solve them together," Jan Dahd told Noor Afzhal.

"Yes, we will come to your aid, and you will come to ours," Noor Afzhal replied.

The informal alliance of the Mohmands and Safis was significant and a direct outgrowth of Jim's work with both tribes. "We need to deal with one malik. I will pick the people to control all the area, for example, the Pech and Chapadara. . . . We need people who are strong and fearsome," Jan Dahd said.

Together, Jim and Jan Dahd made rapid progress in identifying hundreds of Safi tribesmen willing to serve as arbakai in Chowkay. When the Taliban tried to co-opt the process by having their sympathizers in the Chowkay district government take charge of choosing the arbakai, Jan Dahd and Jim worked together to thwart it. Jim wanted to press his advantage and test the Taliban's will directly, face-to-face—in the Dewagal.

THE MOUNTAIN FELL AWAY steeply to the left as our Humvee hugged the road up the valley.

Jim turned to his driver, Pfc. Kyle Redden. "If an IED goes off on us, push right. If we go left, we're done," he said.

We punched through the lower Taliban ambush location and pushed one, then two miles farther up the valley. With a single road leading in and out of the Dewagal, it was an incredibly gutsy move by such a small force. Jim could hardly have done more to invite an attack. The Taliban had plenty of time to prepare to hit us on the way out. Insurgents were particularly active that day because a U.S. Army unit had arrived at the foot of the valley that morning in an effort to meet with local elders but was rebuffed. Another report came in that the insurgents were tracking our vehicles.

A radio call came in telling Jim that his Air Force JTAC (joint tactical air controller), Staff Sgt. Andy Deahn, no longer had communications with some helicopters that had come into the area. Did he want Andy to make comms?

"Negative," Jim said. "I'll call CAS [close air support] if we need it." He almost never called for airpower, preferring to maintain a tactical edge on the ground as long as possible. The aircraft could also scare off the enemy.

Everyone was vigilant. I was scanning the hillside for insurgents using the zoom lens on my camera. I was tracking what looked like a shepherd walking down a path. Just then, I spotted a white flag flying from the corner of a qalat up ahead. White did not signify surrender. On the contrary, it was the Taliban's color—symbolizing their claim to fight for peace.

"A Taliban flag!" I called up. Drew had seen it, too.

Immediately, Jim knew what he was going to do.

"Let's get it," he said, and halted the patrol.

With his Afghan comrades, Abe and Ish, and another U.S. soldier, Jim dismounted and scrambled down the hillside to the qalat. There, Abe climbed onto Jim's back and plucked the Taliban flag off the corner of the qalat wall. They climbed back up to the road and attached it to an antenna in the rear of our Humvee.

Jim led the patrol a short distance farther up the valley. Still the insurgents held their fire. Jim decided he'd pushed far enough, and ordered the patrol to turn around. The return trip down the single valley road was perhaps the riskiest of all, especially after we had stopped and seized the flag. Everyone was braced for an attack, but it never came. Only later, back in Mangwel, would we learn the reason.

We slowly wound down the long road and then drove through the Chowkay bazaar with the Taliban trophy for all to see. Afghans turned their heads and gawked at the white flag flying on the back of a U.S. military Humvee. Jim had made his statement—one far louder than any gunfight.

Almost as quickly as the patrol crossed the Konar River and returned to the qalat in Mangwel, word spread that Commander Jim had ventured unscathed far into the Dewagal. Later, Jim and I walked down the dirt path from the qalat, past the tailor's shack and the baker's house and the village cemetery, to Noor Afzhal's home for dinner.

When we arrived, his son Azmat approached us, excitement shining on his face.

"There were four groups of Taliban in the Dewagal watching you," he said. "They watched you take the flag. But they knew you were Major Gant and you were a friend of Haji Jan Dahd, so they didn't attack," he explained.

The Taliban's hesitation gave Jim a wealth of information, all of it vital to his plan to recruit the Safi tribe. First, it confirmed Jan Dahd's power and influence. It also told him that, months after he had started working with Jan Dahd to raise a tribal force, the insurgents were still unsure how to react. That, in turn, signaled an opportunity, the possibility that Jim and the Safis could eventually win some of the Taliban to their side. From that moment, Jim knew he could press ahead with the Safis as planned—going into a heavily contested area with only a small American and Afghan force and empowering the tribe to safeguard the villages and valleys. It would be risky, but it offered a huge payoff: engaging tens of thousands of people as well as the insurgents and bringing security from the outside in.

More important, six months after the cold, overcast day when he'd moved into Mangwel with the Mohmand tribe, Jim had laid the groundwork to expand to the Safis. The work was painstaking, slow, frustrating, and endlessly complex, but the momentum was unmistakable. Meanwhile, prominent leaders from other important eastern Afghan tribes across the area were reaching out to Jim and expressing interest in launching tribal forces.

What was only a vision on paper in late 2009 was becoming a reality. His strategy was working. Throughout history, foreign forces had successfully engaged the Afghan tribes or ignored them at their peril. Now Jim was advancing on the ground, one tribe at a time.

AS JIM WAS MAKING HEADWAY with the Safi tribe in the spring and summer of 2011, he opened up a major opportunity with another Konar tribe, the Mushwanis.

In late May, five trucks full of armed Afghan security forces sped down a tree-lined commercial street in the eastern Afghan city of Jalalabad and stopped outside the metal gate of a large, four-story home. A young man in sandals and a light blue tunic quickly pulled the gate open, and a sedan quietly drove into a crowded outer courtyard.

Out of the vehicle stepped a well-dressed Afghan man in his mid-forties wearing a black pakol and vest and a white tunic, and sporting a neatly trimmed dark beard. His face was almost expressionless but bore a slight frown. Haji Ayub was the leader of the Mushwani tribe, which dominated several contested districts in northern Konar. He had come to the Jalalabad residence to meet Jim, who waited upstairs. Ayub was the son of Malik Zarin, a former mujahideen commander and previous leader of the tribe, which dominated several contested districts in northern Konar. A tall, strong man, Malik Zarin was uneducated but smart, and respected for his ability to solve disputes impartially. He was in many ways the tribal counterpart in northern Konar to Haji Jan Dahd of the Safis farther south. The two men had known each other for years and vied for power. Both tribal leaders were among the first in Konar to fight Russian occupation forces, battle the Taliban, and come to the aid of the U.S. military after the 2001 invasion. Both also profited from the province's lucrative timber trade, over which they sometimes clashed, and had sizeable militias. Afghan government officials felt threatened by their influence, and the U.S. military tended to view them as warlords. But among Afghans the widespread belief was that between them, the two men could rule all of Konar.

In April 2011, however, disaster struck for the Mushwanis. Zarin and some forty followers were attending a tribal council in Asmar District where Zarin sought to rally opposition to the Taliban and foreign fighters affiliated with Al Qaeda in Konar. A teenage boy wearing a suicide vest reportedly approached Zarin, hugged him, and detonated his vest, killing the tribal chief and a dozen other people, including one of Zarin's sons. The tribal elders urgently summoned Ayub back from Pakistan. Well educated and politically savvy, Ayub was untested but also respected by his people. When he arrived at the Jalalabad residence in May, Ayub was taking charge of the tribe even as he was convulsed with anger and a desire for revenge. He had heard of Jim and his tribal strategy and wanted to meet him.

"Welcome, my brother," Ish told him. "Come this way."

The armed guards took position inside the gate as Ayub and his younger brother followed Ish up a series of narrow staircases and walkways into a formal meeting room. At the age of twenty-five, Ish, with his soft dark eyes and calm, measured voice, was the seasoned diplomat among the three brothers of the Khan family—Ish, Abe, and Imran— who were working with Jim. The Khans also belonged to Ayub's Mushwani tribe and were originally from Konar. Ish's father and mother and more than thirty members of the family lived in the lively Jalalabad home. Ish exuded responsibility, consulted closely with his father and uncles, and was the father of two young sons. Abe, the older brother, was much more impulsive and violent. Abe was a loner prone to strange visions, but he had a tender heart and everyone in his family loved him dearly. Imran was an excellent student and more carefree than his two older brothers. They had studied English while refugees with their family in Pakistan, and together they gave Jim an immeasurable advantage in his dealings with the Pashtuns. He employed each of the brothers skillfully according to their strengths. The meeting with Ayub was critical. Ish would translate for him that day.

Ayub strode across the room, its rich red carpets, golden draperies, and ornately painted ceiling fitting for a family of the Khans' stature. Seated against a row of maroon pillows, wearing a lavender tunic, Jim stood up and lightly pressed his palm on Ayub's chest, a Pashtun gesture of welcoming and respect, while Ayub did the same.

"It is an absolute honor to meet you, thank you for coming," Jim said in Pashto as he shook Ayub's hand.

"The honor is mine. I have heard about you," Ayub said.

"*Keena*, please sit down," Jim said. "I am deeply sorry for your loss. There is nothing I can say, but I believe maybe there is something I can do. Are you familiar with the *mahali* police program?"

"Yes, I know a little," Ayub said.

"The most important thing is that the police are chosen by the people. They defend their home, village, and valley," Jim explained.

Ayub looked at Jim intently. "I can help you with five districts and across the border as well," he said, describing the Mushwani presence in the Konar Province districts of Shigal, Asmar, Naray, Dangam, and Ghaziabad.

He went on: "I have five thousand fighters that are like arbakai securing the border with Pakistan."

Jim knew Ayub was not exaggerating. Indeed, on April 21, a few days after Malik Zarin was assassinated, reports circulated that hundreds of men from an Afghan tribal army, or lashkar, crossed into Pakistan and attacked a Pakistani outpost, killing some forty-four personnel in Lower Dir District, Malik Zarin's former home in Pakistan's Khyber Pakhtunkhwa Province (North-West Frontier Province). Jim confirmed that these were Mushwani tribal fighters loyal to Malik Zarin, and they had actually recaptured the Pakistani military outpost after Pakistani Taliban insurgents overran it. According to Jim's reports, the insurgents had killed several Pakistani military personnel, but the Pakistani military was unable to recover the bodies, so it asked for help from Ayub. Ayub's ability to mobilize the lashkar and retake the outpost was a potent demonstration of Mushwani strength and his own resolve in the wake of his father's killing. Indeed, Ayub had just come from a meeting in Kabul with President Karzai, who was eyeing Ayub for a position in control of Afghan border forces.

The tribe's influence on both sides of the Afghan-Pakistan border held major appeal for Jim. One of the primary goals of his one-tribe-at-a-time strategy was to leverage the tribes to help uproot the insurgent safe havens in Pakistan that were vital to sustaining the Taliban's war in Afghanistan. The Mohmand, Safi, and Mushwani tribes all had large populations on either side of the border.

"The areas that are mine are now yours," Ayub told Jim. He looked him straight in the eye. "Whatever I say to you is a promise, not like other people."

"I don't doubt it," Jim replied, placing his hand on Ayub's knee. "I have fought here many years and I know that the enemy is easy to find—it is friends who are hard to find."

Ayub nodded. "The most important thing now is to take care of the guys who killed my father. We need to get rid of them, or they are going to blow our heads off."

"Has OGA contacted you at all?" Jim asked, using the military acronym for "other government agencies," which referred to the CIA.

"No, you are the first one," Ayub said, his face darkening. His disappointment at the failure of the U.S. government to help him avenge his father's death was palpable. His father had fought the Russians and stalwartly supported the Americans, yet now he was so easily ignored, forgotten? How could the most advanced military in the world come up empty-handed with information about this brazen assassination? "Even a blind donkey can find water," Ayub scoffed, reciting a Pashtun proverb.

Jim took out his white ceramic prayer beads and began running them through one hand.

"I will give you all the information the Americans have on who killed your father," Jim said. "But it will be on paper, and you can't discuss it on the phone."

"Thank you from the bottom of my heart," Ayub said. "This is my destiny, to get rid of them," he said, his voice shaking. "My life is yours," he said. "We will help you however and wherever we can. We will share our intelligence with you—not only about the ones who killed my father, but all of the Al Qaeda and Taliban who are working in Afghanistan. We are with you to the death."


BACK IN MANGWEL a week later in June 2011, Jim gathered Tribe 33 for a team meeting after dinner, as he did almost every night.

In T-shirts and camouflage pants, the soldiers crowded into the tent that served as the crude operations center. They were sweaty and energized. That morning, Taliban insurgents had again attacked the team in the Shalay valley. Jim was riding on the hood of his Humvee after looking for an IED near the Shalay clinic when insurgents opened fire with a Russian machine gun.

"Hold!" he yelled, leaning back on the windshield. He hopped back into the front seat.

In the gunner's turret above, Sonny spotted muzzle flashes from a nearby qalat and returned fire with the M240, pumping out twelve hundred rounds as they drove back and forth under fire. The two vehicles executed a complex figure eight maneuver designed to confuse the insurgents while mounting an aggressive counterattack. Gunning from the second vehicle, Miah shot back with the .50-caliber machine gun at Taliban fighters he saw firing from a mound of dirt. A rocket-propelled grenade exploded about forty yards away, its shell casing flying into the air. After only about fifteen minutes, they had suppressed the ambush. They were getting good.

That evening, Jim reviewed the combat action in detail as usual and invited critiques from his men. But his main message, driven home by his deepening ties with the Safi and Mushwani leaders, was that the tribes held the only key to victory. He knew it, and the Taliban knew it.

"Men, we are getting closer to the tipping point," Jim said. "I feel it's right there. The big fight is coming. In a few weeks, we will be the biggest target for the insurgents in all of Konar. When they figure out we are winning over the Mohmands, the Safis, the Mushwanis, they will fight us with everything they have. "And when that happens, remember," he said. "*Jung de meweey de taskeemawalo zai naday*—war is not passing out candy."

High in the valleys, Jim knew, the Taliban were talking, planning their next moves. As he prepared to expand to more tribes and deal with the Taliban reaction, he badly needed a right-hand man on his team. In his mind, there was one choice for the job: his trusted Special Forces comrade, medic, and gunner, Dan McKone.

# CHAPTER 24

ONE MUGGY MORNING IN early July 2011, Capt. Dan McKone hugged his ten-year-old son, Sean, and boarded a military transport plane in Fort Bragg, North Carolina, on orders for Afghanistan. Going to war for Dan was by then routine—it was his sixth combat tour since he first deployed with Jim as a staff sergeant on Special Forces ODA 316 in 2003. But as Dan joined the single file of uniformed soldiers walking up a metal ramp into the aircraft's dim belly, he was surprised at how strongly he felt the tug of home.

Dan knew that Sean, his only child, was growing up fast and needed his father more than ever. But when Jim asked Dan to join his tribal engagement team in Mangwel, Dan didn't hesitate. Unlikely best friends, as opposite as night and day, the two men had spent years earning each other's trust. Now Jim needed a second-in-command who would never be a yes-man. Dan was it. He agreed to go.

But when Dan arrived at the headquarters of the CJSOTF-A at the U.S. military base in Bagram, he hit a brick wall in the form of its new commander, Col. Mark Schwartz.

"Why does Major Gant assume you are going to work with him?" one of Schwartz's deputies quizzed Dan after summoning him to a meeting at the CJSOTF-A headquarters at Camp Vance.

"I was requested for Mangwel," Dan protested.

Regardless, Schwartz decided to relegate Dan to a desk in a cubicle at headquarters. Dan was stuck.

Schwartz was part of an entirely new chain of command that took charge in Afghanistan during the spring and summer of 2011 and seriously undermined Jim and his mission. Petraeus, Campbell, Miller, Bolduc, Lovelace, and Pelleriti—strong supporters of Jim and his tribal inroads—had all now departed. Jim stayed on, having volunteered for an exceptionally long deployment of two years in Afghanistan.

Schwartz, who replaced Bolduc as head of CJSOTF-A, was no fan of Jim's. He was the same Schwartz who, in late 2009, as operations chief for the CFSOCC-A, had tried to confine Jim to a staff job in Kabul. He had written Jim a blunt email saying there was no intent to place him on a special team conducting tribal engagement. Upon his return to Afghanistan, however, Schwartz found Jim in Konar doing exactly that—with Petraeus's blessing. Schwartz went to visit Jim in Mangwel and acknowledged the success he was having. But he was threatened by Jim's relationship with Petraeus and other senior officers, and wanted to rein Jim in. Back at his headquarters in Bagram, Schwartz made his wariness of the junior officer known.

"Jim is off the reservation! He is out of control," Schwartz fumed when Jim's name came up. Others under Schwartz contemptuously referred to Tribe 33 as "Team Gant."

Schwartz's subordinate and Jim's boss at the time was Lt. Col. Robert Wilson, commander of Special Operations Task Force–East. Wilson was in charge of all Special Operations Forces operating in eastern Afghanistan, including Konar, where Jim was based. Wilson had his own nickname for Jim: "the Scarlet Witch." A Marvel comic book figure, the Scarlet Witch was a mutant who had a wide range of powers but could not always control them. Her hexes and use of "chaos magic" often backfired. Wilson went so far as to introduce Jim to other Army officers as the Scarlet Witch. He would also use the name in a cautionary sense with Jim in conversations.

"Hey, Jim," Wilson would warn. "Don't Scarlet Witch me on this!"

Mostly, however, Wilson acted as though Jim did not exist. Knowing how much Schwartz disliked Jim, Wilson avoided mentioning Jim and his team in briefings and reports, and told Jim so.

Meanwhile, both Wilson's command in Bagram and that of his subordinate in Jalalabad, Maj. Eddie Jimenez, were permeated with a negative attitude toward Jim and his team, according to Sam Schmidt, a civilian intelligence analyst who worked at both commands. Requests

Jim made for intelligence support, including against IED threats, were immediately shot down, he said. "It was unprofessional and personal," said Schmidt, a thirty-three-year-old former Marine intelligence officer who had also deployed to Iraq. Jim asked for specific imagery on IED strikes as well as infrared monitors and security systems that he never received. Schmidt said that, whether intentional or not, the attitude and obstruction heightened risk for Jim and his team. "It could be characterized as wanting him to fail," he said.

When Wilson found out that Dan had been sidelined in Bagram, he did nothing.

"You have lost this battle," Wilson advised Jim. "Leave it alone." He specifically warned Jim against trying to go over his and Schwartz's heads to get Dan released from Bagram.

Jim faced a tough dilemma.

Dan was the comrade he most needed. Whenever a U.S. general officer visited Mangwel, Jim asked for only one thing: send Dan. He needed Dan for many reasons, especially to allow him to focus more on the big picture. As his influence grew, Jim felt stretched thin, even at times overwhelmed, by demands to spend time with his men, the arbakai, and a growing number of tribal leaders. Dan was a smart, seasoned, former noncommissioned officer who could help lead and direct Jim's team. He also spoke Pashto and had worked with the Konar tribes.

But Jim knew that if he pulled strings with the generals to try to force Schwartz to send Dan to Mangwel, it would make Schwartz all the more determined to undercut him.

At the same time, he felt he really had no choice, so he pulled strings. Schwartz got a call from a general officer. A few days later, Wilson sent Jim an email summoning him to Bagram Air Field immediately to see Schwartz.

*This can't be good*, Jim thought.

The 150-mile trip from Mangwel to Bagram covered some of the most dangerous roads in Afghanistan and was known for insurgent ambushes. But Jim had no choice. He loaded up his vehicles and headed out with his men as ordered.

When he arrived at Schwartz's headquarters, a modern office building on the sprawling military base, Jim passed down a sterile hallway lined with the framed plaques commemorating Special Operations members killed in action in Afghanistan. Jim never liked such memorials to the fallen—he had seen too many of them grow shabby with neglect. Then he spotted something on a plaque that angered him: Staff Sgt. Chris Falkel—Bronze Star with "V." Chris, who was Jim's and Dan's teammate from ODA 316, had been awarded the Silver Star, not the lesser Bronze Star, for his valor in the Mari Ghar battle in which he was killed in 2005. Jim went to Schwartz's sergeant major and asked, calmly but firmly, that the plaque be corrected. Then he headed to Schwartz's office.

Dan was outside.

Although it had been more than a year since Jim had seen Dan, they both understood that, given the circumstances, they had to greet each other formally, with a handshake and few words.

"Hey, Dan, great to see you," Jim said.

"You too, sir," Dan said. A smirk crossed both their faces. Dan had not called Jim "sir" since they both joined ODA 316 in 2003.

Then they walked into Schwartz's office and sat down next to each other on a cramped couch, neither of them having any idea what Schwartz would say.

A tall man with a bulbous nose and short-cropped hair, Schwartz sat at a large desk in front of two large computer screens. A 1987 graduate of Idaho State University, Schwartz was a motorcycle buff. At Fort Bragg, he was known as a founding member of a motorcycle club known as the SF Brotherhood. When he took over the 3rd Special Forces Group in 2009, he hung his club vest in his office until one of his enlisted advisors said he should take it down. Schwartz attempted to be intimidating, but Dan and Jim felt he didn't quite pull it off.

After some small talk, Schwartz turned to his main concern: "executive communications."

"Jim, I don't want you having any direct communications with general officers or senators or other senior people without talking with me first," Schwartz said. "Is that clear?"

"Yes, sir," Jim replied.

"Here is Dan. Move out," Schwartz said.

That was it.

Dan and Jim got up, saluted, and walked out of Schwartz's office. They looked at each other.

"What the fuck?" they said at the same time.

Jim was livid that Schwartz had forced him and his men to make the trip to Bagram on bomb-laden roads only to get a lecture on his contacts with generals. Schwartz could easily have called Jim on the phone, sent him an email, or arranged a video conference, and dispatched Dan to Konar by helicopter. But they were both excited that Dan was headed to Mangwel.

"Welcome back to the war. Now get up in the gun!" Jim said, flashing his old gunner a smile. He was overjoyed to have Dan, a warrior and leader, back at his side.

"Fuck it," Dan said, climbing behind the M240 machine gun. "Let's go!"

The episode with Dan was emblematic of Schwartz's tone-deaf leadership. He and his commanders had only a minimal understanding of the environment Jim was working in or what he was trying to achieve. They seemed more concerned with imposing rules than with winning a war.

And Schwartz and Wilson dealt Jim even more serious setbacks. While Jim's previous command had streamlined operations and given him wide leeway, Schwartz and his subordinates attempted to micromanage tactical details. Jim was required to gain the approval of Wilson's command before firing his team mortars—a rule he refused to follow because it would have endangered the lives of his men. The command also attempted to dictate the number and type of soldiers and vehicles on patrols, the use of body armor, and the precise wearing of facial hair. Schwartz was not unique among commanders in imposing strict grooming standards. But coupled with his overarching lack of support for Jim, the rules seemed particularly grating.

For example, in a June 2011 memo, Schwartz specified in excruciating detail that beards worn by Special Forces soldiers in Afghanistan "will not exceed one inch in length . . . the hair above the lip will be trimmed to the upper lip line exposing the lip. The beard itself will not have a 'bush like' appearance but will be kept trimmed. . . . This policy is punitive," the memo said, warning that if even a single member of the team was found to violate it, the entire team could be deprived of the ability to dress and look like Afghans.

The same memo required that when Special Forces teams wore a "non standard uniform" of Afghan clothing, they must wear body armor underneath it, and ideally carry heavy weapons with them. The rule was inane and completely defeated the purpose of wearing Afghan garb to blend in with the population.

On major issues involving Jim, though, the command was often AWOL—unresponsive at best and at worst obstructive to requests. Overly focused on "kinetic" combat operations, Schwartz and his subordinates shortchanged village-level engagement, according to Special Forces officers with extensive knowledge of their decision making. Most important, Schwartz and Wilson did not approve some of Jim's key tribal expansion efforts. They refused to allow him to set up arbakai already approved for Khewa District near Mangwel. They also blocked Jim's efforts to make inroads with Haji Ayub, the newly anointed leader of the powerful Mushwani tribe, whose father had been assassinated. After meeting with Ayub, Jim sent Wilson a detailed memo on the encounter as well as a Mushwani tribal engagement plan. The conventional U.S. Army commander in northern Konar had lost several men from his battalion in recent fighting and was enthusiastic about Jim's idea of recruiting arbakai to quell insurgent strongholds where the Mushwani held sway. But the night before Jim and his team were to drive north to the district of Naray to meet again with Ayub and the U.S. commander, Wilson ordered Jim not to go. "You are not authorized to go to Naray or to speak with any Afghans about expansion," Wilson said. Jim's emails to Wilson seeking an explanation went unanswered. It was a huge missed opportunity. Jim's frustration over the decision grew as Ayub rose in national prominence in the months to come.


ONE LATE SUMMER AFTERNOON soon after Jim brought Dan to Mangwel, a rainstorm struck the village. Many overcast, windy days had come and gone without so much as a drop of rain falling from the sky. But that day, with tremendous thunder and high winds, the rain finally came down in sheets, as if the sky at last gave birth. We all felt relieved. In the Tribe 33 qalat, wind whipped at the tents and gun covers on the trucks. I ran up into one of the guard towers, soaking wet, and looked out at the mountains surrounding the village with misty silhouettes of gray. Then the clouds broke open and a rainbow arced over the wet green fields—the first rainbow I had seen in Afghanistan.

That afternoon as we dried off after the rain, Jim reflected on promises—both those he made and unspoken promises that he believed must be held sacred by those who have the arrogance to send men to war.

Jim had kept his promise to return to Noor Afzhal. And the tribe was keeping its promise to him—as surely as the feisty arbakai manned the towers day in and day out, squabbling or calm, keeping us safe. Jim had kept his promise to protect his men. And his men had shown him great loyalty. He had kept his promise to himself—to strive to win, or at least to leave everything he had on the battlefield.

"I only want three things," he told me. "I want a battle, a war that means something. I want worthy men who are willing to fight and die. I want good commanders, who will support me with everything they have."

But by then, he knew, those commanders were gone—replaced in his view by straw men, men without honor.

Jim's face darkened. "Remember this: whoever does me in will be wearing a U.S. Army uniform, with a Special Forces tab."

While Jim's U.S. commanders were failing him, his enemies better understood the impact he was having on the ground. One after another, the Taliban came calling.

# CHAPTER 25

NIQ MOHAMMED STOOD OUTSIDE the tall blue metal gate of the Tribe 33 qalat in Mangwel one early September day in 2011, knowing his next step could cost him his life.

The burly thirty-seven-year-old, with a broad black beard, large nose and deep-set eyes, cast a glance at the dozen Afghan elders who had come with him to Mangwel from their contested village cluster of Kawer down the road. They were counting on him.

Niq had been struggling with what to do for six months, since the Taliban executed his friend Gujar. The shifty leader of the tribal police in Kawer was shot four times in the back of the head. The February 2011 incident led to the collapse of the fledgling tribal force in Kawer. As a result, Kawer stayed under the influence of insurgents, while security improved in Mangwel and other nearby communities with arbakai.

The road that twisted through Kawer, nicknamed Zombieland, remained the most dangerous place for improvised explosive devices and other attacks for Jim and his men. In response, Jim had shunned Kawer and its people, cutting them off from wells and other development projects offered to surrounding villages. His patrols through the area with scores of Mangwel arbakai were large, guarded, and hostile. Adopting a Taliban tactic, he distributed night letters in Kawer warning the people that they were being watched constantly by "eyes in the sky," drones. He wanted to retake Kawer, but he knew the people themselves had to stand up to make that happen. So he kept up the pressure.

A big man with a booming voice, Niq was feared and respected in

Kawer. Originally from Shalay valley, his grandfather had moved to Kawer and bought land to farm. His father had fought the Russians, who killed Niq's brother. In recent years, he had served on the local government shura, or council, and ran the bus station. He was viewed by many as the only man who could lead a new group of tribal police and bring security to the community. There was only one problem: Niq was former Taliban.

Jim had received intelligence reports that indicated Niq had been involved with an insurgent cell that emplaced the roadside bombs and carried out attacks on U.S. troops. Jim had been advised by Noor Afzhal that he must deal with Niq in one of two ways: kill him or put him in charge of the arbakai in Kawer.

Niq, for his part, had heard much about Jim and the security and projects he brought to Mangwel and surrounding villages. He decided to turn against the insurgents and ally with the American. Now he would find out what Jim had decided.

"*Zu*. Let's go," Niq said to the group of elders, his voice loud and hoarse. He rapped his knuckles on the metal gate.

One of the Mangwel arbakai, with an AK-47 slung over his shoulder, swung open a small door cut in the gate. Niq paused for a second, then walked through, followed by the elders. He tossed over his shoulder the cream-colored Afghan blanket that he wore over his ochre tunic and strode into the arbakai tent, which served as a meeting room.

Moments later, Jim came in, and after a terse greeting sat down across from Niq and looked him in the eye.

"So the people of Kawer want local police now?" he asked.

"Yes," Niq replied.

"Are you sure?" Jim pressed.

"Yes," Niq said. "We already have ten men who want to join."

"Insurgents have used Kawer to move through," Jim continued. "Don't say you don't know about that, because you do. I will not go ahead with this unless someone stands up and says they will not let the enemy operate in that area. Being 'Taliban' is not the issue. I am concerned with people who are trying to kill Americans, to kill Afghan security forces, and to stop development from coming," Jim said. He was drawing a distinction that was difficult for many Americans to grasp, but one that meant everything to these Pashtun tribesmen. In their minds, "Taliban" at this juncture in history translated loosely as fellow

Pashtuns who were not happy with the corrupt Kabul government and foreign military intervention. That unhappy group was a very large subset of the tribal people and sympathized with the Taliban. All Taliban were not necessarily active insurgents. Often families contained some adult males who were serving in the Taliban and others who worked for the government. Tribal leaders including Noor Afzhal sometimes referred to Jim, in a positive way, as "Taliban."

"So what are we going to do?" Jim asked.

"Give me a map," Niq said.

One of Jim's soldiers handed him a map of Kawer, and Niq spread it on the floor and traced over it with coarse, thick fingers.

"We can put in checkpoints in these hills, and patrol from our houses like the arbakai do in Mangwel," Niq said. "I will secure Kawer and chase the bad guys to Maya." He vowed to push insurgents out to a remote highland area beyond the top of the Shalay valley on the Pakistan border.

"I am tired of everyone saying we are bad in Kawer," Niq went on. "We can't tolerate it. I want some respect. I want you and elders and government people to come there and drink tea, and for roads to be paved, wells dug, and retaining walls built."

Jim watched Niq intently.

"You want what I want," he said.

"All the people say I am a bad guy. But if we don't do this now, no one will. I am on your side. If the Taliban shoot me, just bury my body," Niq said.

Jim knew then. Niq was the commander he was looking for. But he held off on making an agreement.

"I've been waiting a long time for someone to come from Kawer and say they wanted local police," he said. "Now, show me you mean what you say."

One morning later that week, dozens of arbakai from Mangwel joined Jim and his soldiers for a show of force into Kawer. Wearing chest racks of ammunition and their black uniform vests, the arbakai patrolled single file along raised dirt paths between the fields of corn and vegetables to Kawer, and then swarmed through the village to Niq's qalat. There, the arbakai stood guard while Niq and Jim met with more Kawer elders, and then Niq led Jim to his proposed locations for checkpoints.

Jim was satisfied.

"We will move ahead," Jim told Niq. "I and my men will fight along-side you."

Within days, Jim and his men had armed and trained fifty Afghan Local Police for Kawer. Niq launched the arbakai on aggressive patrols and positioned them in the high ground above the Zombieland road. Whenever insurgent activity was reported, Niq and his men responded immediately, without Jim calling first. The atmosphere in Kawer changed dramatically almost overnight. Village men began descending on Niq's qalat to ask him if they, too, could become arbakai and defend Kawer. "Everyone is coming to me. Now the people want a hundred Afghan Local Police!" Niq said with a gruff, booming laugh.

Niq was quickly proving himself one of Jim's most capable arbakai commanders.

THE TALIBAN REACTION TO Jim's inroads with the Konar tribes was broad and deep, underscoring the power of his strategy.

Jim never missed an opportunity to invite the Taliban to either talk or fight. "I have food in one hand, a knife in the other," he would say.

Jim believed reintegrating the Taliban was vital to any lasting peace in Afghanistan. He also realized that the formal reintegration program run by the Afghan government and U.S. military was slow, bureaucratic, and corrupt. As a result, it was destined to fail, because it could only appeal to a narrow subset of the Taliban. It required the former Taliban to hand over their weapons and publicly renounce the insurgency, yet it allowed the government to arrest them at any time. So, working quietly through the tribes, Jim reached out to the Taliban using his own, more sophisticated, nuanced, and ultimately flexible approach—one that was riskier and more time-consuming but which offered more options for both him and them. For example, short of public reintegration, a Taliban member could become a source of intelligence or could even conduct "red-on-red," or surrogate force, operations against other insurgents. In many cases Jim's approach was not one approved by his command, but now that outreach was paying off.

The most positive shift, and the one Jim sought from the start, was to push some Taliban to switch sides and work with him as arbakai. Niq and others epitomized this trend.

Switching sides could be incredibly dangerous, though, and required some minimum level of protection from retaliation. For some Taliban, breaking with the insurgents proved much more complicated. That was the case for Mohammed Jalil, a middle-aged malik of the Mohmand tribe who worked for the government but was also closely tied to the Taliban, as well as for Obeidullah, a young Taliban commander who belonged to the Safi tribe across the river in Chowkay.

It was early September 2011 when Jalil walked into the meeting room at Noor Afzhal's house one afternoon and sat down across from Jim. The Muslim holy month of fasting known as Ramadan had just ended. Mangwel had a festive, relaxed atmosphere as families feasted and visited friends and relatives. Boys in colorful new cotton tunics and pants raced through the village laughing and playing with plastic toy AK-47s. The village was more secure, and there was a tangible sense of normalcy that hadn't existed only a few months before.

Despite the celebratory mood, Jalil looked gaunt and glum. His large, powerful figure was stooped, and he had dark circles underneath his eyes. As it turned out, Jalil had just been temporarily released from jail for the holiday.

Jalil and his family belonged to the Mohmand tribe, and Noor Afzhal welcomed him as such. Officially, Jalil was deputy director of the local government shura, but his ties with the Taliban ran as thick as blood. Two of his brothers, Abu Hamza and Momen, were midlevel commanders of the Afghan Taliban. Jalil was a passive supporter of the Taliban, part of what Jim called the Taliban "underground." Jalil and his family had roots in the Shalay and neighboring Walay valleys. There was an ongoing feud between Jalil's brothers and rival insurgents from the Pakistani Taliban led by Jim's nemesis Abu Hamam. Overlying all of these relationships, however, were the tribal imperatives of Pashtunwali.

Afghans living in the high rugged valleys were isolated from the settled towns below. With no forces to protect them, they had little choice but to provide Taliban fighters with food, water, shelter, and refuge if they needed it, or face beatings or other retribution, and Jalil's family was no different. As a rumor spread in the late spring of 2011 that Jim and his team were going to set up arbakai in the Shalay and Walay, pressure on the local people intensified. Jalil went to Shalay to fetch his son's bride for their wedding and found himself surrounded by Hamam's

Taliban fighters. They accused Jalil of supporting the arbakai plan and threatened to cut off the head of the bride's father in retaliation. It was a blow to Jalil's namoos, or honor, because the bride belonged to his son. At the time, Jalil went to see Jim, complaining about Hamam and the absence of any security forces in the Shalay.

"When I am up in the Shalay, they say I am working for the government. When I am down here, they say I am Taliban!" Jalil fumed. "I just want my family to live safely with no one bothering us."

Nevertheless, Jalil faced continual questioning from American and Afghan forces because of his Taliban ties. In August, during a search of his home and fields, Afghan soldiers found two rifles and a rocket-propelled grenade booster, and arrested him. Afterward, a large group of elders from the Walay and Shalay valleys came to see Jim in Mangwel to ask for his help in releasing Jalil. They appealed to Jim's tribal affiliation to win his support.

"You are Mohmand. I am Mohmand. You are a man. I am a man. We guarantee Mohammed Jalil will not do anything bad if you get him released," said Malik Mirwais, from Shalay.

"I will do my best," Jim said. As usual, he also asked the elders to let the Taliban commanders in the Shalay know he wanted to meet with them.

After Jalil was freed from jail for the holiday, he immediately went to see Jim. Jalil was downcast after his time in the miserable Afghan prison in Asadabad where he was being held, and wanted to get out. Most of all, though, he felt a deep sense of dishonor because the Afghan soldiers had taken his weapons. He wanted them back.

"This is peghor," he said, using the word that means "shame" in Pashto. "Maybe a hundred years from now someone will talk about how they took my weapons and I didn't do anything. I can't tolerate that. A man can kill or be killed for that. So it's not that those weapons are priceless—it's about peghor."

"I understand completely," Jim told him. Soon after, Jim went to Asadabad to try to track down Jalil's weapon, but in vain.

On a chilly day in late December, after a fresh blanket of snow had fallen on the mountains surrounding Mangwel, Jim and Abe got into a pickup truck and went to visit Jalil in prison. A few days earlier, Jalil's brother, Taliban commander Abu Hamza, had been shot and killed leaving a mosque in Peshawar, Pakistan, and that was one reason for Jim's trip. The prison lay north of Asadabad, past the mouth of the Pech River valley, and was often attacked by insurgents.

A warden led Jim through the high prison walls overlooked by guard towers and into his office. After a short wait, Jalil was brought in, his hands and feet shackled, his spirit broken.

"Take them off," Jim told the guard, who removed the shackles. "Brother," he told Jalil, "it is good to see you."

Jalil's face lit up. He kissed Jim on the cheek.

"Commander Jim," he said. "No one has come to see me, not even my own sons, but you come? My life is yours from now until the day I die."

"I will help your family until you are released," Jim told Jalil, who had a six-month term. He said he would give Jalil's son about 20,000 Pakistani rupees a month.

A guard came in with green tea in glasses on a tray. But Jalil refused to drink it—it was one small way that he could show strength and gain a modicum of honor in front of Jim.

Once again, Jim saw a chance to open a dialogue with the Taliban.

"Abu Hamza, your brother, is dead," Jim said. "You are in prison. The pressure is on your family. I want to speak with your brother Momen. I'm not going to kill him or capture him."

"He won't listen to me, but I will try," Jalil said.

Jalil was in a vise between the Afghan government and his Taliban brothers that he was unable to escape. But the relationship Jim forged with him created yet another inroad into the Taliban—one that months later would bear fruit.

ACROSS THE KONAR RIVER in Chowkay, Taliban commanders from the powerful Safi tribe also decided to cooperate with Jim.

One was a low-level Taliban commander named Obeidullah, called Obeid for short, who contacted Jim in the fall of 2011.

A soft-spoken nineteen-year-old too young to grow a full beard, Obeid was part of a new generation of Afghans who joined the insurgency after the Taliban regime fell in 2001. He was drawn to fight for economic rather than religious reasons. Obeid's father, Per Mohammed, was a subcommander to Safi tribal leader Jan Dahd and fought with him against the Russians as well as against the Taliban. He was killed trying to stop Taliban forces from crossing the river into Konar in a major battle at the Kama Bridge in 1996. Per Mohammed shot three Taliban trucks with rocket-propelled grenades but was then killed by a volley from a Russian PKM machine gun. Given how his father died, it was telling that Obeid later felt compelled to join the Taliban.

Obeid lived near the Konar River in the village of Barabat, where his mother, two brothers, and two sisters farmed corn, wheat, and rice on a small plot of land. Obeid loved swimming in the Konar River in the hot summer months. To provide fuel for his family to burn, he would cut reeds and grass growing on river shoals and tow it on a wooden raft to the banks.

One summer morning in 2009, when Obeid was sixteen years old, his cell phone rang. A strange man's voice greeted him. "I am Haqyar, I fought alongside your father," the man said. "I hear you are a good swimmer."

The hair rose on Obeid's neck. He knew about Haqyar—he was a former mujahideen, about forty years old, who had joined the Taliban.

"I will pay you to bring something across the river," Haqyar continued.

A few days later at around midnight, Haqyar met Obeid by the river and gave him five AK-47s, two rocket-propelled grenades, and a Russian machine gun. Obeid put them on a jala, or raft, and covered them with a tarp. He had never swum across the river at night before and was a little scared. But he waded into the rushing water and together with a friend nudged the raft across using sticks because they had no rope. They hoisted the raft onto the opposite bank and, soaking wet, picked up the weapons and carried them to a house in the Taliban-held valley of Badel. When they arrived a man handed Obeid 30,000 Pakistani rupees, enough to feed his family for a couple of months.

"I was so happy. I had a job," Obeid recalled thinking.

For two months, he continued making the shipments, and word spread of the brave young man who swam the Konar for the Taliban.

Then his cell phone rang again. This time it was Taliban commander Abu Hamam himself.

"I have heard of you," Hamam said. "Meet me in Maya."

Obeid walked up a red-dirt path high into the Shalay valley and beyond into the insurgent territory of Maya near the Pakistan border. There, some of Hamam's fighters took him to a small house where they had lost comrades in a firefight not long before. Hamam walked in. Obeid was impressed by his strong build, long hair, and large beard.

"Come with me and be my friend," Hamam told him, and offered him a Russian machine gun.

Obeid decided on the spot to leave school and join Hamam and his men.

"When I saw that weapon, I was happy. I just wanted to have the gun," Obeid recalled.

But when he returned home to gather some things, his brothers were furious and beat him.

"The Taliban killed your father, and now you are going to join them?" his older brother yelled.

Obeid left, angry, and spent the next few months in the mountains with Hamam. He admired Hamam's cleverness and strength, learning he had experience in the Pakistani military and also had studied karate for ten years. Obeid went with Hamam to the Shalay valley when Hamam and his men were shooting mortars at COP Penich, but mainly his jobs were to cook and gather wood.

"Let me go with you!" Obeid pleaded.

"You are too young," Hamam responded.

After five months, Obeid's desire to fight led him to join another, more senior Pakistani Taliban commander, Abdul Wali. Abdul Wali was also a member of the Safi tribe that resided in Pakistan. They met at Abdul Wali's home in the village of Kandaro in the Mohmand Agency of Pakistan's Federally Administered Tribal Areas (FATA). Obeid initially liked Abdul Wali. He was tall with a thin beard, about twenty-seven years old, and well educated, according to Obeid. Obeid would stay in Pakistan for two years, fighting against the Pakistani military. He was paid, 10,000 or 20,000 rupees at a time, and sent money home to his family.

But when Abdul Wali started pressing him to return to Afghanistan to kidnap Afghans for ransom, Obeid refused. He stole away, saying he was going to see his family in Afghanistan, and never returned.

"It was the first time I went home," Obeid recalled. "I told my brother that I didn't want to be Taliban anymore, and my family was so happy." From then on, his family had to support itself by farming alone.

In November 2011, Obeid went to Safi tribal leader Jan Dahd and his son Jan Shah and asked them to help him start a new life. Obeid brought with him nine of his fighters, and they turned over all their weapons to the government. Then Obeid asked to meet Jim, and they sat down in Jan Shah's safe house, a qalat in Asadabad.

"I can help you," Obeid told Jim. "We can kill the people who need to get killed, and arrest those we need to arrest."

Obeid and Jim began talking about establishing Afghan Local Police with Safi tribesmen in his home village of Barabat. As a former Taliban, Obeid had the advantage of detailed knowledge of who the insurgents were, where they lived, and how they operated—and would be effective at targeting their networks. With Obeid flipping, the local police could lead other Taliban fighters to follow suit. He also knew which tribesmen he could trust to defend the village and surrounding area. Obeid's close ties with the strong Safi tribal leadership and Jan Dahd amplified the potential for reintegrating the Taliban.

Obeid started providing Jim with highly credible intelligence on the insurgency. A few weeks later, Jim gave Obeid an AK-47 so that he could better protect himself. Obeid would need it.

One day Obeid's phone rang. He answered, and immediately recognized Abdul Wali's staccato voice.

"I am going to kill you," Abdul Wali said, and hung up. A chill ran down Obeid's spine. It would not take long for Abdul Wali to make good on the threat.

One afternoon a few days later, Obeid and three friends climbed into a station wagon taxi—the kind ubiquitous in Afghan towns—on their way to a meeting outside of Barabat. The taxi slowed as it passed through a crowded market, with wooden carts piled with tomatoes and skinned sheep carcasses hanging on hooks from bamboo stalls. Out of the corner of his eye, Obeid noticed some suspicious men.

"Stop, turn around!" he told the driver.

Seconds later, the men opened fire on the taxi with machine guns and rocket-propelled grenades from behind a wall on the road. The gunfire shattered all the windows. As the taxi sped away, Obeid started shooting back from his door with the AK-47 Jim had given him. But it was too late: one of Obeid's companions was dead, and two others were wounded.

After Obeid's narrow escape, Jim lobbied his command harder to employ Obeid in starting up local police in Barabat and surrounding areas.

As usual, the command's response was slow. Some inside the command objected to Jim's discussions with Obeid because he was a former insurgent, ignoring the possibility that once he turned he could be of tremendous value.

Meanwhile, Obeid was exposed, vulnerable. "If I don't get the ALP, it will be really tough for me to live," he said. He eventually was given command of a small ALP force, which proved effective at cutting off Taliban movements and supply routes. The Taliban intensified attacks on Obeid, using ambushes, IEDs, and even shooting at his house. Knowing he was a marked man, Obeid allegedly killed a key tribal leader with connections to Abdul Wali—a man involved in targeting him in the taxi. In response, Abdul Wali put pressure on local Afghan government officials, who arrested Obeid for the killing and jailed him in Jalalabad.

ABDUL WALI HAD WANTED to exact revenge on Obeid, but a far bigger problem was Jim. Like other hard-core insurgents, Abdul Wali understood that Jim's work with the tribes was a very serious threat. He had been targeting Jim with suicide bombers, IEDs, and an assassin named Khetak.

Then one day in August, Jim got word that yet another Taliban wanted to meet him. When he learned who it was, he was intrigued and excited. It was Khetak.

A thousand calculations rushed through Jim's mind. Was it an ambush? What was Khetak's agenda? Who else was behind this?

Jim knew from classified sources that Khetak was wanted for gunning down several men in cold blood, including a police chief. Khetak was still believed to be a paid killer working directly for Abdul Wali and was on the U.S. terrorist watch list.

Yet instinctively Jim felt he had to agree to a meeting. He could never report such an encounter to his command because it was unauthorized, but it was too lucrative an opportunity to influence the insurgent leadership to pass up. What clinched it for him, though, was his implicit trust in the tribal intermediary Khetak went through to set up the meeting: Noor Afzhal's son Asif.

Asif and Khetak were from different tribes—Asif was a Mohmand and Khetak a Safi. But they were peers and shared an interest in dog fighting. Asif had heard Jim ask again and again in meetings to talk with the Taliban, and understood his goal of reintegrating the insurgents.

Jim mitigated some risk by having the meeting take place at Noor Afzhal's house in Mangwel, and took a slew of other tactical precautions. It was the ultimate test of his trust in the tribe's protection. Khetak made similar calculations. Still, each man was prepared for the other to try to shoot him on the spot.

On the morning of the meeting, Jim brought Noor Afzhal and his son Azmat to the Tribe 33 qalat for their safety. Then he and Ish walked down the dirt path toward Noor Afzhal's home. They had already discussed contingencies. It was silent except for their footsteps and the rhythmic whistle of the village wheat mill. Jim took the pulse of Mangwel as they walked. Did the children come out? Were the normal people around? They stepped through the tall wooden gate of Noor Afzhal's qalat. The guard dog was chained and curled up in the corner as usual. Raza Gul came out and greeted him, and then Asif did the same. Jim read his brother's face. He had seen Asif upset, on edge, and he could see that he wasn't now. It was all right. Jim took a deep breath, strode quickly into the meeting room, and embraced Khetak.

"*Salaam aleikum*, peace be upon you," Jim said.

"*Waleikum salaam*," Khetak replied in a low voice. Each man accorded the other the respect due a fellow warrior—though on opposite sides of the war.

Heavyset and balding, Khetak was thirty-eight years old and wore a black tunic and red kandari cap. Jim knew some of Khetak's background from intelligence reports. A Taliban leader based in Chowkay, Khetak shot and killed the district police chief in 2008 and then fled to Pakistan to work for Abdul Wali. He returned to Chowkay in early 2011 and spoke about turning in his weapons. Then government forces raided his house near Chowkay market, and he fought back with machine guns and rocket-propelled grenades.

Khetak had also heard much of this American Special Forces officer, Commander Jim, who dressed and lived like an Afghan and was working with the tribes.

"*Keena*," Jim said. He deliberately broke the space between them, sitting cross-legged on the floor within easy arm's reach of Khetak. "Brother, it is good you are here. We will talk," Jim said.

"Yes, there is much to talk about," Khetak said, his voice strong and confident.

After an exchange of pleasantries, Khetak got down to business. Unsurprisingly for Jim, he launched into a salvo about feuds, revenge, and honor—things that coursed through the blood of Afghan men.

Khetak was Safi, but he detested Safi tribal leader Jan Dahd and his son, provincial intelligence deputy chief Jan Shah. He claimed that the trouble had started seven years earlier, when Jan Dahd had his eye on a girl to become a bride for one of his sons, but Khetak's brother married her instead. Soon afterward, his brother was gunned down as he worked shoveling in an irrigation ditch. Jan Dahd was responsible, Khetak said. "Jan Dahd has killed many people in Chowkay, and no one can say anything because they are scared of him. If you put him in charge of the local police, he will have even more power," Khetak said. "My brothers and I will have to leave."

Jim took out a string of prayer beads, and began running them through his fingers with feigned calm. The possibilities afforded him at that moment were stark, bewildering.

*An assassin who might be trying to kill me wants my help.*

*He has direct access to a top insurgent commander.*

*His worst enemy is my friend and ally.*

Jim was honor-bound to side with Jan Dahd. If he crossed the Safi tribal leader, both he and Khetak would be dead men. And Jan Dahd had proven loyal to him, so Jim had no reason to betray him. If anything, Khetak's protests about Jan Dahd confirmed to Jim that he had chosen the right man to lead the local police in Chowkay, someone who was feared even by a Taliban killer such as Khetak. Who else could control the lawless area?

And yet there just might be a way to assist Khetak as well. It would have to await their next meeting.

"Before we go on, there is something I want you to know," Jim said. "I will never run out of bombs or bullets, I will never tire of fighting. Neither will you, my brother. We can go on this way forever, or we can meet face-to-face and try to deal with the issues. I will do my best to help you. Just be patient. We will meet again."

Khetak studied Jim's visage.

"I have one more thing to say to you," Khetak said, recognizing a fighter like himself. "There is a Pashto proverb. The people who live by the river, the lowland people, they all dance to the same music. In Pashto, that means they will say yes to anything. But the people who live in the mountains will not.

"We Safis are mountain people. You should come with me to the mountains and become a Taliban commander."

Jim let out a laugh. The irony was almost too much for him.

"If we did that, every American commander would come after us. Do you think it was bad for Osama bin Laden?" he joked.

Then, just as quickly, his smile disappeared. His face grew distant as Khetak's invitation sank in.

"Do not ask me that again, my brother," Jim said, patting Khetak's knee with his hand. "For if you do, the next time I might just say yes."

SEVERAL DAYS LATER, JIM and Khetak met for a second time at Noor Afzhal's house.

Both men were cagy enough to be wary without showing it, and joked like old friends—with a peculiar humor that came from knowing they could kill the other as naturally as they breathed in and out.

"I have a nice gift for you," Jim said, smiling. From a leather holster, he pulled out a 9 mm pistol. "I don't ever carry a weapon that is not loaded," he added. To show Khetak he was telling the truth, he dropped the magazine. Then he pulled back the charging handle with a sharp metal clack and ejected the bullet from the chamber.

Jim then casually picked the bullet off the carpet and reloaded the pistol, slapping the butt to insert the full magazine and chambering a round.

"So it is loaded. It is good," he said, handing it to Khetak.

Khetak took the pistol in his hand, admiring it.

"And here is my picture," Jim went on, handing Khetak a large printed photograph of himself in Afghan clothes. "So you have a weapon and my picture. Shit, whoever wants me, now they can come and get me," Jim said.

It was the supreme gesture of trust, as well as a challenge—and Khetak knew it.

"When are we going to the mountains?" Jim asked. "I want to meet Abdul Wali. Set it up, I will go. But you have to guarantee my safety. You only," he said.

"A lot of Americans try to meet me, but I have not met anyone. Asif asked me to come, and I trust him. Now I trust you, and you can trust me," Khetak said.

"Do you know how many reports I have sent up about you?" Jim asked. "*Sifr.* Zero."

"Governments, they make war. People make peace. That is the truth," Khetak said.

Jim then explained how he intended to help Khetak. He was going to use his close relations with Jan Dahd and Jan Shah to try to mend the rift between them and Khetak. If that failed, Jim would let Khetak know in advance so he could leave the area before Afghan Local Police were established under Jan Dahd and his men.

Khetak's face grew cloudy. He would have nothing of it.

"I will tell you right now. I am going to get rid of his son [Jan Shah]. He killed my brother. He is a bad person. I am going to kill him, and in our next meeting, I will tell you about it."

The die was cast. Jim had to protect Jan Shah.

OVER THE NEXT SEVERAL weeks, Jim moved ahead with the plan to raise three hundred Afghan Local Police in Chowkay under the leadership of Jan Dahd. An official "validation shura" was held, with American and Afghan senior officers flying in to endorse it.

Word got back to Jim that Khetak was very angry and seeking instructions from the Taliban on how to deal with Jim.

Jim went to Asif. "What's going on with Khetak?" he asked.

"Forget about him," Asif said, clicking his tongue and brushing the air with his hand.

That confirmed Jim's suspicions. He asked Abe to get Khetak on the phone.

"Don't ever come fucking around here," Jim told Khetak. "If I ever see you again, I will kill you."

"No," Khetak replied. "I will kill you."

WHILE SOME TALIBAN BROKE ranks to join Jim, intelligence reports proliferated that die-hard Taliban commanders such as Abu Hamam and Abdul Wali were stepping up attacks against Jim and his team.

On July 30, four men in a Hilux pickup truck left the village of Sarkay in Khas Kunar District and headed toward Mangwel. Two of the men were wearing flowing blue burkhas, the cloaks that covered Afghan women from head to toe except for a small opening for their eyes. One of them had a suicide vest under the burkha. Two others were armed with pistols and wore chest racks full of ammunition. Abdul Wali, angered at the protection Jim received from the Mohmands in Mangwel, had reportedly brought them in specifically to target Jim. They had been training for three months in the use of suicide vests, interrogation resistance, and other tactics. The men drove back and forth along the road that lead to Mangwel all morning, waiting for Jim. But when he did not arrive, they headed back through the Khas Kunar bazaar toward the White Mosque Bridge. A suspicious villager had alerted the police, who stopped the truck at the bridge. The two men wearing burkhas jumped out. One of them, trained to never be captured alive, blew himself up. The blast stripped away the burkha, revealing a corpse with a large black beard and bushy eyebrows dressed in bloody white clothing. The second ran and was shot in the chest and groin by the police; he died. The other two men were detained in the truck.

In August, reports surfaced of two other efforts to kill Jim with suicide bombers, including one ordered by Taliban commander Abu Hamam.

One day in mid-August, a suicide bomber was reported to have infiltrated Mangwel. Jim, Azmat, and several arbakai immediately grabbed rifles and headed into the village on foot while Americans and Afghans reinforced the guard towers of the Tribe 33 qalat. Justin watched from one tower through the scope of a sniper rifle as arbakai fanned out through Mangwel. Jim detained three men by the road and headed on to Noor Afzhal's home to check on him. What happened next surprised even Jim. Mangwel's mullah began broadcasting from the village mosque, saying there was a possible threat to the village and asking the people to report any strangers to the arbakai. Tribespeople poured out of their homes looking for attackers but found none, signaling either that it was a false alarm or that the bomber slipped away. Either way, Mangwel was safe.

Roadside bombs were another prime means of attack. But Jim was so skilled at finding and dismantling the improvised explosive devices (IEDs) that they rarely escaped his notice. Still, it was a constant cat-and-mouse game, and the insurgents weren't giving up.

One clear September day we headed for the Shalay valley on a mission to clear a reported IED. We rolled out of the Tribe 33 qalat at about two o'clock in two vehicles. Jim was in the lead in his Humvee, called Vehicle One. Dan followed in command of a mine-resistant armored vehicle, or RG-31, called Vehicle Two. Pvt. Kyle Redden was driving the RG-31. Miah and Staff Sgt. Andy Deahn, the Air Force JTAC, were manning the RG's guns. Ish, Mike, and I were in the back of the RG. Jim almost never used the RGs because the big armored vehicles were intimidating to villagers and they offered limited visibility to the soldiers inside, but he had Dan take one that day because of the elevated IED threat. It was the right call.

Jim's Humvee swerved onto the dangerous and twisting stretch of road known as Zombieland. A short distance behind, Dan's RG made the turn with Kyle at the wheel.

Dan ordered Miah and Andy to test their guns, a routine combat drill at the start of missions.

"Test fire!" yelled Miah. He fired off some rounds of the .50-caliber machine gun into the mountainside. Standing in the open roof hatch above me, Andy pumped out several rounds with the M240 machine gun.

The RG made a sharp left turn and then descended onto a flat stretch of road lined with ironwood trees. Just then, a huge explosion rocked the RG, blasting the vehicle with shrapnel and surrounding it in a cloud of dust and debris. It was a direct hit on our vehicle.

FROM THE LEAD HUMVEE, Jim heard the explosion and immediately knew Dan's vehicle behind him had been hit.

*Dan . . . Ann . . . Ish!*

A wave of dread swept over him. It was the purest emotion he had ever felt in his life. Three of the people he loved most in the world were in that vehicle.

*They're gone*, Jim thought. Everything was moving in slow motion.

"Turn around!" Jim told his driver. "Gun it. Go! Go! Go!"

As the Humvee turned, he saw the big cloud of smoke.

"Two, Two, Two, this is One. Over!" Jim radioed, trying to get a response from Dan in Vehicle Two. "Two, this is One!"

No answer.

IN THE RG, THE force of the blast left us dazed.

"Shit!" Miah yelled.

"Is everybody okay?" Dan called out.

"Yeah!" Ish, Miah, and I yelled together.

"We're good!" said Mike. An Iraq war veteran, Mike managed the Tribe 33 operations center tent and had not been on a mission for months. He was in the RG to get dropped off at COP Penich, and from there catch a helicopter to go home on leave, while the rest of us went on to the Shalay valley. Mike was glad Jim had not let him ride in the luggage trailer the RG was pulling.

"Andy?" Dan called to the Air Force sergeant in the gunner's hatch. "Andy?"

"Yeah . . . I'm all right!" came Andy's voice from above. Exposed in the turret, Andy had borne the brunt of the blast. It had blown off his hat and gear, leaving him stunned and disoriented. There was a ten-foot-long branch caught on the .50-caliber machine gun. But, miraculously, he survived unscathed.

Dan didn't miss a beat.

"Looking for the triggerman! Looking for the triggerman!" he called. "Gunners, get scanning!"

"Get out of this cornfield!" Miah yelled, warning about the cover provided by the ripe cornstalks in the field next to the road.

"Push! Push!" Dan ordered.

At the wheel, Kyle kept the RG steady and moving forward, as Jim had trained him to.

IT WAS LESS THAN two minutes, but it seemed like an eternity for Jim until he heard Dan's familiar voice.

"Be advised, looking for the triggerman. No positive ID. Over," Dan radioed. "Be advised, everybody is okay," he said.

Jim's Humvee passed us and headed straight to the blast site. Jim and

Abe jumped out and started walking through the cornfields, looking for the triggerman.

At the same time, scores of arbakai began converging from all around. Azmat led one group from Mangwel, and Niq and his men came from Kawer. Some of the arbakai carried rocket-propelled grenade launchers on their shoulders. They scoured the fields and darted up on ridges overlooking the area. It was a powerful show of force. Whoever emplaced the IED had chosen an area just outside of Kawer where there were no arbakai, taking advantage of the gap.

After searching everywhere for the triggerman, Jim and Abe walked to a shop down the road and within view of the IED location. Outside, the shopkeeper was lying on his cot under the trees.

"You are now my enemy," Jim said, his eyes wild and intense.

"I didn't see it. I don't know anything," the shopkeeper protested. Then, to his great misfortune, he smirked.

In one swift move, Jim pulled out his knife, swept the shopkeeper's feet out from under him, pushed him down on the cot, and put the knife to his throat.

"I will cut your fucking head off!" he screamed, his face crimson and spit spewing from his mouth. "The IED went off a hundred meters away from this shop. You can't tell me you didn't know about it! I will kill you, motherfucker. I will kill everyone here!"

A crowd of Afghans was watching. Jim released the shopkeeper, leaving him shaken, and stormed away past them. The Afghans could have dismissed Jim as a crazy American soldier—but they all knew him. He had drank tea with them and bought things in that shop.

For Jim, the psychological blow of not having spotted the IED was profound. It was only the third IED that had gone off behind him in his military career—but those were the ones that kept him up at night. It turned out the IED was powerful and highly unusual. It consisted of about a hundred pounds of homemade explosives placed not in the road but in a tree and detonated by remote control. The top of the tree was blown off. Still, Jim felt a crushing sense of failure mixed with anger.

Jim's rage was all the more intense because he could not help but see the incident through the lens of Pashtunwali, as an unforgivable blow to his honor, to his namoos. They had attacked his family, his wife.

"I feel as though you have been raped in our home," Jim told me later. "And I could not protect you."

A few days later, Jim and Abe returned to the scene after dark with a can of gasoline and set the remnants of the blown-up tree on fire. The red flames flew into the night sky, the sparks rising until they disappeared high above the valley.

# CHAPTER 26

EARLY ONE MORNING IN Mangwel, Jim dreamed that he awakened in his bed, rolled over, and opened his eyes. The windowless shipping container that served as his room was dark but for a sliver of sunlight coming through a crack above the heavy wooden door. Each day that stream of light beckoned to him from across the room, reminding him of where he was and the enormous task before him. Often he slept fitfully and woke up weary, his body, mind, and soul exhausted after months in combat.

Voices chastised him. *Get up! Get up!*

All he had to do was make it to the door and look out. Then he would see the arbakai in his guard tower at the far corner of the qalat, smell the smoke from Salim's kitchen, and catch sight of the mountains, and he would once again find the strength he needed.

Jim pulled on a white Afghan tunic and pants and slipped on his sandals. Then he pushed open the door and stepped outside.

Right away, he noticed the tower was empty.

*The arbakai must be changing shifts.*

It was oddly quiet. Jim's footsteps crunched loudly as he walked from his room down the gravel path beside the qalat wall. On the other side of the wall was the makeshift shooting range. That was silent, too, but it was early in the day for training.

He walked farther down the path. He could not even hear the distant whirring of the village flour mill. Then he turned into the courtyard.

No one.

A wave of panic swept over him. He ran down to the metal door of the qalat, swung it open, and stepped through.

The scene in front of him was ghastly.

In the tree across from the qalat gate, a small boy in Afghan clothes hung lifeless from a rope. His feet were bare, his arms and legs limp. His head drooped to one side and his eyes were closed. Jim recognized the figure immediately and rushed toward it.

As if hearing the footsteps, Little Malik raised his head and opened his eyes wide.

"Why did you do this to us?" he pleaded in Pashto, looking at Jim. *"Wailay?* Why?"

Horrified, Jim turned away. Then he saw them, a short distance down the path that led from the qalat to the center of the village. Lying in the dirt were three beheaded, bloodied bodies. He walked closer. Then let out a scream of agony and desperation. They were the bodies of Noor Afzhal, Asif, and Azmat. Mangwel was empty, deserted.

Jim cried out again and sat up in his bed, sweating. *Not that dream.*

Troubled, he got up, made a cup of coffee, and checked in on the operations center tent. Then he walked out to the armored vehicle in the qalat and began his daily 5:00 a.m. guard shift, using the sight of the gun mounted on a truck to scan the hills surrounding the qalat. But even as he scoured the area through the gun sight, images from his dream kept assaulting his mind. He trusted his dreams. The possibility that he could bring such harm to Little Malik and his family was extremely disturbing. He tried to focus but he could not shake the vision off. Later that day, he and Abe went to see Noor Afzhal alone in his house and told him.

Noor Afzhal put down his glass of green tea, looked at Jim intently, and then spoke as father to son.

"Do not tell anyone about your dream," he said. "I know what you must do. Come with me."

Picking up his cane, the tribal elder climbed into Jim's pickup truck and directed them down a bumpy dirt road through the village of Chamaray toward the Konar River. When they reached the river's edge Noor Afzhal told them to stop, and they got out. The sun glistened on the swiftly moving water and the breeze blew softly through the reeds on the grassy green bank.

"Go," Noor Afzhal told Jim, motioning to a path that led beside the river toward a long, stone retaining wall. "Tell the river your dream, and ask the river to carry it away.

"Do not rush," he added. "The war is not going anywhere. We are not going anywhere."

Jim walked alone along the water's edge. Then he sat down on the wall and gazed toward the cool, flowing water, his mind lost in thought. His eyes traveled to the far side of the river. Just then, a large U.S. military convoy started to pass on the road that hugged the opposite side. He turned away, not wanting to spoil his conversation with the river, which to him felt almost spiritual. He closed his eyes and recounted aloud his nightmare, casting it into the current.

One evening after that, Jim and I ate dinner with Noor Afzhal. His youngest son, Raza Gul, spoke more of Pashtun beliefs about dreams, seeking to reassure his brother.

"I had a dream last night that you were taking a boat across the river, and I jumped in to get you," he said. "The good dreams are at four in the morning. They will come true.

"But," he warned, "if you dream during the day it will be the devil's dreams."

Raza Gul's words were reminiscent of another man, another time, and another betrayal: "The dreamers of the day are dangerous men," wrote T. E. Lawrence. "For they may act their dreams with open eyes, to make it possible."

For several weeks after he went to the river, the demons left Jim alone.


AS JIM'S SENSE OF foreboding over the mission deepened, he reacted by pushing himself and his team even harder. As Obama's clock ticked down and the withdrawal of combat troops was well under way, Jim felt time gripping him like a vise. The Taliban were converging on Mangwel, as were other tribal elders, but Jim's command was hell-bent on extracting itself from Afghanistan. Jim and everyone on his team were tired. The unbearable stress he felt rubbed off on everyone around him. Dan, who had recently arrived at Mangwel, noticed how Jim was constantly on edge, and mentioned it to me. I was relieved Dan had arrived to become Jim's right-hand man, his second in command. Jim, too, knew he needed backup. His nerves were frayed. He felt incredibly exhausted, yet found it impossible to relax. The only way he could sleep at all was with the heavy use of sleeping pills, usually mixed with a little liquor.

One night, after taking the pills, Jim put on a headlamp and walked out the door of our room into the darkness. He often went to check on his men one last time before going to bed, so I didn't think much about it. Several minutes passed. Then the door swung open again and Jim stumbled in, a dazed look on his face. He walked unsteadily toward the bed and crawled in. Then he fell asleep, with his headlamp still on.

What I didn't find out until later was that Jim had gone to the operations center on the qalat. Inside, two of his men sat working and Jim checked his email. But as his sleeping medication kicked in, Jim unexpectedly fell into a stupor. As he got up to leave he stopped and pulled his AK-47 rifle out of the wooden rack by the door where he always kept it at the ready. Then he picked up the rifle and, before his men could stop him, he put the barrel in his mouth and pulled the trigger. As usual, Jim had kept the rifle unloaded in the operations tent and so there was no bullet in the chamber. Then he set the rifle back down and with an unsteady gait walked out the door.

By sunup the next morning, word of the incident had spread from man to man around the camp. The morning cool had all but dissipated when Dan walked up to Jim and me as we stood talking outside the operations center. Dan looked the most serious I had ever seen him.

"Jim, I'm going to say this in front of Ann," Dan started, his face stern. "Someone came up and told me you put an AK in your mouth last night. It made him uneasy, but he wouldn't tell you. Brother, I'm your friend. I am here because you want me to be here. So I am telling you."

Jim gave Dan a look that was cold, emotionless. He had little doubt he had done what Dan said, although he had no recollection of the incident.

"If you want to send up an email, just send it," Jim said, staring at Dan. "I don't give a fuck."

"Please listen," Dan said. "I need you to understand."

"Why wouldn't they tell me?" Jim snapped. It pissed him off that *his* men—the ones he'd spent months leading, loving, and protecting, the ones for whom not long before he'd plunged his hand into a pile of dirt and yanked out the wires of an IED—would turn to Dan first.

Dan looked straight at him. "You are one intimidating motherfucker. That's why," he shot back. "The pressure here can be cut with a knife. The guys are coming to me and saying they can't deal with it much longer. You are so angry. You are not the same Jim Gant I have known all these years. I am not sure I want to stay here with you anymore if some things don't change."

Jim said nothing. He turned and walked away, leaving Dan and me standing there. Then he grabbed his AK-47 and left the qalat, walking through Mangwel and into Kawer. For more than an hour he sat next to an irrigation ditch that overlooked Zombieland, thinking. Dan's words had devastated him. He was unsure what to do. Should he contact his command and tell them he had enough and wanted to return to the United States? Should he tell Dan to pack up and go back to Bagram? He trusted Dan and knew his words had merit. But Jim also believed Dan did not fully grasp—and perhaps never would understand—just how difficult it had been to push the mission as far as he had with an inexperienced infantry team.

That evening, Jim asked Dan to come to our room in the shipping container. After dinner, Dan knocked on the door. Jim pulled it open, and the two men sat down cross-legged facing each other, as they had so many times before.

"Dan, I needed to hear that, as a person, as a commander, and, most importantly, as your friend," Jim began. "Your friendship means more to me than anything you do here. You have earned that a thousand times over with me. You have nothing to prove to me. The bottom line is that I don't want you to leave, but I understand if you do. I am sorry I have let you down. I have no excuses for you."

Dan's response was immediate. "Jim, I don't want to leave. I just want some things to change."

"I know. Brother, I know," Jim said. His heart was broken, and Dan and he both knew it.

Jim spoke to his team about the incident and assured them he was not a threat. But he offered no apology.

For my part, I hadn't known what Jim did with the AK-47, but I was not surprised. A few days earlier, I had walked into our room and found Jim lying faceup on the floor in a daze with a 9 mm pistol in his hand. I took the pistol and put it back in its usual place, stuck between the mattress and wooden frame of our bed. It was the first time that had happened, and I was concerned, but I didn't believe Jim was a danger to himself or others. He sometimes did crazy things. After Dan spoke up, I wondered whether I was too close to the craziness.

LATE ONE NIGHT JUST before Christmas in 2011, Jim was working in the operations center in Mangwel when FRAGO 27 showed up in his official military in-box.

It was a fragmentary order issued by Wilson, a multipage document updating the assignments for Special Forces units in eastern Afghanistan. Buried in the fine print was an abrupt change of mission for Jim's team: Tribe 33 was to close down its base in Mangwel no later than January 15, 2012.

Jim was shocked. No one in his chain of command had consulted or even informed him in advance about the withdrawal plan. The order gave his team about three weeks to leave—no time to prepare the tribe. It described the move as part of the overarching U.S. military transition to the Afghan government and security forces in preparation for the withdrawal of most American forces by 2014. The team was ordered to move into the Khas Kunar District Center for a few months in an effort to shore up the local government before departing completely by May or June. On that timeline, Jim believed the plan was premature and nothing less than abandonment.

At the time FRAGO 27 came down I was back in Washington, writing and conducting interviews and doing my best to support Jim from seven thousand miles away. We spoke at least once every day.

We agreed that as a strategy, pulling out of the Mohmand tribal area and leaving behind the Afghans who had most steadfastly supported the arbakai program from the very first—when the risk was greatest—made no sense. It reflected a catastrophic misunderstanding of the importance of the hard-won relationship with the tribe and the advantages of maintaining that tie. The Mohmands and Mangwel had set the example that other areas and tribes wanted to follow. The arbakai in Mangwel and the rest of the district were the most powerful security force in the area. Jim's bond with the tribe was what created the potential for expanding the arbakai into other areas and winning over former Taliban. Reaping those benefits required a long-term commitment. He knew he could not remain in Mangwel forever, but his team had been in the village just ten months.

Jim and I worked together on a memo that urged Wilson to postpone shutting down the Mangwel base, arguing that it could undermine security in the area and pointing out that the district government was ineffective and corrupt. The Afghan official who was to be empowered by Jim for security, a district police chief named Jungee, was embezzling funds and giving the insurgents intelligence on Jim and his team. The six Afghan National Police stationed near Mangwel were reporting on Jim's movements. Jim told Wilson that. Still, Wilson refused to slow the departure from Mangwel, but he did allow more time for the team to move to the district center.

"I want the ALP to be fully integrated into the District security plan and to be self-sustaining prior to your departure," Wilson said. "We are supposed to be working with the Afghans." In fact, Wilson's own headquarters was about to return to Fort Bragg, and the short, rigid timeline was largely based on his desire to realign units—like so many plastic soldiers in a sandbox—before he left. He stated that Jim's team and another team would leave and not be replaced. But in an Orwellian flourish, he insisted that this did not represent abandonment. "We are not withdrawing!" he wrote. "We are not abandoning them!"

Jim's sense of betrayal over the decision was intense. He felt he had been a fool to trust that his command or the U.S. government would back up its promises to the villages and tribes with a long-term commitment. In his view, his current command no longer wanted to "win"—if it ever had. He blamed himself for naively believing otherwise. After all, he had predicted this very thing on the last page of his paper "One Tribe at a Time." Under the heading "What Scares Me Most," he wrote:

On a personal note, my gravest concern is that a Tribal Engagement strategy in some form will indeed be adopted and implemented, but that the U.S. may eventually again abandon Afghanistan—and the tribes to whom we have promised long-term support will be left to be massacred by a vengeful Taliban.

This is by far the worst outcome we could have.
It is immoral and unethical to ask a tribe to help us and

promise them support and then leave them to defend themselves on their own. If our forces do withdraw from Afghanistan, we should decide now to arm the tribes who support us with enough weapons and ammunition to survive after we leave.

Jim's anguish over the betrayal was doubly strong. Based on early assurances he had received that top commanders backed his strategy, he believed Special Forces would remain on the ground in Afghanistan in significant numbers for many years. So he had personally recruited the tribes, rallied his men behind the mission, and persuaded Taliban to switch sides—all at the risk of their lives. He had pledged his loyalty to them, and to a degree they were all fighting for him. They trusted him. Now an order from above was making him break that trust. Worst of all, he felt he was being forced to abandon his own family and the only place that to him truly felt like home.

Jim dreaded breaking the news to Noor Afzhal. For two days he struggled with what he would say. Then on Christmas Eve, he asked Noor Afzhal, Asif, and Azmat to come meet with him, Dan, Ish, and Abe at the qalat the next morning.

It was sunny but cold and Noor Afzhal could see his breath as he walked toward the tent wearing a heavy coat and vest. Jim came out and hugged him. Asif took hold of Jim's right hand with his left—the one mangled by the threshing machine. Only a few weeks earlier, Jim had experienced one of the proudest moments in his life when Asif had nonchalantly offered him his disfigured hand for the first time. Jim felt a pang in his chest.

They entered one of the qalat tents and sat down on the red carpet.

*There is no good way to do this.*

"You all know how much I love you," he began. "You are my family. I want you to listen to what I am telling you, and tell me honestly what you believe."

Jim pulled from his pocket a copy of FRAGO 27.

"The U.S. military is ordering me to leave Mangwel. They gave me no warning," he said. He read the order verbatim to Noor Afzhal.

"So now I am asking you all. The U.S. military is ordering me to move away from Mangwel. Should I leave Afghanistan and go back to the United States, or stay? Father, you have always given me your best advice. Tell me what would be best for you and your people," Jim said.

Noor Afzhal's face dropped. His first thought was that the tribe had somehow let Jim down. All he wanted was for Jim to leave with his honor. But as he listened to Jim he understood that it was an order, that the tribe had done nothing wrong.

Noor Afzhal looked at Jim. His eyes were resigned but his shoulders as determined as ever. He was to the point, and spoke from the heart.

"I want you to stay, as close as you can," he said. "If you must leave Mangwel, stay in Khas Kunar. If you leave Khas Kunar, stay in the province. If you leave Konar, stay in Afghanistan."

Asif was next. His eyes were clear and piercing but not angry. They were serious and very sad. Strangely, it was the same look Asif had had a few weeks before, when he and Jim were at the district cricket championship in Mangwel. Scores of people were playing and laughing around them, and Jim grabbed Asif's hand.

"We've come a long way, brother," he said.

"Yes, we have," Asif replied, and then for a reason neither of them fully understood, he added, "I will miss you when you go."

Now Asif looked both sad and distraught.

"You should stay," he told Jim.

And so it went around the room. No one, including Dan, Abe, or Ish, believed Jim should leave.

The next day, Jim patrolled down to Chamaray and sat down over tea for a frank talk with his arbakai commanders: Azmat, the former Taliban Niq, and the Chamaray ALP commander Mohammed Ghul.

"Make no mistake, we are leaving," Jim told them.

"Jim," said the burly Niq, "you are the only American who has spoken of our future in an honest manner. I understand now. We will only have what we have when you leave. After that we will be on our own."

"Yes, Niq, you are right," Jim replied.

Niq was quiet, his face solemn, as the truth sank in.

Then Azmat spoke. "Jim until you came, we were not together. Now we can talk. We have guns that are legal. What you have asked of us is not easy, but it is our duty as Pashtuns. We can protect ourselves. If the Taliban comes and kills ten, we will kill one hundred."

Then Niq spoke again, his eyes steady and voice strong. "Jim, what

you have done for us is without words. You look Afghan, yes. You understand our language, yes. You follow Pashtunwali, yes. But you are a great
warrior with a great heart. I would happily fight and die alongside you.
It is my duty. But," he added, "it is not your duty to be here. You are here
because you want to be. Please know that if you are killed here it will
bring great, great dishonor to all of us."

Jim smiled and took Niq's hand, and for a moment his heart felt
relieved. "Your words mean more than you know," Jim said. Once again,
Jim could count on the Pashtuns. Having lost faith in his U.S. military commanders, Jim was basing his decisions on the needs of the tribe, his tribe.

Toward his command, he felt only anger. FRAGO 27 was a slap in the
face. It afforded him no respect, no honor. He felt used and tortured, like
a caged beast, prodded with sticks and only occasionally let out to kill.
Tormented by guilt, he could no longer keep the monsters and demons
that taunted him out of his mind; he had lost control of that space.

"Ann, I feel like I'm just going to break," Jim told me one night on
the phone. "I feel like something in my head is going to snap, and I won't
be able to make sense of anything anymore."

I could tell his mental state was precarious, that the situation was
threatening to overwhelm him. After a year and a half in the war zone,
with his life and those of his men constantly at risk, and now the under-
cutting of his mission by his command, Jim was struggling to hold
everything together.

"I'm switched on all the time. My emotions are frayed. I am so tired.
And the deceit on top of all of this . . . ," he said.

We agreed that I would return to Afghanistan earlier than planned,
and meanwhile I encouraged him to pace himself and focus on the
essentials.

After several sleepless nights, he knew what he had to do: fight, on
his terms. He would fight to get all his Americans home alive. And he
would fight to get Abe, Ish, and Imran and his old interpreter from
ODA 316, Khalid, out of Afghanistan. We were extremely concerned
about the dangers the Afghans faced as a result of working so closely
with Jim and other Special Forces teams. All four men and their families had faced specific death and kidnapping threats from Taliban
insurgents. The Taliban called the two Khan brothers "infidel Abe"
and "infidel Ish" in intelligence intercepts. Khalid had been wounded
in battle. Abe had acted courageously in the firefight in which his com-

rade Staff Sgt. Robert Miller was gunned down and for which Miller posthumously received the Medal of Honor. We believed that, along with other U.S. military interpreters, these Afghans had sacrificed more and risked more for the U.S. mission than many of their compatriots and that they should not be left behind to fend for themselves by the United States. Jim had felt the same about his interpreter from Iraq, Mack. Jim had helped Mack and his new bride, Amal, escape Iraq for the United States in 2007, and brought them to live in his home in Fayetteville for nearly a year. Mack and Amal since had two children and were on their way to becoming American citizens. Mack embodied what Jim had fought for in Iraq, just as Abe, Ish, Imran, and Khalid did in Afghanistan.

Finally, Jim decided he would ready his Afghan tribal allies for a total loss of U.S. support. He would prepare them to defend themselves, by stockpiling weapons and ammunition for the tribesmen and killing or capturing as many insurgents as possible. It was the only way he could ever leave with his honor intact.

"SENATOR McCAIN, AS WE pull out of these places, the Afghan government—at least in some places, I would argue in many parts of Konar—is not prepared," I said, cradling my cell phone as I typed at my computer. "They cannot pay their people. They cannot supply them with fuel. They just can't do it."

"I agree with you, Ann."

"Well, so what's our obligation here?"

There was a pause on the phone line. I looked out the window of my house in Bethesda into the gray December morning. I always enjoyed interviewing Senator McCain. He did not dodge questions, and he spent a lot of time downrange, seeing things for himself. McCain and two other senators had traveled together to Mangwel in July 2011 and were deeply impressed by the relationship Jim and his small team had forged with the tribe. Petraeus had told McCain and the other senators that Jim was the best counterinsurgent in the Army. Before McCain left Mangwel, Jim asked him one question: "Sir, are we still trying to win?" McCain had looked troubled. "I knew when Petraeus was here. Now, I just don't know."

I was curious to hear what McCain would say now.

"Well, I think our obligation is to fulfill the commitments we have made," McCain told me. But, he said, President Obama overruled the recommendations of his military leaders to keep troops in Afghanistan longer. As a result, he said, he worried that Afghanistan would "become a cockpit of competing influences" from Iran, India, and Pakistan, and "deteriorate into a situation not unlike after the Russian withdrawal . . . It does not bode well for the people of Afghanistan."

Then he added, "I am glad you are interested in this issue, Ann. It needs to be talked about. A realistic assessment of this situation needs to be rendered by someone."

"Senator," I went on. "I know you have been on the ground a lot, and I have, too. I have covered the wars."

"I know you have, yes."

"I believe there is some morality in war. I believe that we have asked people to risk their lives for us in a very profound way, and to not at least make sure that they can fend for themselves when we leave is—"

"A tragedy," McCain interjected, "and a stain on the honor of the United States of America. And I'd like to remind you, Ann, there was another time in Vietnam where we left some people—some of them were called Hmong, as you might recall—and we were able to get a bunch of them to the United States, but a hell of a lot of them were slaughtered by the Vietnamese."

The image sent shivers down my spine.

"Sir, I have a very specific personal request for you. There are some very brave Afghans who have been working with Major Gant—and he has sacrificed an incredible amount in his own career, with many years away . . ."

"He's an amazing man, yes," McCain said.

"He asked me if I would ask you if you could possibly try to help get out of Afghanistan a couple of the Afghans who have worked side by side with him, fought side by side with him, who you may have met while you were there. That is the only thing he will ever ask anyone for."

"I would be honored to do so."

SNOW LAY THICK ON the Hindu Kush Mountains in mid-January 2012, as Jim's Afghan mechanic, Shafiq, steered our car along a rocky, twisting road above the Konar River toward Mangwel.

We passed a man cloaked in a woolen blanket herding a flock of sheep, and descended into the lowlands, where barren fields stretched out under a cold, gray sky.

I sat in the back covered with a blue burkha, and Jim rode in the front passenger seat in Afghan clothes with his AK-47 resting against his legs. The burkha had felt suffocating when I last made the journey in July, but now it afforded a welcome extra layer of warmth. It made me think of a poem by the Pashto poet Khushal Khan Khattak called "The Coming of Winter":

> *When Libra travels from the sun, then does winter come.*
> *The world, once weak with summer's heat, grows strong again;*
> *Man eats with joy and finds the taste of water sweet;*
> *Lovers embrace again, arms and lips entwined.*
> *The warrior welcomes now his coat; the horse, his winter trappings;*
> *The one feels not his armor; nor the other his saddle's weight.*
> *From Swat the falcon now returns, like traveled a yogi coming home;*
> *And in the radiant moonlight hours comes the heron screaming in*
> *the sky.*

My arrival in Mangwel this time was bittersweet, with all the comings and goings.

Jim's first team of soldiers from the Kansas-based 1st Infantry Division had finished their yearlong deployment and departed before Christmas. They had bonded with the arbakai and villagers and together improved life for the people in Mangwel. To a man, they believed in the mission and it had changed them profoundly. Miah, Sonny, and Kyle planned to alter their Army careers and try out for Special Forces or Civil Affairs, branches of the military that partner closely with foreign people. Drew and others were leaning toward quitting the Army, knowing no future deployment could match this one. Mitch loved Mangwel and had begged his command to let him stay. Everyone was sad to see them go. As he dropped them off at the helicopter landing zone, Jim grabbed each one of them, hugged them, and whispered in their ear: "Be great. Meet your potential. I won't forget what you did here. Strength and honor." Then he walked away and did not look back. For Jim, it was a relief to again send all his men back to their families alive, but he missed them badly.

After days with only a handful of U.S. soldiers at the qalat, a new group of infantrymen arrived from the 2nd Battalion, 3rd Infantry Regiment, part of a Stryker brigade based at Fort Lewis, Washington. They came just in time to begin shutting down Mangwel and moving to COP Penich. As a result, they had little motivation to learn about the tribal people. For Jim, starting from scratch to break in a new team was particularly disheartening—along with the knowledge that it would be the last group of men he would ever command in combat. The nominal leader of the soldiers, 1st Lt. Thomas Roberts, was a 2010 West Point graduate on his first deployment, and his inexperience led immediately to insecurities and friction as his men gravitated toward Jim's leadership. The first red flag was that Roberts refused to sign Jim's initial counseling packet. "I am uncomfortable with some things you said," the bespectacled Roberts told him as the team prepared for weapons training at Mangwel. "I don't know if you meant it or it was just a speech." That infuriated Jim, but for the first time ever he let it pass. He had no confidence his command would back him up if he sent Roberts packing, and he lacked the time and energy to deal with Roberts. He expected that some mentoring would eventually bring Roberts around. So Roberts moved back to Penich with about half of the new team of soldiers, while the rest stayed in Mangwel with Jim.

The atmosphere on the qalat felt more subdued. The morning after I arrived I sat outside the kitchen drinking milk tea with Sher Ali, a young, boastful arbakai who wore sleeveless shirts with ammunition belts crisscrossed over his chest.

"Jim came a year ago and brought a big change here. No one shot at us because we had good security in Mangwel. The Taliban were scared of us," Sher Ali said. "I am very sad he will leave. No one will support us like Jim supports us," he said. "Everyone is worried the Taliban will take revenge."

Hakim Jan, an older arbakai with a long tanned face and thick eyebrows who resembled Anthony Quinn, agreed.

"Before we came to this qalat we had a lot of bad guys in Kawer and Zombieland, and the Taliban could move freely through Mangwel and Chamaray," he said as he stood in the guard tower with his AK-47 over his shoulder. "Now if that happened someone would call us because we have a relationship.

"If Jim leaves and the Taliban attack us, we don't know what will happen. We have a small ANP checkpoint but it is not strong. The ANP guys are calling the Taliban," he went on. "We are very unhappy he is leaving."

Squinting in the winter sun, Hakim Jan looked out over Mangwel and proposed a tribal solution. "If Jim wants to stay, we have a lot of elders in the village. We will send them to Kabul to talk with the American generals," he said.

As a farewell for the arbakai, Jim invited musicians from Jalalabad to play one evening in a tent on the qalat. Everyone crowded into the tent as the long-haired musicians warmed up their *rabab* (lute), accordion, keyboards, and handheld drums. Younger arbakai taught a few of the new soldiers to dance in the Pashtun style, holding out both arms and shimmying their shoulders, while Abe tossed wads of Pakistani rupees in their direction. Salim, the cross-eyed cook, belted out a forlorn love song, and Chevy, the orphan boy worker whom Jim had taken in, danced until his face was red. Noor Afzhal reclined on a pillow next to me and Jim. But as the night wore on and the musicians sang a long ballad about fighting the Russians, Jim retired to our room.

When I joined him, he was lying on the bed looking at the ceiling. I sat down next to him, and he took hold of my hand.

"We are going to leave. I am going to leave these people. And when I do, I will take with me the hope I have given them," he said. "What I have accomplished here will be as insignificant as a mosquito landing on a still pond in the early morning. It will be like when you stand on an ocean beach with violent, crashing waves, pick up a handful of sand, and toss it little by little into the water. You rub your fingers together and the last grains fall away until there is only a single, tiny grain between your fingers. It will be like that. Nothing I have done here will be sustained. I have made no difference."

"You have shown them what is possible, but now they must do it themselves," I said. "You have done so much. What more could you give them?"

"That is an easy answer," he said. "My life."

The next morning, it was chilly and overcast. Clouds hung low over Mangwel, obscuring the tops of the mountains. It was raining lightly in the early afternoon when Jim came to me and asked me to go with him on a walk. He wrapped a thin, cream-colored wool blanket around my shoulders over my Afghan dress, then took an identical wrap for himself. He told his men we were leaving the qalat, and we set off toward the village.

I delighted in our strolls through Mangwel, although I never knew when or where we were going. I had learned not to ask. It was part of our agreement after the IED strike on my vehicle. Jim understood that, as a war correspondent and writer, I wanted to go everywhere to document unfolding events. He knew I was not afraid. But although I had been on countless combat patrols with infantry units in Afghanistan and Iraq, my situation in Mangwel was different. I was under the protection of a man who loved me, who held my life dearer than his own. He was responsible. So he decided where I could go—especially in the wilds of Konar.

We set off across some empty fields on the edge of the village, instead of heading toward the center of Mangwel. Jim walked slightly ahead of me, as usual. The rain was making small puddles in the rocky, sandy soil, and I stepped over them.

"You know that when we met, I was not looking for you," he began. "But I fell for you very quickly. One of the reasons was that I felt you, and your care for me."

We began to climb a narrow dirt path that led in switchbacks up a large hill overlooking Mangwel.

"Then when you came here and I saw your bravery, and how you risked your life to be with me, I fell more deeply in love with you," he said. "I trust you."

I knew that for Jim and men like him, trust was an act of extreme courage. It was not until I had walked in his steps in Afghanistan and become a true comrade that he could trust me completely. It reassured him that I would still love him after witnessing the person he was in war, and that I would be there when he got off the plane. After eighteen months in Afghanistan, the constant swirl of lies, deception, and outright treachery by both Afghans and Americans made trust seem at once impossibly naive, laughingly quaint—and more important than ever.

I wanted to hold his hand, but I could not. Only a few people were out in the rain, and they were far below us in the village. But we knew they were watching. To hold his hand publicly would have been extremely shameful for us and for the tribe. He knew that, and walked on a few steps in front of me.

The trail brought us up the side of a large hill to a flat patch of earth from which we could see all of Mangwel stretched out before us. The fragrant smoke of ironwood fires wafted from some of the mud qalats, which disappeared in the distance into a soft, misty tree line.

Jim squatted down in the Afghan style, looking out at Mangwel, and I did, too. Raindrops were falling on our faces. Our hands were bundled in the blankets against the cool air.

"You can think about this if you need to," he said, looking at me. "You do not have to answer me right now."

I nodded.

"Ann, will you marry me?" he said.

I looked into his eyes and smiled.

His face was very serious, but his eyes smiled back.

"Yes," I said simply, "I would be honored to be your wife. I believe in you and us. You make me happy."

He reached into his pocket and pulled out a clear box holding an engraved gold ring on a pillow of red velvet. He put the ring on my finger and closed his hand around mine. My eyes filled with tears. We wanted to embrace, but we could not.

At that moment, a single shot rang out in the distance and echoed across the valley. It was not aimed at us, but it required a response. We stood up to leave, and Jim walked a short way farther up the path. He raised his AK-47 rifle and fired two rounds toward the mountain behind us. Then he paused, listening and scanning the village below.

"Are you all right?" he asked, leading the way back down the path.

"Yes," I said, following behind.

FORTY-EIGHT HOURS LATER, JIM was riding through Zombieland at the head of a slow-moving patrol including a crane and two Afghan cargo trucks transporting gear from Mangwel to COP Penich.

He was irritated about the move. The short timeline forced his team to be out on the road far more than usual, exposing them to increased risk from roadside bombs. Jim had gone to extremes in recent days to clear the route, even walking it on foot. Still, everyone by then knew about the move, including the Taliban. Jim's departure from Mangwel gave the insurgents a huge incentive to strike so that they could claim to have pushed him out.

Knowing the threat was high, Jim was riding on the hood of the lead Humvee to better spot IEDs. The patrol twisted through Kawer, secured by Niq's force of local police. It then entered an area called Mygon that lay beyond Niq's control.

Jim's Humvee pulled into a turn where the road hugged a rocky face on the right and dropped off into fields on the left.

*Right side road, Chris*— Jim waved his arm in a signal to his driver, Spec. Chris Clement.

Jim glanced to his left. He noticed movement in an unusual place in the field and was just about to react when—*boom!*

Chris gripped the steering wheel as the left side of the Humvee flipped up and came crashing down.

"Contact! Contact! Contact!" he shouted.

A blast of dirt hit Jim's gunner, Pfc. Jonathan Salyer, and machine-gun ammunition flew out of the can into his face.

"Get ready for the follow-on attack," yelled Salyer, a twenty-year-old from Columbia City, Indiana.

Chris pulled forward in the Humvee and stopped.

"Where the fuck is Jim?" he shouted.

Jim was gone.


ABE BURST THROUGH THE wooden door in the qalat wall and ran toward me. I was sitting outside our room, writing.

He had heard the boom and called Ish, who was on the patrol.

"Jim's vehicle hit an IED," he told me. "He was on the hood."

My heart stood still.

"Is he okay?" I asked frantically.

"I don't know," he said.

"Stay here. I've got to go."

I went into our room, grabbed my camera and notebook, and quickly followed Abe to the courtyard.

Noor Afzhal was waving his cane, shouting at Abe.

"Get going, get out of here!" he yelled, urging Abe to go to Jim's aid with some of the arbakai.

Abe was shaking.

"Not yet," he told the elder firmly. "My brothers are out there. But Ish told me to wait."

Abe turned to the dozen Afghan workers in the qalat and told them to take cover in the mortar pit. He realized there could be a two-pronged attack—with Jim and his men hit outside, the Taliban could strike the undermanned qalat. Tribe 33 had rehearsed the defense drill many times, and Abe knew the qalat had to be secure. He also knew Jim had a brand-new team of U.S. soldiers.

Together with the arbakai, the soldiers were climbing into the guard towers to reinforce security.

I ran to the tower beside the front gate and clambered up the wooden ladder to the top of the wall, making sure not to trip in my flowing Afghan clothes.

I peered out the tower down at the road, desperate to see something, anything that would let me know Jim was all right.

DAN WAS COMMANDING THE vehicle about three hundred yards behind when he heard the blast and saw the cloud of the explosion hit Jim's vehicle. He pushed forward and passed the big bomb crater on the left side of the road. When he didn't hear Jim on the radio, Dan ordered his driver to back up while he dismounted.

Then he heard a faint voice and saw him.

On the side of the road up against a wall of rock, Jim was lying face-down. He had been thrown over the road by the blast and landed on his left side.

Dan ran over.

"Jim!"

Jim was breathing.

"Hey, Jim!"

Jim moved a little and then tried to push himself up. Dan grabbed his arm and helped him onto his feet. For a moment Jim was dazed and unsteady as he watched the world spin before his eyes. His head was throbbing and he couldn't understand what Dan was saying. Then everything came into focus.

"Are you all right?" Dan shouted again, shaking him.

"I'm good," Jim said.

"What do you got?" Dan said.

Shrapnel was stinging Jim's left arm, leg, and head. But he had to lead.

"Charlie Mike—keep the patrol moving!" Jim said, using Army

lingo for "continue mission." "You need to push up to Penich and get this stuff downloaded. I will secure the scene." Halting the patrol, already slowed by the unarmed Afghan trucks, would be even more dangerous.

Soon Niq's fifty arbakai and about fifty more led by Azmat were swarming the area.

Jim told Ish to let Abe roll with more reinforcements. Abe drove to the blast site as fast as he could. Together he and Jim found the IED's long, thin command wire and traced it into the field. They found a pair of sandals and excrement left by the triggerman, who must have been watching the road for at least a day, waiting for precisely the moment to strike.

The Taliban wanted Jim dead.

UP IN THE TOWER, the radio crackled.

Then I heard Jim's voice.

"Sustained significant damage to the lead vehicle—break—one pax with superficial wounds—break—everyone is good to go," he told a soldier in the Tribe 33 operations center. "We have about a hundred ALP in the area—break—they are going to search five qalats. Request no CAS [close air support], no medevac, and I am in contact with the ground battlespace owner. How copy?"

A soldier in the operations center repeated back what Jim said, but left out some details.

"Okay, close, but no cigar," Jim said.

I burst out laughing, relieved and amused. The worse things got, the more calmly Jim spoke on the radio. He prided himself on that.

Jim made the soldier get it right.

"Roger, good copy," he said at last. "Get that up to Bushman 33, time now," he said, referring to the call sign for the Special Operations Task Force–East, commanded by Wilson.

On Jim's orders, Dan and the rest of the team completed the mission to Penich, while he and the arbakai pursued the insurgents. When Jim finally returned to the qalat, he was bleeding where shrapnel had pierced his arm, leg, and head. He had a bad headache and his ears were ringing. But he downplayed the injuries, not wanting to be forced to leave his men for medical treatment.

The next morning, he called his team together for an after-action review on the IED strike. Apart from appraising his team's performance, he wanted to show his men that he was all right.

"This is war. This is what you came here to do," he told them. "If not, you are fucked up."

He was also concerned about the atmosphere at COP Penich and complaints some of the soldiers who had remained there were making about Lt. Roberts.

"Be loyal to one another," Jim said. "There are serious loyalty issues at Penich."

Then he pulled on a shoulder harness loaded with ammunition and grenades, picked up his M4 carbine, and climbed back on the hood for another mission.

But back at the Bagram SOTF-E command, news of the IED hitting Jim was greeted with more disparaging remarks.

"Finally," one soldier remarked, "the Great One got his."


IT WAS DUSK ON January 18, two days after the IED strike, when Jim and his team packed up the vehicles and prepared to leave the qalat for the last time.

Concerned about a Taliban attack, Jim made sure Noor Afzhal was at his home. Earlier in the day, he visited the tribal elder and told him he would see him soon. He also had gotten word from the son of Mohammed Jalil, the Taliban sympathizer whom he had visited in jail, on the possible IED trigger puller, as well as a warning that another IED was in place.

Back at the qalat, Jim hugged Asif, Azmat, and the arbakai, who were busy loading up goods from the camp, and walked toward his vehicle.

A small figure ran toward Jim and embraced him. It was the arbakai Umara Khan. He was sobbing openly. He would not let go.

Jim looked into those intense green eyes, and saw the heartbreak of an entire people. At that moment, he knew, it was over.

A few days later, Noor Afzhal stood in the deserted, gutted qalat and felt numb, in shock. It had happened so quickly.

The scene reminded him of his return to Mangwel from Pakistan many years before, finding it a desolate ghost town. But though his people had suffered foreign invasion and occupation, massacre, and exile, they never forgot their tribal homeland. Now, he thought, they had to stay united.

He felt like crying, but he knew he had to be strong.

"It is not in his hands," Noor Afzhal told the tribesmen gathered around him. "It is his command making him do it," he said. "He is still our brother, and we will see him."

# CHAPTER 27

SITTING IN OUR ROOM at COP Penich before a board covered with butcher paper, Jim used colored pens to draw an elaborate map of Konar. Having lost faith in his U.S. military superiors and their commitment to the war, he continued to push forward his own tribal expansion plans. The Safi tribe led by Haji Jan Dahd was a critical building block for the security of Konar. A foothold with the Safis in Chowkay District would allow for further expansion north into the dangerous Pech valley and beyond. To the south, it would create a buffer for the Mohmand tribe. Jim was plotting his own course to empower Noor Afzhal, Jan Dahd, and other tribal leaders so that they could protect their people from whatever came their way—Taliban retaliation or the full-scale Afghan civil war that he believed was coming. He wanted to make sure Niq, Azmat, Noor Mohammed, and other ALP commanders who had sided with him had the supplies they needed to defend their communities. From now on, he would ignore whatever rules got in the way of carrying out the mission as he envisioned it. His commanders fell into a by now familiar pattern—they let him keep operating because, as wary as they were of Jim, they realized that no other Special Forces team leader on the ground had anything close to his experience, knowledge, and connections in Konar. He used that fact to his advantage.

"I'm dangerous now," he told me, "because I don't have anything to lose."

The four hundred Afghan Local Police that Jim worked with in Khas Kunar were already emerging as the biggest and most robust security force in the district. Jim set out to make them even stronger. In violation of U.S. military rules, he gave them heavy weapons, ammunition, and fuel for their vehicles that the Afghan government was not supplying. He also fixed broken vehicles and weapons for them. Without the fuel Jim's team provided, the arbakai would not have been able to keep their vehicles running reliably. The weapons he handed out included scores of repaired AK-47 rifles, eight rebuilt PKM machine guns, and half a dozen RPGs, together with tens of thousands of rounds of ammunition. Without such weapons, the ALP could be outgunned by the Taliban. The ALP began to prove their aggressiveness and strength well beyond village defense.

Meanwhile, yet another change of command created an unexpected opening for Jim. In February Wilson left and was replaced by Lt. Col. William Linn, an intelligent but by-the-book officer who had never served in Afghanistan. In Jalalabad, Maj. Kent Solheim replaced Maj. Eddie Jimenez. Solheim, a thirty-nine-year-old combat veteran from Oregon City, Oregon, had earned a Silver Star for valor in Iraq during a battle in the height of the fighting there. Solheim saved the lives of several comrades, but his wounds cost him a leg. Jim met with Solheim in early February in his office at Camp Dyer, the Special Forces headquarters at Jalalabad.

"Schwartz really dislikes you. He's said a lot of bad things about you," Solheim told Jim. Schwartz's disparaging comments in meetings had tainted Linn's attitude, he said. Linn was known as a black-and-white thinker. "Linn told me, 'Jim Gant is not operating in my battlespace. I can't have someone who doesn't know his left and right limits,'" Solheim said.

But Solheim told Jim that he had realized a lot of information Jimenez and others gave him was flat wrong. Keeping an open mind, he asked Jim to provide his own assessment of the situation in Konar. Impressed by Jim's knowledge, Solheim peppered him with questions, which Jim answered in an exhaustive four-hour briefing and a series of memos. Jim explained why the Afghan Local Police initiative had worked so well in Mangwel and surrounding areas, and outlined for Solheim his expansion plan, starting with the Safi tribe in Chowkay. Solheim encouraged all his Special Forces team leaders to learn from Jim's approach.

Two days later, on February 8, Solheim was on a patrol near the town of Pashad when the RG-31 mine-resistant vehicle he was riding in was hit by a massive IED. The explosion flipped the vehicle on its back end and tore off a large chunk of the front. The blast was so great that it blew one of the tires through a qalat wall about forty yards away. As soon as he heard about the IED, Noor Mohammed and some twenty of his arbakai raced to the scene and helped secure the scene, which was in his area of operations, followed soon after by Jim.

Jim approached Solheim, relieved to see he and his men were unhurt.

But Solheim's first remarks to Jim had nothing to do with the IED. Instead he praised Noor Mohammed and his men.

"I love ALP," Solheim said. Then he surveyed the wrecked armored vehicle. "They can have my leg," he added, "but they can't kill me."

"I hear you, man. Fuck them," Jim said, then added, "It's good to see you."

What Jim was thinking, but didn't say, was that Solheim was fortunate to be alive. He had put himself and his men at unnecessary risk by patrolling the area without alerting Jim; had he done so, Jim and the ALP could have cleared the route and probably prevented the attack. The area had long been dangerous—it was the same place where Jim's team hit the IED in 2003 that cost Luke Murray his leg. Still, Jim was glad Solheim saw the rapid ALP response to the IED.

Solheim was sold on the effectiveness of the ALP and the importance of expansion. Both he and Linn were eager to have more Special Forces teams living in rural villages and working with ALP, as Tribe 33 did in Mangwel. But other teams were having great difficulty moving into villages—including the ODA that was then living on Forward Operating Base Fortress, a U.S. military base in Chowkay. Together, Solheim and Jim discussed a plan by which Jim would lead a team to embed with the powerful Safi tribe in a village in Chowkay, train and mentor three hundred ALP there, and then hand off the location to another team.

"In sixty days, we'll have three hundred ALP and an embed site," Jim told Solheim.

There was only one obstacle: Schwartz had to sign off on it.

Schwartz's criticism of Jim had intensified and grown more open. In February, after Jim evacuated his driver, Chris, for treatment for back pain as a result of the January 16 IED strike, Schwartz implied that Jim was slow in acting and should "take better care of his men." After the extreme sacrifices Jim had made to protect his soldiers, the verbal attack was particularly galling. Yet despite Schwartz's seeming concerns about Jim and his men, he agreed to give them a mission no other Special Forces team had been able to execute.

Solheim gave Jim the news, telling him in an email, "Pack your bags." He was to move into Chowkay as quickly as possible.

This was it. After twenty months in combat, he had landed the most complex, challenging, and risky mission of his life. The Safis were the strongest and most dissident tribe in Konar, and Chowkay was a highly contested district. We would be living in a small qalat in the middle of the tribal lands. Mangwel had proved that Jim's tribal engagement strategy could work; Chowkay would be the ultimate test.

# CHAPTER 28

METHODICALLY PACING INSIDE THE dimly lit wooden B-hut at COP Penich, Jim wore a tan baseball cap with a ghairat honor patch, his lower lip full of Afghan dip.

"We may not all make it out," he said. He spat into an empty water bottle, screwed the top back on, and slapped the bottle against his hand for emphasis. "There's going to be some fighting," he went on, but "that doesn't mean we're failing." In fact, if the team *didn't* have to fight to get in there, he said, then "something is wrong."

Jim had picked eleven other Americans, eight Afghans, and X for the Chowkay mission. I was X in his operational plans. The new team was designated "Tribe 34."

It was February 21, 2012, the day violent Afghan protests had erupted outside Bagram Air Field in reaction to the accidental U.S. military burnings of the Koran. In the nationwide unrest to follow, two American military officers working as advisors at the Ministry of Interior would be shot in the back of the head by an Afghan co-worker, a total of thirty people would be killed, and two hundred would be wounded. Stateside support for the war in Afghanistan would fall to a new low.

While protestors in Jalalabad burned an effigy of President Obama and screamed "Death to America, death to Obama, death to Karzai," we planned to exit the secured gates of COP Penich the next morning. Then we would head ten miles northeast, into the Taliban-controlled capillary valleys that descended to the Konar River.

American commanders would soon put all troops in Afghanistan on lockdown, but Jim pressed forward. Our final destination? A mud-brick Safi tribal qalat in a small village in Chowkay. The village was in the direct line of fire for Taliban holding the mountains above, highlands with well-worn insurgent routes through Konar and across the border into Pakistan.

Despite the paranoia NATO forces now had about our Afghan allies, Jim knew we didn't have the luxury to wait until religious passions simmered down. Before he left to become director of the CIA, Petraeus, then the head of all U.S. forces in Afghanistan, had given Jim the green light to engage the Safi tribe in May 2011. Jim pitched his plan in Mangwel when Petraeus visited to see how his village security initiative was working on the ground.

"What are you waiting for, Jim? Go," Petraeus said, right in front of Jim's immediate chain of command. Jim got his second mission, but it had taken a critical nine months to get the order to move to the valley.

With those months burned by bureaucracy and Jim's two-year tour approaching its end, he had no time to spare. And the Taliban knew we were coming.

Jim warned me and the rest of his team again and again before we left Penich that we would get hit by insurgents, and hit hard. The qalat in Chowkay was tiny, so Jim had carefully chosen which of his new infantry soldiers to take, selecting the most squared-away and gung ho among them. Jim decided early on that I could go, and I never hesitated, despite the heightened risk. I wanted to go. I was by then a core and relatively experienced member of his team. I provided valuable information and perspective, and had demonstrated my ability to navigate the Pashtun tribal culture and create valuable relationships with Afghan men as well as women. I knew the Safi leaders. And my presence as a woman would send a unique message of Jim's trust in the tribe. I believed in the mission and wanted to document this critical phase. Mobilizing the fierce Safis was an indispensable part of the one-tribe-at-a-time strategy Jim had developed over his four-plus years in combat. He wasn't going to pass up this chance to prove the plan would work.

The Safis had dominated the oft-contested Konar Province for centuries. An uncompromising and war-driven tribe, they were at the center of the last major tribal uprising against the central government in 1947, the first to fight the Soviets in the Konar in the 1980s and the first to stand up to the Taliban there in the 1990s. It had taken years, dating back to 2003, for Jim to build his relationship with the Safi elder, Haji Jan Dahd.

Jan Dahd was a shrewd leader and a warrior, brutal and yet generous. Also, he was a devout Muslim and staunch defender of Pashtunwali. Jim had been told by higher-ups that Jan Dahd was Taliban. They feared him because he would not roll over for anyone.

One of the truest statements Jim ever heard about the cagy tribal leader was, "Haji Jan Dahd is not Taliban, he owns his own Taliban!" Jan Dahd was the only man Jim ever met whose clout transcended the Afghan government, the Taliban, and U.S. forces—he was two parts tribal leader and one part warlord. Jim believed he could trust him.

Moving in with the Safi tribe marked the biggest challenge of Jim's life—the last opportunity, as he saw it, to turn the tide of the war in eastern Afghanistan. But was Jan Dahd the man he said he was, with the power and will to take on the Taliban?

Jim didn't have the same deep relationship and history with Jan Dahd and other Safi elders that he had with the Mohmand tribe in Mangwel. The Mohmands had made Jim one of their own, a tribal malik. All he had with the Safis was an honorable and fearsome reputation, enough to ensure safe passage into the valley. He knew we'd be pummeled the moment we kicked up Chowkay dust. The question was, would the Safis back him now, when the bullets started flying? Jim was betting all of our lives that they would.

To add to his burden, he also had to execute the strategy without strong supporters in the chain of command, and without the men he had meticulously groomed for the mission. He'd had his new infantry soldiers for less than two months, and he was exhausted.

"Where we are going is not safe," Jim explained to his unseasoned infantry team. Raw fear and excitement filled the room. In just the previous few months, he told them, insurgents in Chowkay had staged scores of attacks with mortars, rockets, machine guns, recoilless rifles, and IEDs, including several catastrophic hits on U.S. forces. Jim threw down the after-action reports from two major attacks in which the Taliban overran U.S. forces in the vicinity in recent years, and invited the team to read them.

The Afghans among us included Abe, Ish, the mechanic Shafiq, and five silent mercenaries whom Jim hired without approval from higher command—breaking yet another Army rule. Jim must have sensed our hesitancy trusting heavily armed Afghans we'd never seen before.

"The mercenaries are very well trained and loyal to us," Jim said. "They are mean, evil motherfuckers." They would bolster our initial protection. It was not unusual for senior tribal leaders to have body-guards, and so the mercenaries would also add to Jim's status as a man not to be trifled with.

But Jim's real focus as he spoke was not on the enemy but on the tribe we would live with. For the nervous young men in front of him, this was tribal engagement 101.

"We will be living out of vehicles and under ponchos. We will sur-vive on our wits and our relationships," he said. "We can make no mistakes—not one . . . I am betting on the tribal connections I have built in Chowkay to protect us," he said. "You had better hope I'm right."

Under a bright sun and almost cloudless blue sky, our patrol of eleven American soldiers and eight Afghans rolled out the gate of COP Penich the next morning. We crossed the Konar River at the White Mosque Bridge and entered Chowkay District and the territory of the Safi tribe. The road passed farmland and orchards lining the river valley and then twisted as it rose into foothills. On one steep hillside stood a large and isolated compound with a large patio overlooking the river. It was the home of Jan Dahd. On an adjacent hillside, across from an old stone fort, lay the village of Chinaray.

Jim stopped, got out of the vehicle, and walked up to Jan Dahd's qalat, where the tribal elder was sitting on his porch.

"I am here. Is it all right if my men and I move into the village?" he said, in an act of deference to the Safi elder. The village of Chinaray was built on his land.

"My brother, you are welcome," Jan Dahd said.

Our patrol stopped next to a narrow dirt road that led up the hillside into the village, and I got out. I was wearing a dark green U.S. military uniform and boots with my hair tucked up in a baseball cap because Jim did not want anyone who was watching to see a female moving into the Afghan qalat that day.

We proceeded to a mud and brick qalat overlooking the Konar River. The tiny structure—the outer wall was about a hundred feet long and thirty feet wide—had the advantage of being a short walk from Jan Dahd's home. But it was located in even lower ground than I imagined—backed up against steep, rocky mountains on its northern side. Two peaks, whose names in Pashto meant "Martyr Mountain" and "Dishka Mountain," rose directly behind us. Along the mountain ridgelines, small walls of rock created natural fortified fighting positions for the Taliban. They would be able to rain down fire directly into the compound.

Jim got eighteen tribal fighters from Jan Dahd straightaway. They worked alongside Tribe 34 in nine-man shifts. So that made it twenty-six Afghans and twelve Americans against God knew how many Taliban in the mountains.

As the team hustled to move in, I took Jim's advice and sat down for green tea with the two ALP commanders, Mohammed Sadiq and Abdul Wali, in a corner room of the qalat. Although they seemed a bit surprised at an American woman in their midst, they treated me respectfully and did not hesitate to speak with me. Abdul Wali had large round eyes and a gentle smile. He originally came from the Badel valley, an insurgent stronghold just up the road, and had moved to Chinaray a few years before. The father of three daughters and three sons, he worked building houses with sand and rock before he joined the ALP. "Now this is my only income," he said as his four-year-old son, Ishmael, played nearby.

Sadiq sported a pakol pushed back to expose a fringe of curly black hair over his forehead, giving him the look of a French artist. Like Abdul Wali, he was from the Badel valley, where he used to be a shepherd. He bought a small four-jerib plot of land in Chinaray and farmed wheat, corn and vegetables. Sadiq was the more senior ALP commander, and carried himself as such. As a teenager, he had fought with Jan Dahd against the communist regime of Afghan president Mohammed Najibullah, and had vivid memories of dogs eating the dead bodies of Afghan soldiers strewn across a field after a big battle in Khewa District, Nangarhar Province.

Jim's plan was first to establish a foothold in the qalat at Chinaray, and from there recruit hundreds of Safi tribal police in a span of about three weeks. He would then push those police up to observation posts to secure the high ground behind the qalat, the main road, and eventually the surrounding district.

When we arrived in Chowkay there were only thirty-nine ALP out of the three hundred that had been approved, an illustration of the anemic results of the Special Forces team tasked with this mission before us. The team had proven unable to move into a village from behind the walls of FOB Fortress.

In the first hours and days, we depended completely on the tribesmen living with us in the qalat to protect us. But Jim knew once the Taliban started attacking, that force would not be enough to keep us from being pinned down. To prevent that, he rapidly trained groups of about fifty Safi tribesmen—spending only one day enrolling, training, and equipping each group, not the three weeks mandated by the U.S. military. Every Pashtun tribesman worth his salt knew how to shoot an AK-47, so Jim jettisoned that part of the training program, as well as the parts that were supposed to teach them about "personal hygiene" and the Afghan constitution. Jim's attitude was, *Jesus, this is a war and we're telling the greatest living warriors how to fight!*

After training the first group, Jim gathered the new ALP and their commanders outside the qalat. The local police stood in two rows of twenty, their rifles slung in tandem over crisp khaki uniforms. Sadiq and Abdul Wali stood between the rows, with Jim at the top, looking out at the police and beyond to an old stone fort and river. The police trained their eyes on Jim, their faces proud and intent.

"What you are doing matters," Jim began. "What you need to do is very simple. Protect yourself. Protect your qalat. Protect your village. Protect your valley. Protect your tribe.

"Do not misuse the power that you have. You have guns, you have power now. And pretty soon there will be more local police than any other force in this area. Do not misuse that. You are powerful now. Your commanders, the local people, your tribal leaders, myself, and the government have shown you special trust. Do not break that trust.

"Tell everyone that you come into contact with, I did not come here to fight. I came here to help the people. But if someone wants to fucking fight, they know where I am.

"Last thing: if you get into trouble and you need us, we will be there." Jim went to shake each of the arbakai's hands, his white Afghan tunic billowing in the wind.

The first group of about fifty was assigned to man two observation posts on Martyr and Dishka mountains. Amir Mohammed, a Safi tribal commander of twenty of the new police, was the only one to raise a concern. The thirty-five-year-old commander with a long auburn beard stepped forward to speak with Jim.

"If you want to fight the enemy, you can't fight them with AK-47s. You have to have PKM machine guns and RPGs," said Mohammed, who would be responsible for Martyr Mountain. Jim agreed. One way or the other, Amir Mohammed would have the firepower he needed.

To get the guns necessary for the tribal police, Jim broke other rules. He and Ish wrested hundreds of AK-47s from corrupt local Afghan government officials who sought to control the guns. They handed out the rifles to the tribesmen, giving them back a symbol of honor. Next, he asked his command for PKM machine guns for the force to defend the high ground. Then they scraped and borrowed to provide food, water, blankets, tents, and other critical gear for the Afghans.

Jim constantly drilled into the team the need to befriend the Safi tribesmen—to eat, drink tea, talk, and play cards with them. Wearing a pakol and Afghan tunic, his beard long and his face almost haggard from fatigue, Jim appealed to us to have faith in his approach in Chowkay and hold the Afghans close.

"What I am telling you is: trust me," he said, surveying the sober faces of the Americans and Afghans crowded into the operations center. "You want security tonight? Then eat with the local police. I am telling you, it's not their mortar positions, machine guns, or RPGs. None of that shit is going to matter. It's helping people with their everyday lives, it's talking to people, it's treating people well," he said.

Despite the lockdown imposed on U.S. troops earlier that day because of spreading anti-American protests, he said, some of the team was going to violate the rule and leave the qalat. "In fact, I am leaving the qalat right now," he announced. "I am going to see Haji Jan Dahd to shoot the shit after dinner."

The very next day, Jim came to me as I was sitting bundled against the cold in a corner of the qalat in my makeshift chair—a piece of scrap lumber I'd set on two bricks. I shared a windowless, mud-walled room in the qalat with Jim, Abe, Ish, and Shafiq. It was one of four connected rooms lining the side of the qalat closest to the mountains. The rooms faced onto a central courtyard. We slept under thick Afghan blankets on Army cots and hung our coats and rifles on long nails I hammered into the wall. But since the room was dark and I loved the sunshine, I spent most of my days outdoors. I chose to work in that particular corner because it was one of the few spots in the courtyard sheltered from possible gunfire from the mountains behind us. Unlike the only other protected corner, it was also not in a direct line of fire from the qalat gate.

"Let's go for a walk," Jim said. I put my notebooks and computer inside and followed him out the qalat gate. The only weapon he carried was his 9 mm pistol in a shoulder holster. We walked down the dirt road, now covered with gravel, and crossed the main paved road, with me staying slightly behind him. After passing the old stone fort, we strolled down a path toward the river and the other side of Chinaray. The sky spread out before us above a range of snow-capped blue mountains on the other side of the valley, as the Konar River snaked silver beneath them. The village seemed poor even by Afghan standards, its small qalats made of stone and surrounded by rocky, barren plots of land with a few withered trees. A couple of Afghan men standing nearby who were building a stone wall saw Jim and he waved at them.

"*Tsengayay, jore, wror,* how are you, brother?" he greeted them, and stopped to chat. "I see how hard you are working," Jim said in Pashto. "I want you to know I am not here to make your life more difficult. I am here to help you."

"We know who you are," they said. "And thank you. Our lives are very hard, and our days are long. We will do all we can to help you. Please come and have tea with us."

We moved on, and came upon a group of boys and girls. Jim sent me ahead to speak with them. I did, and before long their mothers and sisters began coming shyly out of nearby qalats to talk with me, too. Even a pregnant woman approached me, but she hid behind a rock so the men in the distance could not see her. They had never spoken with an American before. I asked the children about their families and school.

As dusk fell we walked to the school and visited the home of one of the teachers, speaking with him and his relatives until it grew dark. Just then I noticed an Afghan taxi pull up on the main road above us and stop, letting out three men. My heart started beating faster. The biggest threat we faced living in the qalat was a suicide bomber on foot or with a vehicle. I had been traveling in Afghanistan long enough to know that it took no time for word of our walk to spread—one phone call to the insurgents and we were done. I stepped a little faster as we crossed the road in front of the taxi headlights and headed toward the qalat. Just then, a few of the ALP came out to meet us and escort us back.

I felt relieved as we entered the gate to the qalat. I was rarely rattled that way, but both Jim and I knew that our solo walk through Chinaray was risky. How risky, we were not sure. We had several things going for us—our walk was completely unexpected by the people, they already knew who we were, and they undoubtedly knew of Jim's relationship with Jan Dahd. Yet our reception in Chinaray remained uncertain. Perhaps the only way to know for sure was simply to go, as Jim did. In doing so, he reaped huge gains. He had once again demonstrated the ultimate trust in the Afghans, walking among them with his *khuza*, his wife, tacitly entrusting her safety to them. He also wanted to show them that he was one of them and they had nothing to fear from him.

The next morning, Jim pulled his usual guard duty, from four to six in the morning, after a late night with Abe, Ish, and Jan Shah, Jan Dahd's son. If his father was the most influential tribal leader in central Konar, the son was the deadliest. Jan Shah's knowledge of the insurgents and their activities was second to none. At around nine-thirty later that morning, Jim woke from a short nap to the sound of gunfire. Incoming rounds were hitting the qalat from a Taliban PKC machine gun. Next he heard the return fire from an American-made M240 machine gun he'd given the Afghan Local Police.

*Fuck, here we go*, he thought.

Everyone knew what to do. He'd detailed the qalat defense plan, and there had been several talk-throughs and walk-throughs.

Jim glanced at me for an instant, saying nothing. He swung out of bed, pulled on his ammunition harness, and threw back the door. He wore a tattered gray sweatshirt over traditional Afghan pants.

I grabbed my camera and flicked on the switch to start filming from the doorway.

Fully in control, but quick and instinctive, Jim had already begun to tick off his internal combat to-do list. He moved across the courtyard, pointing at the qalat's blue metal door, ignoring the incoming fire.

"Secure the gate!" he yelled.

The rounds kicked up dust and worked their way toward the operations center and checkpoint just outside the gate. The gate and the walls had to be protected at all costs from possible suicide bombers.

Afghan tribal police, Jan Dahd's men, moved to reinforce the entrance to the gate and the road leading to the qalat. Just outside at a guard station, the commander Abdul Wali unloaded two hundred rounds at the mountainside with Tribe 34's U.S. military M240 machine gun. He'd only learned how to fire it the day before. Abdul Wali then handed off the gun to his fellow commander, Sadiq, and rushed into the qalat to support our counterattack. Standing exposed in the middle of the courtyard, he fired off more volleys with his AK-47. Outside the qalat, several other of Jim's newly minted arbakai took positions and fired into the hillside.

Jim listened to the volume and intensity of fire. He heard the Afghans shooting back—but not his heavy guns, which were supposed to be manned by the Americans.

*Where are my guys?*

Jim's U.S. soldiers—many of them facing their first gun battle—deliberately moved into position according to Jim's qalat defense plan. Their mantra, repeated over and over by Jim during training, was to give the enemy a "momentary impression of superior firepower." Or, in his words: "Let loose on his ass with everything you have." He needed those few seconds—not more—to assess the landscape of the battle and how to counterattack.

Sgt. 1st Class Tony Carter, a weapons sergeant from the National Guard's 19th Special Forces Group, scrambled up a wooden ladder and moved along the wall on the qalat roof toward his .50-caliber machine gun. Rounds zinged past him—close.

*I'm going to get shot in the head*, he thought.

But Tony kept moving. With minimal cover, he returned fire toward the muzzle flashes he saw coming from Martyr Mountain. Dan, who had been directing the arbakai barrage, rushed up the ladder behind Tony, covering for him in his last stretch to his .50-caliber station.

Pfc. Jonathan Bartlett, a brawny high school football player from Springfield, Massachusetts, jumped on another .50-caliber machine gun in one of the Humvees parked in the courtyard. Bartlett opened up on the ridgeline, his gun smoking.

Jim had made it past the incoming fire to the crude operations center, a tiny, mud-walled room with a couple of tables mounted with computers. Spec. Fernando Ruiz, a twenty-two-year-old from Los Angeles, was yelling into the radio in what sounded like Spanish—whatever it was, it was incomprehensible. Jim grabbed the radio and calmly told his commanders what he wanted them to know. In his experience, staff officers got in the way of the fighting more than they helped.

"Iron 34, this is Tribe 34," Jim said. "We are under enemy fire. I request no CAS, no QRF, no medevac. Break. I will update you as necessary. Tribe 34 out."

But as he kept listening to the outgoing fire, he knew something was still wrong. He could only hear three machine guns—they had six.

*Where the fuck are my goddamned guns?*

Jim ran back out and across the courtyard.

Both M240s on his Humvee were idle. So was the .50-caliber machine gun on the RG-31, a mine-resistant armored vehicle.

Jim's gunner at the time, a young soldier on his first deployment, had not even climbed into the turret.

Abe rushed up, wearing his helmet and carrying his AK-47. "Do you want me on the M240?"

"Fuck yeah!" Jim replied.

Within seconds, Abe was laying down heavy fire on the enemy positions. Next, Sgt. 1st Class Fernando Gonzales, the team's skilled intelligence expert, started shooting another machine gun mounted on the RG-31. The escalation of fire was enough to suppress the Taliban shooters and end the gun battle—for now. But what about tomorrow?

At the after-action meeting that night, Jim and the team reviewed what had gone right and what had gone wrong. The Afghan police and their commanders had fought back without hesitation. Sadiq had quickly reinforced the qalat gate and road with ALP in case of a complex attack, such as a suicide bomber. But some of the inexperienced U.S. soldiers had reacted too slowly and cautiously. Ruiz worked incredibly hard and was unusually competent, but the first incoming rounds of his life had temporarily rattled him. Jim's Humvee gunner had seemed confused about what he was supposed to do and didn't fire his weapon. Shaky their first time under fire, they had hesitated. It took far too long for the soldiers to get the RG-31's machine gun up and running. It was all Jim could do not to start ripping into his guys. But they were all he had. There was one solution to that problem: more combat.

Jim explained that the delayed response was not the right message to send the Taliban. He then laid out an aggressive two-vehicle patrol the next day into the Dewagal valley, the nearby territory of the Taliban commander who ordered the attack. To leave the Taliban attack unanswered would have been dangerous, dishonorable, at odds with Pashtunwali. Jim handpicked the men to accompany him.

Sadiq, wearing a dark green U.S. uniform and a black balaclava in order to hide his identity from the Taliban—he knew some of them—rode in the back of Jim's Humvee. Jim's vehicle was by then composed of almost all Afghans—Abe, Shafiq as the driver, Sadiq, and one of the mercenaries.

The only other American was Jim's gunner. The patrol passed an ambush zone nicknamed "Shark's Tooth" and continued another mile up the valley. When the insurgents opened fire, Sadiq unleashed a two-hundred-round burst from the machine gun, leaving the barrel smoking. The rest of the team laid down consistent and heavy fire. But again, Jim's American gunner hesitated.

Back in our room Jim told me about the incident. He had so few men that his options were limited; what did I think?

I was normally one to give people a chance to improve. But this time I said, "Take him off the gun. He's going to get someone killed."

Dan agreed.

While the rest of the U.S. military made it their priority to prevent killings of U.S. soldiers by Afghan forces, known as "green-on-blue attacks," our lives were now completely in the hands of Afghans.

Jim had earned their trust, and now—forged in battle—they owned ours.

# CHAPTER 29

AT A TEAM MEETING that night, Jim told us the three main insurgent groups in the area, including Afghan and Pakistani Taliban, were said to be collaborating to attack the qalat. The word was that they sought to destroy the alliance between Jan Dahd and Jim's team. Maulawi Basir, the insurgent commander from the Dewagal valley who had kidnapped Linda Norgrove in 2010, was reported to have two suicide bombers with motorcycles targeting Jim.

"Any one of them would be a problem for us, but now they are all coming after us," he said. "It is only a matter of time before we start taking casualties.

"If they want to fight, we will fight," he went on. "I think they will attack tomorrow. It is Friday, so no ALP will be here training. The Badel valley elders will be at a meeting at Haji Jan Dahd's, so they can say they are not responsible."

One of the most critical and fascinating pieces of this puzzle was Sadiq. Much of Sadiq's family was Taliban from the Badel valley. He had sources within the Taliban who began to report planned attacks on the qalat.

Early the next morning, Jim was up talking for two hours over the radio with the U.S. Air Force operator of a Predator drone that was monitoring the area surrounding the qalat. Jim knew from Sadiq that the fighters had left the Badel valley and moved in their direction, but he didn't know exactly where they were. The Predator scanned the possible insurgents' hideouts and infiltration routes from the Badel valley, but the camera spotted no insurgents on the move.

At around 10:00 a.m., Jim and Ish walked to Jan Dahd's qalat for the meeting with the Safi tribal elders from the Badel valley. Before he left, Jim turned to Sadiq.

"No matter what happens, if the qalat is attacked, protect Ann. She is my namoos," Jim said.

Sadiq's face turned serious.

"No, she is *my* namoos," Sadiq said, patting his chest with his hand.

Jim nodded, speechless. He could ask nothing more of Sadiq. He and Ish left on foot.

Half an hour later, I was sitting outside in the corner of the qalat writing when machine gun rounds started flying in again, coming this time from both Martyr and Dishka mountains behind the qalat.

Bartlett and Pfc. Richard Lerma were the first to jump up in their truck and start hammering the ridgeline with machine guns—faster, more confident, and more deadly than ever. With so few of us, each member of the team knew his job was critical.

Abdul Wali stood beside them firing well-aimed rounds with his AK-47, then rushed out the gate to supervise the ALP.

Shafiq ran across the courtyard and grabbed an automatic weapon, while Abe got on the rear gun of Jim's Humvee. A mercenary named Sahib Zada rushed to help Abe reload.

Dan took charge overall. "Stay in position!" he yelled, and ran to the operations center, where Ruiz was effectively manning the radios. The team's air controller, Air Force Senior Master Sgt. Wesley Brooks, had his headset on and was trying to contact any aircraft in the vicinity.

Insurgent fire tore through the dirt barriers the team had just built in front of the operations center, and shot through the hood of Dan's Humvee. An RPG landed in the qalat next door.

Outside by the ALP guard shack near the gate, rounds were coming in thick. Sgt. Danny Bird, a twenty-one-year-old from Albuquerque, New Mexico, ran to the shack and made sure the ALP were in position. Then Bird sprinted to the Mk-19 grenade launcher placed between the qalat gate and the main road and fired off several rounds.

Dan rushed out the qalat gate and backed up the ALP by firing his favorite old M14 rifle. Then Dan ran to Danny and helped him reload the grenade launcher—but it jammed. Dan tossed Danny an Mk-48 machine gun.

Suddenly, an enemy bullet hit Abdul Wali in the lower right leg.

"I'm shot!" he yelled to one of the mercenaries, thirty-four-year-old Basir. Then he kept firing, blood running down his leg, until he ran out of ammunition.

One of his men, Mahmud Dwaher, wrapped his commander's leg in a tourniquet made of cloth. Then he picked him up and carried him pig-gyback under fire. Dwaher rushed into the qalat, up the stairs, and into one of the rooms, where he laid Abdul Wali on a cot. I ran in behind them. Dwaher rolled up his pant legs and showed me the wound.

I ran out the door. "Abdul Wali's been shot!" I yelled.

"Doctor!" Dwaher cried, coming out behind me.

Staff Sgt. Ed Martin, a medic from the National Guard's 19th Spe-cial Forces Group, rushed in and began treatment. I got Abdul Wali some water and went back out.

Dan came back into the qalat. As he passed through the gate he sud-denly felt his leg burning.

*What the fuck?* he thought. He had been shot in the leg, but he could still move, so he decided to ignore it. As a medic, he wanted to make sure Abdul Wali was all right.

Across the way at Jan Dahd's compound, Jim had been meeting with the Badel elders.

"I'd like to provide development projects and help you, but if you want to fight, we'll fight," he told them. Just then, gunfire rang out in the direction of the qalat.

The Special Forces team from FOB Fortress was also at the meeting with Badel elders. However, they did not fire back when the shooting started. So Ish, annoyed at the passive response, climbed up to the M240 machine gun on the back of one of their parked vehicles and started returning fire—ignoring the rule that Afghan interpreters for the U.S. military were not allowed to carry weapons. Jim considered that rule immoral and always made sure his interpreters were well armed.

Jim and Ish ran back to the qalat and arrived after the firefight had died down. A short while later, Dan and Ish left for FOB Fortress to transfer Abdul Wali, who was on a stretcher in the back of a pickup truck, to a U.S. military field hospital in Asadabad. Once Dan arrived at the U.S. base wearing an ALP jacket, however, the U.S. medic mistook him for an Afghan and started speaking with Ish instead.

"He's the interpreter," Dan interjected brusquely. "I'm the U.S. soldier."

The medic initially resisted evacuating Abdul Wali by helicopter for Asadabad, arguing that ALP were not allowed such treatment. Incensed, Dan wouldn't take no for an answer. He finally prevailed and Abdul Wali was flown out on a helicopter. Later back at the qalat, Dan, with Ed's help, treated his own leg, which had been grazed by a bullet.

THE RAPID PROGRESS JIM was making with the Safi tribe was the result of many months of invested effort. But the cost to the U.S. military of the Chowkay operation—in terms of manpower, weapons and other equipment, and money—could hardly have been lower. The qalat was austere. We were living largely off the local economy. Jim's command did not advance him any additional funds for the mission to Chowkay. The upshot was that Jim had to borrow heavily from local Afghans while at times he and Dan spent their own money. We were not paying large sums for security forces, as many other Special Forces teams did.

Indeed, Jim was already concerned about making the handoff in April to the incoming Special Forces team. The team leader, Capt. Randy Fleming, had emailed him asking questions about the gym and chow hall. Clearly Fleming had no concept of living in an Afghan qalat, let alone the overall mission. We'd been in Chowkay just three weeks, and Jim was trying to gain more time.

Meanwhile, our requests for the most basic needs to keep us alive and the mission moving forward were routinely denied. For example, Jim asked for some resources to build observation posts on high ground: money to hire donkeys to help ferry supplies, tents and blankets for the ALP who would man the posts, ammunition, and, most important, heavy weapons. But still no funds arrived—so he had to pay out of the limited existing budget, or again borrow from Ish or other Afghans. The team also had to scrounge at the nearby U.S. military base, FOB Fortress, for packaged military meals, fuel, and water. Even then, the command complained that the team was consuming too much water. All we had was bottled drinking water and a small water tank built in an open-air wooden stall. We washed ourselves in buckets. Given the threat to the qalat, Jim asked his command for ten thousand rounds of additional AK-47 ammunition for training the ALP who guarded the qalat. That request was denied also.

On March 9, Jim, Dan, Fernando, Abe, a few other team members, and I made a trip back to COP Penich to gather critical supplies such as water, ammunition, and two M240 machine guns. We drove in pickup trucks down a dusty dirt road on the opposite side of the river from Chowkay. Lt. Roberts had been asked to make sure that the supplies were ready to pick up. Jim decided to have me stay at Penich for a few days, for my safety.

But when Jim arrived at Penich, he received some disturbing news. Abe and Ish's brother Imran, who had stayed at Penich working for Roberts, told him that Roberts had been openly hostile to requests from Tribe 34. One day Imran had approached Roberts to pass on a simple message.

"Sir, Capt. McKone needs you to call him," Imran had told Roberts.

"Fuck them," had been Roberts's response.

A few other soldiers at Penich were griping about the requests from Tribe 34. "They think they are the only guys at war," another soldier said, according to Imran.

In a way, the simmering resentment was understandable because the splitting of Jim's team between Chowkay and Penich divided the men who were eager to fight and those who were not. "The rules Jim puts out may not be actual Army doctrine, but they are actual warrior doctrine. If you are a natural soldier, it's an easy transition to make," said Bartlett, who had been itching to go to Chowkay and had already proven his skill as a gunner on Dan's truck. "But if you joined the Army because there were no jobs in Michigan, or because you wanted to impress a chick, it would be a harder transition." Several of Jim's men were under consideration for awards for valor, unlike the less aggressive soldiers who had stayed behind with Roberts.

Before Jim left, he pulled Roberts aside and asked him to respond to what Imran had said. It was Roberts's chance to broach his concerns directly with Jim, man-to-man. But Roberts shied away from the opportunity.

"Is there anything you need to say to me or anything I need to know?" Jim asked him.

"No, sir," Roberts replied.

Then Jim called Roberts and the rest of the Penich team together outside on the muddy ground where the vehicles were parked, and dressed them down.

"Listen here, motherfuckers," Jim told them. "You have leadership issues, loyalty issues, and teamwork issues. Three things are wrong, and it starts with you, Lieutenant," he said, looking at Roberts, who stared back blankly. "You had better get your shit together. When we call and ask for something, you better have it ready," he said. "If you can't do the job, let me know and I will get you out of here. Your buddies over there are fighting; we are fighting for our lives. We need you to work as a team. Does anyone have any questions?"

Silence.

Jim walked away. One of the soldiers, Sgt. Adrian Cone, approached him. "Thank you, sir," he said. "It is miserable here. No one is working together. Everyone needed to hear that."

Two days later on March 11, twenty of the newly recruited Chowkay arbakai set off on the dirt path that led through Chinaray and up the steep valley to Martyr's and Dishka's peaks. A few donkeys followed them laden with food, water, tents, and blankets, as well as weapons and ammunition. Jim and his men watched over them as they moved out.

*They are going to get hit*, he thought.

Jim had taken extensive precautions to ensure that other American units in the area did not mistakenly attack the Afghan force, but he was sure the Taliban would strike.

The arbakai were going to build observation posts on the two mountains. There they would absorb insurgent attacks otherwise directed at the qalat. Jim knew the arbakai would not survive without heavy weapons and ample ammunition. But his requests for extra ammunition and PKM machine guns still had not been approved. That left him one option: he gave the arbakai two U.S. military M240 machine guns along with thousands of rounds of ammunition. It was unorthodox, but it would keep them and us alive.

# CHAPTER 30

AS JIM AND HIS comrades, Afghan and American, were fighting to expand tribal security in Chowkay, back in the relative safety of COP Penich, Lt. Roberts took out a pen and began printing in small, neat letters: "I, Thomas Christopher Roberts, want to make the following statement under oath."

Things were not right with the world, according to Roberts. He wanted to be in Afghanistan, he wrote, to "conduct operations that benefit the people and destroy terrorism." However, he claimed indignantly, he was being asked to do things that were "immoral or illegal."

Roberts was not sure of his facts, but in six scattershot paragraphs, the newly minted West Point lieutenant lodged a number of suspicions, allegations, and complaints.

One of the problems, he said, was that Dan had asked him to give the ALP fuel. This was a common practice by Special Forces teams across Afghanistan, but Roberts believed it was wrong, so he refused to do it. Another concern, he said, was about reporting fifteen workers present on COP Penich when at the time there were only three. Roberts did not realize that those numbers also covered the workers at Chowkay.

Then Roberts turned to Jim. He claimed that Jim had failed to report his own injuries after the January 16 IED so that "his concussion would not be means for him to leave the country." Roberts also stated that he had once detected alcohol on Jim's breath—it was in the evening of February 16, the same day that Jim, Dan, and a small team including nine arbakai climbed twenty-five hundred feet into the Shalay valley and captured a high-value Taliban target. "I did not smell alcohol on him before this time, but this was the first occasion I have been around him after heavily perspiring," Roberts wrote. Jim was "walking abnormally because of the dismounted movement" and was offered pain medications by the camp medic. Roberts said he "suspected Major Gant of being intoxicated and under the influence of pain medication."

Roberts noted that I was at COP Penich but had not been put on the daily situation report, or SITREP. "No one knew where she came from, only that she lives with Major Gant and he refers to her as his wife," he said.

For whatever reason, Roberts had decided not to raise his concerns directly with Jim. But it was clear from what he wrote and confided in others that he feared getting in trouble and decided instead to point the finger. "I cannot work in this environment any longer without being asked to do something immoral or illegal."

Shortly after that, Roberts quietly dispatched his sworn statement up the chain of command. It landed at the Bagram headquarters of Special Operations Task Force–East, headed by Linn, who was then Jim's boss. Within seconds, the information went viral, causing an almost apoplectic reaction by the military hierarchy, according to a senior member of the command.

Immediately an email flew in from Col. Richard Kim, the U.S. Army commander and battlespace owner in charge of Konar and surrounding provinces. The message was sent with an exclamation point, indicating high priority. "We have to get this guy off the battlefield," Kim wrote, according to the senior member of Linn's command. Minutes later, Schwartz called, followed by Brig. Gen. Christopher Haas, who had taken over in July 2011 from Brig. Gen. Miller as head of CFSOCC-A. Next came a call from the office of Gen. John Allen, who had replaced Petraeus as the top U.S. commander in Afghanistan.

The accusations against Jim arrived at a particularly tense time for the Special Operations commands of Haas and Schwartz. On March 11, the same day that Roberts made his sworn statement, Army Staff Sgt. Robert Bales was detained at his base in the southern province of Kandahar for allegedly slaughtering sixteen Afghan civilians in a methodical killing spree. Bales was a member of the same Fort Lewis–based conventional Army battalion as Roberts, and was similarly assigned to work with Special Forces teams conducting village stability operations. The incident sparked more anti-American demonstrations, leading the U.S. military to again suspend combat operations in Afghanistan. Internally, it was a leadership crisis for Haas and Schwartz, whose commands Bales fell under.

Based upon Roberts's statement, Linn decided to conduct a "probable cause" search of Jim's living quarters and those of his teams at both COP Penich and the qalat at Chowkay.

ON MARCH 12, SOLHEIM flew into Chowkay for a planned visit to see the progress Jim and his team were achieving with local security forces in the Safi tribal area. His helicopter churned up a cloud of dust as it set down in a newly built landing zone guarded by arbakai just outside Jan Dahd's compound.

The team had rushed to flatten the ground and cover it with gravel to create the landing zone, anticipating high-level visits soon by Haas and Allen.

Jim greeted Solheim as soon as he got off the chopper. "What's up, Jim?" Solheim asked casually.

"Hey, what's up, man?" Jim replied.

Solheim knew Jim was in trouble, but instead of pulling his fellow major aside and giving him a heads-up, he chose to say nothing. It was a stab in the back that would later surprise and anger Jim. Instead, he and his senior enlisted man, Sgt. Maj. Brian McCafferty, appeared relaxed and positive as they walked over to have tea with Jan Dahd, and then spoke with several Afghan officials who were involved with the local police and who stopped by the compound.

For Jim, Dan, and the rest of the team, it was a proud day. The fact that a U.S. military helicopter was now landing at the home of the tribal leader was a sign of recognition for the Afghans—both Jan Dahd and the Safi tribesmen working to protect their area—and underscored the importance of their efforts. And as far as Jim knew, no other Special Forces team in eastern Afghanistan had embedded twice in Afghan villages and lived in qalats guarded only by themselves and their arbakai. The rapid momentum achieved with the Safi tribe was resounding proof that Jim's strategy was working.

Later Jim led Solheim across the hillside to the qalat and introduced him to his men. They ate lunch with Sadiq and Abdul Wali, who was back from the hospital in Asadabad and had a cast on his wounded leg. "I got shot in the leg, too!" Solheim joked, pulling off his artificial limb and laughing.

Solheim appeared genuinely impressed by what he saw during the three-hour visit. "You are living the life," he told Jim.

Solheim had agreed to hold a ceremony at the qalat to present Jim and Dan with Purple Heart decorations for their recent combat wounds. But shortly before his departure, Solheim excused himself to make a phone call. When the time came to give out the Purple Hearts, he prevaricated. "I am prepared, right now, to present you and Dan with your Purple Hearts," he said. "But just know we will do the ceremony again, because Lieutenant Colonel Linn wants to come out and give it to you. He will fly out in the next couple of weeks," he lied. "So we can do it twice, or we can do it once."

Jim had turned down several Purple Hearts and did not really want the latest one. When he dropped me off at COP Penich, he had commented that he was reluctant to accept the award, knowing other recipients had suffered far more crippling wounds.

"Fuck that, let's just wait," Jim replied.

Then he walked Solheim back to the secured landing zone, shook his hand, and watched his helicopter take off.


AT NINE-THIRTY THE NEXT morning at Bagram, Brig. Gen. Haas quietly launched an Article 15-6 investigation "into the alleged misconduct of MAJ Gant, CPT McKone, and by Operational Detachment Alpha 3430(G) personnel and augmentees at VSP Chowkay and VSSA Penich." Haas appointed Lt. Col. Robert Kirila to conduct the investigation.

A slight man with short-cropped hair, Kirila had just arrived in Afghanistan to become Schwartz's deputy. The forty-three-year-old from Norfolk, Virginia, had graduated from the University of Richmond with a bachelor's degree in Spanish and later commanded a battalion in the 7th Special Forces Group, which focused on Latin America. Once when he was the company commander in charge of Special Forces training at Robin Sage he forced the students and cadre to work on July 4, even though training was finished, according to a former Special

Forces officer familiar with the incident. A drinking party ensued and several members of his command were punished.

Schwartz, meanwhile, drafted two memos based on Roberts's statement that outlined the alleged wrongdoing of Jim and Dan.

"I have received a report of serious allegations regarding your conduct as the ODA 3430G Team Leader," he wrote about Jim. ODA 3430G was the formal designation for Jim's team, whose call sign was "Tribe 34." The G stood for "Gant." "The allegations include, but are not limited to alcohol consumption, misuse of pain medication, misappropriation of government funds, misuse of fuel, falsifying documents and a potential inappropriate relationship with Ms. Ann Tyson."

At his headquarters in Jalalabad, Solheim rallied a team and prepared to move against Jim. Solheim told other officers he felt betrayed by Jim. He assigned Sgt. 1st Class Markus Eckart and Staff Sgt. Bryant Brown to conduct a "health and welfare" inspection of Jim's room—the shipping container that had been moved with the rest of his team's equipment from Mangwel to Penich. Then he called on Capt. Fleming, the leader of ODA 3412 who had already been assigned to take over for Jim in Chowkay, and ordered him to have his men armed and ready.

At 2:50 a.m. Kirila left his quarters in darkness and boarded an aircraft for Jalalabad. At 8:45 a.m. on March 14, Kirila, Solheim, and the rest of the team flew by helicopter to COP Penich.

# CHAPTER 31

THE FOUR DAYS I spent at Penich waiting to return to Chowkay were dreary ones. Back behind the outpost's barb-wired walls and dirt-filled Hesco barriers, I felt completely removed from Afghanistan. Only the mountains reminded me of where I was, so I spent hours each day writing while sitting on a wooden chair outdoors, where I could see snow-capped peaks in all directions.

Imran was even more miserable than I was, having been left behind as the youngest brother while Abe and Ish moved with Jim to Chowkay. Shunned by Roberts, he spent hours each day alone in a large wooden hut he had shared with his brothers at Penich. One day I went to the door and knocked. "Come in," a quiet voice called. The room was only dimly lit, and Imran was sitting on the floor with his back resting against the wall, doing nothing. His eyes looked hollow and sad.

"I feel like I am in prison," he told me. I decided to try to cheer him up. Every day, when we were feeling especially stir-crazy among the monotonous routines of the camp, we would go out for a walk together. One windy afternoon, we played basketball with Chevy, the eleven-year-old orphan boy who had come with us from Mangwel. Another day, we took a long stroll in a drizzling rain, pacing round and round the perimeter of the camp, like caged zoo animals. We avoided the chow hall, with its blaring televisions and bland, canned food. One day, as a treat, Salim, the cross-eyed cook, boiled a chicken in a heavy cast-iron pressure cooker—the same type of cookers that were sometimes used to make IEDs. "This chicken came from my house—not the market," he told us proudly. We drank the soup heartily. In the evening, we played cards. I ended up staying in Ish's small back room, which attached to Imran's, at night because the shipping container was so terribly lonely.

Imran wore his heart on his sleeve, and we talked about his furtive relationship with his fiancée, who lived in Jalalabad. Given the cultural taboo on dating, Imran rarely saw her, but they spoke on the phone. I gave him some advice. In return, Imran translated a Pashto love poem for me. He also gave me an Afghan cell phone so Jim could call me. Little did I know then how critical that phone would be.

ON THE MORNING OF March 14, shortly before 9:00 a.m., I received an urgent call from Jim.

"Ann, Solheim is flying into Penich. You need to stay in the room," Jim said. "He will be there in about twenty minutes."

"Okay. I've been staying in Ish's room, so I'll be there," I told him.

"Whatever you do, move now," Jim said. "I've got to go. I love you."

"I love you, too."

Something about his tone and Solheim's unexpected visit made me nervous. As a precaution, I gathered all my notebooks and computer and hid them in Ish's room. Then Imran helped me replace the padlock on the door of the shipping container, leaving my backpack and cameras inside. I went to Ish's back room and closed the door.

Imran was in the larger connected room. About ten minutes later, after Solheim's helicopter landed, Imran went out to see what was going on. He came back, a shocked look on his face.

"Ma'am, they broke the lock on your room," he said in a hushed voice. "They are in there taking pictures. They have gloves on."

My whole body tensed.

*This is it. The black helicopter has landed,* I thought.

Jim had warned me about this moment many times. *Remember this: whoever does me in will be wearing a U.S. Army uniform, with a Special Forces tab.*

Now I had to break the news to him.

Just then, there was a knock on the door. Imran quickly stepped back into the main room.

ANN SCOTT TYSON

"Everyone turn over your cell phones. Now!" announced a stern male voice. It was Navy SEAL chief Kasey Heiland, the American advisor to the district government, who had recently arrived on the camp. The cigar-smoking SEAL had only been on one mission with Jim, a live-fire training event near Mangwel. Every inch of the walls of the hut he and others lived in was plastered with pictures of naked women.

I stayed in the room and kept my Afghan cell phone, hiding it under some clothes. The moment after I heard Heiland leave, I called Jim.

"They broke the lock and are searching our room," I said in a whisper.

There was a pause on the line.

"It's over," he said. "You know that, right?"

"Yes, I know."

"It will be okay. Just tell me what happens. I have to talk with Dan."

"Okay."

I hung up the phone.


JIM CALLED DAN INTO our room on the Chowkay qalat.

"They are coming after us, brother."

Like Jim, Dan was shocked and angry, but not necessarily surprised.

Then Jim called COP Penich. Roberts answered the phone.

"I have to talk to Kent right now," Jim said.

"Just a minute," Roberts said. After a few seconds, he came back on the line. "Sir, he will call you back."


A HUNDRED CALCULATIONS RACED through my mind.

*My passport, backup drives, and cameras are in that room. I have to get them. There is no reason to hide now. I have nothing to lose.*

I steeled myself and walked out of the room and down the gravel path to our shipping container. Sgt. Maj. McCafferty stood in front of the door, blocking my way.

"We are conducting a health and welfare inspection of the camp," he said, not moving.

"My bags are in that room," I said.

"We are conducting an inspection."

"I am getting my things."

I stepped forward and he let me pass. Inside the room, two uniformed men with gloves were searching, taking photographs, and tearing the room apart. They had gone through my clothes, which were under the bed, and flipped through photo albums including a sentimental one of me growing up—eating blackberries from a pail with my brother as a five-year-old in Seattle, playing with my sisters as a third-grader in Ireland—that I had given to Jim before he left for Afghanistan.

"This is just like something out of a movie!" one said to the other.

I slung my backpack and cameras over my shoulders and walked out, incredulous that they did not stop me.

Then I went back to Imran's hut. A soldier standing outside told me everyone was supposed to stay out of their rooms. I ignored him and went in anyway.

I phoned Jim again and told him what I had seen. We agreed that neither of us would answer any questions.

Then I went back to gather more of my things. This time I ran straight into Solheim at the door of the shipping container.

"Hello. How are you?" I said, and shook his hand.

"Nice to meet you," Solheim said. "You can pack your bags. We are going to be taking you to Jalalabad later this afternoon."

"All right," I said, and went to get one of my duffel bags from under the bed.

Soon afterward, I noticed the inspectors had left our room and moved on.

I called Jim again.

"They left the room. Do you want me to get anything for you?"

"No, it's too late," he said, knowing the room had already been photographed. His computer was there, his medical records, his writings. There was a bag of empty alcohol bottles in the room that had accumulated over months; Jim never openly drank in Afghanistan, out of sensitivity for the Muslim prohibition on alcohol use, and did not throw the bottles in the trash, where Afghans might see them. In addition, there were controlled medications, including pain pills, a small quantity of steroids, and some sleeping pills. Two photographs of Jim and me were on the wall. One showed us cleaning weapons after training. In the other, I was posing playfully with Jim's M4 carbine and AK-47 rifles.

Below the photograph, I had written in Greek the admonition of Spartan wives as their husbands left for war: Η ταν ή επίτας—"Return with your shield or on it."

"I have to get out of here—I am not going to Jalalabad to be questioned by them," I told Jim. "They are not getting anything from me."

"I've got it. But you need to wait until they leave. I am going to pack some things and send Abe to get you."

"Okay."

I got off the line. I told Imran I needed Chevy to bring me some large black plastic trash bags. A minute later, Chevy appeared with the bags. In Ish's room, I packed all my critical gear into three of the bags. Then I waited.

AS THE INVESTIGATORS DESCENDED, Abe and Shafiq were far north of Chowkay in a poor mountain village in Konar's Asmar District, where Abe was born, on an unusual mission.

A few days before, Abe had misplaced the golden Cape Hatteras lighthouse necklace that Jim and I had given him after his oldest brother died—either that or it had been stolen, along with a bundle of his money. Few could match Abe's bravery or passion for fighting, but he was forgetful and was always losing things. Distraught because he knew the lighthouse meant so much to us, he was determined to find it. Like many Afghans, Abe could be highly superstitious. So he and Shafiq, guided by one of the arbakai from Chowkay, were seeking out a certain old mullah, Mullah Baba, who was famous for finding lost objects.

"The mullah can look at a child's fingernail, and see things in it as if he were watching on a television," Abe had told me.

Abe steered Shafiq's old Toyota Corolla high up the mountain until the road ended. Then they got out and hiked through a wood of tall pines for another fifteen minutes to the village. Finally, they reached the mullah's mud house, set on a terraced slope.

The mullah's family was poor but friendly and immediately offered their visitors a simple meal of corn bread and milk tea, served with raisins and dried chickpeas. Then Mullah Baba entered the room, his dark eyes offset by a flowing white beard.

"I lost the money and I lost this lighthouse necklace," Abe explained.

"It was with my clothes in a bag and someone took the bag. I don't care about the money, but I want the lighthouse."

The mullah looked around the room, his eyes glossing over Abe and his companions and focusing in strange places.

"You know, I cannot find things all the time," he said. "When the genies are here, I can ask them. They often come early in the morning, or at midnight. But they are not here now."

Abe was crestfallen. Mullah Baba offered him more tea.

Just then Abe's phone rang. It was his brother, Ish.

"Ibrahim, you need to get back here!"

Abe had rarely heard Ish sound so desperate.

"What is wrong?" Abe asked, listening over the phone for gunfire. "Did they get attacked? Was someone killed?"

"Get back here. I will tell you," Ish said, not wanting to say more on a phone that could be monitored.

Abe, Shafiq, and the arbakai excused themselves and ran back down the trail. But when they got to Shafiq's Toyota they found its battery dead. They had to borrow a solar-powered battery from Mullah Baba to start the car.

Abe drove like a madman back to Chowkay and the qalat.

AT ABOUT 11:30 A.M., Solheim called Jim from Penich.

"Jim, I am headed your way," he said. "We need to talk."

By then, Jim had packed up a large plastic trunk with his and my belongings. When Abe and Shafiq pulled in, Jim rushed to meet them and briefly told them the news.

Jim knew they had only a few minutes. Ish had called the Afghan National Police (ANP) at the White Mosque Bridge and asked if they had seen the U.S. convoy approaching.

"Yes, I can see it coming now," the officer said.

"Abe, brother, this is bad," Jim said, his voice low and tense. "No one is our friend. They stabbed us in the back. Now take this shit, and get the fuck out of here. Move toward Ann. Ann is at Penich. Get Ann to safety."

Jim was blunt for a reason. There could be no gray area, no doubt in Abe's mind what his task was now.

He looked at Abe, whose face told him everything.

"I love you, brother," Jim said.

"I love you, too, sir," Abe said. He understood. This was a mission—a mission for Jim, for me, for all of us.

Abe turned his back and walked down the road. He picked up his AK-47 and 9 mm pistol, climbed into a white ALP truck, and drove off. With him were Shafiq, the mercenary named Assad, and two arbakai, who rode in the back. Abe drove through the Chowkay bazaar, crossed the Konar River at the Hakimabad Bridge, and took the small back roads toward Penich in order not to cross paths with Solheim's convoy.

Jim called me one last time.

"Abe is on the way. He will protect you with his life. I love you," he said.

"I love you," was all I had time to say. Then he was gone.


AT THAT MOMENT, THERE was nothing more I could do for Jim and the mission we both believed in and risked our lives for—nothing but to try to escape the U.S. military and disappear into Afghanistan.

I shut down emotionally. I had to act, quickly.

"Imran, get Salim and Chevy!"

I was standing behind one of the wooden huts with the black plastic garbage bags full of my gear.

Salim came around the corner, looking worried. No more than five feet tall, Salim carried himself with confidence—especially since Jim had made him an arbakai and given him an AK-47 just before we left Penich for Chowkay. He also had poise and was an excellent horseman; as a tour guide on the beach in Karachi, he rode a white steed decked out with garlands, brandishing a sword as the horse reared. I had spent hours talking and joking with Salim in the kitchen at Mangwel. His mother had suffered from a long illness, and Jim had paid for her medical care in Pakistan. She had passed away only a few weeks earlier. But this was the most serious I had ever seen Salim.

Chevy was not far behind. A bit of a troublemaker, Chevy used to career around the qalat in Mangwel in an all-terrain vehicle until Jim threatened to boot him. His mother was ill, too, and Jim paid for a surgery for her. All the soldiers liked to wrestle with Chevy, who was like a little brother to them. Now he looked at me questioningly.

"Take these bags for me," I said, handing them the garbage bags with my critical gear. All of my notebooks, my passport, my cameras, and my computer were inside. But I could not risk being seen carrying anything as I tried to slip away.

Imran explained that they should take the bags out the small back gate of the fenced Special Forces compound to the edge of COP Penich and drop them there. They did.

Next, I sent Imran to make a reconnaissance of the main gate.

He came back to the room, breathless.

"They put two Special Forces guys to watch the gate," he said.

Our first plan, to walk out the main gate to meet Abe, wouldn't work. We needed a vehicle.

Just then my phone rang. It was Abe.

"I am outside on the road," he said.

"Stay there. I am coming," I said.

Imran and I walked out the back gate of the Special Forces compound and down a dusty road that ran between the compound wall and the perimeter of COP Penich. There, sitting on the side of the road, were the black garbage bags with all my gear.

Imran walked swiftly ahead of me to the hut of the Afghan Security Guard (ASG) commander who was also a friend of Jim's—they used to eat dinner together—and was conveniently in charge of manning the guard towers at Penich. A few minutes later, the commander rolled up in a dark blue pickup truck. Imran gave him the garbage bags and told him to drive them out to Abe. He put them in the back and took off.

We stood there waiting. Chevy and Salim came and joined us. It seemed like forever, but the commander finally returned.

I was next.

"Take care, ma'am. I will see you soon," Imran said.

Chevy hopped in the front passenger seat next to the beefy ASG commander, who was aptly nicknamed "Engineer." I climbed in the backseat and crouched down in the footwell, covering myself with a scarf. Salim climbed in the back and sat down near me. Looking straight ahead, expressionless, Salim picked up a crumpled pink hat ornamented with plastic flowers that was in the truck and placed it over my head to further conceal me. The truck lurched and bounced over the dirt and gravel road toward the main gate. I held my breath.

# CHAPTER 32

AT THE QALAT IN Chowkay, Jim hastily gathered Tribe 34 in the window-less, mud-brick operations center. He had changed out of his Afghan clothes and was wearing his dark green uniform.

"Men, they are coming to get me," he told the team huddled round him. "Take care of each other. Take care of the Afghans. Accomplish the mission. Whatever happens here, it will be all right," he said.

Everyone was silent, somber.

"Whatever I did, it was to keep you guys alive," he said. "I took care of you.

"Now, I am giving you an order: do not lie about anything. I will go to prison for you. I will kill for you. I will die for you. No matter what, just tell the truth."

Dan spoke up, his voice strong and unwavering.

"Brother, I will go to jail with you. I am with you," he said.

Jim looked at Dan—his medic, his best gunner, his close comrade for nearly a decade—and, surrounding him, his last team.

"It was my life's privilege to serve with you," he said. "Strength and honor, men."

Jim left the room and walked across the courtyard and out the front gate to wait for Solheim.

At about noon, two heavily armored vehicles and a mine-resistant personnel carrier came grinding up the narrow road to the qalat. Jim went out to meet them. An entire ODA of Special Forces soldiers, led by Capt. Fleming, piled out wearing helmets and body armor with their weapons up. They followed Jim in, fanned out, and took up position in the qalat. Solheim and McCafferty were with them, as were Kirila and Lt. Col. John Corley—all outfitted in their full combat gear.

Solheim walked over to Jim.

"Jim, it's over," he said.

"All right, man," Jim replied. "Let's do this."

Jim went into our room in the back of the qalat, followed by Solheim, Kirila, and Corley.

Kirila spoke first.

"Major Gant, you do not have to answer any question or say anything. Anything you say or do may be used as evidence against you in a criminal trial—"

"Sir, I am not going to make a statement," Jim interjected.

"You are being suspended from your duties," Kirila continued tersely. "An Article 15-6 is being opened, and you are being investigated for multiple offenses. We are going to take you out of here."

He showed Jim the memo signed by Schwartz directing his suspension from duties and listing the allegations against him. Jim acknowledged receiving it. Then Kirila took out an order forbidding Jim to have any contact with me. Jim refused to sign it.

"Is there anything I need to know for our safety?" Kirila asked.

"Yes, sir," Jim said. "Abort criteria has been met—time now."

Jim knew on a visceral level—far more than Kirila or Solheim could even imagine—what a gravely dangerous situation they had created. The U.S. command could not have devised a better way to sabotage the Chowkay mission and alliance with the Safi tribe than by pulling Jim out in this way. It was another demonstration of their inability to grasp the importance of genuine relationships with the Afghans. If the Afghans sensed Jim was threatened, they could have turned on these new Americans in a second. Sadiq could have called his Taliban relatives: "Hit them! Put an IED in the road!" It even crossed Jim's mind that if he wanted to, he could ensure that none of them made it out alive.

*I have to keep everyone calm. The less I say, the better,* he thought.

Just then, explosions sounded up the mountainside. One of the observation posts manned by the arbakai was under attack. The gunfire quickly intensified. Jim picked up his M4 carbine and started moving toward the door.

"Sir, I am still the ground force commander. Do you want me to do anything?"

"No," Kirila said.

*These guys don't even see me as stable. They see me as a danger,* Jim thought.

The sudden realization was like a blinding slap in the face. As if punch-drunk, Jim slowly lifted up his M4 carbine and pushed a button to release the magazine, which crashed to the floor and lay there. He pulled back the charging handle with a sharp clack to clear the rifle, put it on safe, then set it on the floor. He grabbed his 9 mm pistol and did the same thing. Reaching into his pocket, he pulled out his cell phone and tossed it on the bed next to Kirila. Then he ripped the Special Forces tab off the shoulder of his uniform.

He slumped down on Ish's cot, angry and overwhelmed. A wave of utter exhaustion swept over him.

"We are going to take you to BAF [Bagram Air Field]. There is a helo landing at two-thirty," Kirila said. "Pack your shit."

Kirila walked out of the room and went to the operations center, where he summoned Dan.

Dan walked in.

"Who are you?" Kirila asked him.

"I am Captain McKone," Dan replied.

Kirila read Dan his rights, then asked curtly: "Do you have a statement?"

"No, sir," Dan said, looking at Kirila's compact face, which struck him as resembling that of a weasel. *Jesus, what a fucking tool!*

"Captain McKone, you are being suspended from your duties."

Back in his room, Jim began packing. Ish came in, looking miserable. He sat on his cot, and tears started rolling down his face. Jim hugged him, then began giving Ish his U.S. military gear—the chest rack that held his ammunition, his boots. As he packed, Jim was filled with resentment toward the commanders who were acting against him, as well as toward the new team that he knew was not invested in the mission. But he could not allow them to be overrun.

"Ish, there is one thing you must do for me," Jim said.

"Anything, sir," Ish said.

"Introduce the new team to Haji Jan Dahd and Haji Jan Shah and ask them to take care of these guys," he said. At the moment Jan Dahd was in Jalalabad and his son was in Asadabad, so Jim had no chance to make the request himself.

Then Ish called Jan Dahd's younger son, Zia Ul Haq, a tall, beardless, and delicate-faced man who was also an ALP commander. Zia Ul Haq met them in the courtyard, and Jim took him aside.

"Brother, I am so sorry. You did nothing wrong. Tell your father I will never forget what he did for me," Jim said. "The only thing I ask is, treat these guys the way you treated me. Help them. They don't know anything," he said.

Zia Ul Haq nodded.

There was still a half hour until Jim and Dan would be taken away. Rumors of what was happening were spreading around the qalat, sending shocked expressions across the faces of the Afghans.

Ish overheard someone say that the Americans were going to handcuff Jim. Incensed, he walked up to Sgt. Maj. McCafferty.

"Hey, if you handcuff Jim it will be ugly," Ish warned.

"No, we won't handcuff him," McCafferty said. Lying, he added, "He is fine. We won't do anything to him. He will be back."

Jim asked Ish to bring Sadiq, Zia Ul Haq, and the mercenaries to talk with him. He squatted on his toes in the courtyard, with Dan by his side, and the Afghans gathered around him, forming a horseshoe.

"It is over," he said. "I am sorry it ended this way. I love you and I care about you. Keep fighting," he said. "And please take care of these guys. They don't have anything to do with it."

After a moment of stunned silence, sobs started coming from the Afghans, and Jim and Dan started crying, too. Within a minute, the entire group was openly weeping.

"We are losing a brother," said Sadiq, tears running down his face. "We will never forget you."

As this was happening, Solheim was meeting with Jim's soldiers and noncommissioned officers in the room of the team's air controller, Air Force Sr. Master Sgt. Brooks, trying to reassure them. It wasn't working.

"What is going to happen with security?" asked Fernando, the intelligence sergeant.

"We understand it is a turbulent time for you," Solheim said with a consoling tone.

"You don't understand," Fernando shot back. "These people are like our brothers. You don't know all the work we did to get these guys so close to us. Major Gant and Captain McKone are the nucleus," he said.

"The new ODA is here to replace them. They have your security," Solheim said, speaking louder now.

Fernando was nonplussed.

"It's not the same," Fernando said. "We are in danger right now, because these people are watching this go down."

"Sergeant, you need to drive on," Solheim said. The room was silent. After a few more exchanges, Solheim wrapped up the meeting and left the room. But as he walked into the courtyard, he was stunned by the sight of the Afghans weeping with Jim. Also surprised were the soldiers on the armed ODA that had taken up positions inside the qalat, waiting to relieve Jim and Dan. They lowered their weapons and stared.

For several minutes, Jim, Dan, and the Afghans cried together, hugging one another. Then Dan left the group and walked down the stairs to the arbakai room to see Abdul Wali. He found the arbakai commander lying on the floor under a blanket, still on duty even as he nursed his wounded leg. His deep-set eyes were tired, but they lit up for a moment when Dan appeared.

"I am sorry," Dan said in Pashto. "I hope you will be all right. They are making us leave," he said.

Tears rolled down Abdul Wali's cheeks as he reached out to hug Dan, shattering whatever stoicism Dan had left in the face of what he saw as a horrible injustice. Dan cried harder, tears of anger and sadness. His heart was bleeding.

Jim came into the room, speechless with grief. He hugged Abdul Wali, too, and kissed him. Abdul Wali couldn't fathom why this was happening. As they were leaving, he plaintively raised both of his hands in the air, asking, *"Wailay?* Why?"

When Jim and Dan returned to the courtyard, the infantry soldiers of Tribe 34 surrounded them, asking them to sign their metal weapons' magazines and cloth lambda patches with marking pens and take photographs.

"Sir, this is fucked up," said Dan's gunner, Bartlett.

"You guys have to do the best you can with what you have here," Dan told him.

It was time to go. Dan and Jim walked out of the qalat toward the waiting vehicles. Jim was keenly aware that the rifle and pistol he wore had no magazines, a glaring sign of dishonor.

Outside the gate, the dozen arbakai who guarded the qalat were standing in two rows in formation, with Sadiq and Zia Ul Haq between them. All of them were crying. Jim shook each man's hand, and embraced him, and then turned to go.

Solheim was watching, with a pained look on his face.

"You have had quite an impact on these guys," he told Jim. "They are really going to miss you."

"Yeah," Jim said.

Last of all, Jim said goodbye to Ish. He grabbed his face with both hands and kissed him.

"I love you," Jim said.

"My brother, please," Ish said. His voice trailed off into nothing.

Dan and Jim rode enclosed in separate armored vehicles that carried them down the road, through a gate, and inside the thick walls of FOB Fortress. Dan was let out next to a helicopter landing zone. He had been ordered not to speak with Jim, and watched as Jim and Kirila climbed out of the other vehicle some distance away.

The faint whirring of helicopter blades sounded down the valley. It was a sound that had always got Dan's adrenaline pumping. But this afternoon, as the thumping of the blades grew louder, Dan felt numb, drained. The UH-60 Black Hawk helicopter circled and dropped down onto the landing zone in a cloud of dust. A helmeted crew chief got out. Dan watched as Jim picked up his bags and started walking toward the chopper, his hair and clothes blown back by the rotor wash—out of earshot, almost out of sight.

Dan felt a pang of anguish. *This is unreal. How do I show Jim: I am here, let's not give up, we accomplished the mission?* In an instant, it came to him.

In one swift movement, Dan threw back his shoulders and came to attention. Then he raised his hand in a crisp salute and held it, his eyes riveted on Jim, who was about halfway to the chopper. It was Jim's practice when leaving places to never look back. But at that moment, something made him turn around. His heart froze.

*Dan!*

Jim dropped his bags. With his weapon slung over his shoulder, he raised his hand, and for several seconds he returned Dan's salute.

# CHAPTER 33

"ANN, IT'S OKAY NOW."

Chevy was calling me from the front seat of the truck driven by the big-bearded ASG commander.

"You can sit up!"

I hesitated for a minute, and the truck came to a stop. Just as I was getting up from the footwell, the door opened. It was Abe. We were outside COP Penich on the road, at a spot where Abe knew the cameras positioned on the camp could not see us.

"Come on!" he said, taking my arm.

We ran together to the white ALP pickup truck. One of the arbakai from Chowkay riding in the back with an AK-47 flashed me a big grin. I climbed into the backseat and Abe took off driving, with Shafiq beside him.

I felt giddy. I was escaping the Americans, surrounded and protected by Afghans. The sense of safety and freedom I felt was almost indescribable.

Behind us, Azmat drove another ALP truck filled with arbakai from Mangwel.

Shafiq's radio crackled. It was Azmat, speaking to him in rapid-fire Pashto.

"There are big Army guys at the police checkpoint in Mangwel!" Abe told me. "The only way to bypass them is to go through Kawer," he said.

A few minutes later we wound into Kawer, and found Niq Mohammed's arbakai lining the roads. Then Niq himself appeared, smiling broadly. His face alone was immensely reassuring to me. The irony that this former Taliban commander would fill me with a sense of security was so striking it made me laugh.

We stopped briefly so Niq could climb into the back of Azmat's truck and escort us through the area. Members of three arbakai forces were now part of my getaway.

To skirt the U.S. patrol in Mangwel we drove along a dirt road nicknamed "Sleepy Hollow" that hugged the Konar River, and then passed through Chamaray. From that point, I knew no one would stop us. The trip itself became a blur, as my mind, free from worry about the escape, was not yet able to fathom our catastrophic loss.

We continued along the river toward the partly collapsed, V-shaped bridge called Zire Baba. From there, it was only another twenty miles on the hour-long journey to the thronging city of Jalalabad.

AT ABOUT FOUR O'CLOCK, Imran heard a knock on the door of his room at COP Penich. It was Dan.

Imran rushed to him and they hugged each other, crying a little.

"How in the hell did you get Ann out of here?" Dan asked.

Imran started telling him the story, when suddenly Solheim and McCafferty burst in the door. They walked straight back to Ish's room.

"Is Miss Tyson there?" Solheim asked.

"Who?" Imran said.

McCafferty pushed open the door to Ish's room. It was empty.

"Oops!" he said. "She's gone."

"Where is she?" Solheim demanded.

"I have no clue," Imran said. "She was talking with her sponsor, I think."

"Did you see her sponsor?" Solheim asked.

"No, I didn't."

Solheim threw an accusing glance at Dan.

"Don't look at me!" Dan said. He knew he should shut up, but he couldn't resist. "You know Jim—if he's good at one thing, it's making things happen. He is always two steps ahead of you."

With a look of irritation, Solheim turned and left. Imran later heard he and McCafferty had gone to check the Penich surveillance cameras to try to find out whether I was still on the camp. They found nothing.

"Get some chai going," Dan said after they left. "I am going to pack my stuff and come back."

In an hour or so, Dan returned. He and Imran spoke and drank chai into the evening as they waited for the helicopter that would take Dan to the base at Jalalabad.

"They can think whatever they want," Dan said of his commanders. "But you know, and I know, and the people we worked with know, we have been honest with our country and tried our level best to win this war that has gone on for eleven years.

"When we infilled in Chowkay, they didn't give us a dime. My guys were freezing; the Afghans in the tents were freezing. Our lives depended on the guys up in the OP [observation post] at night when it was raining," he said. "What we did in Chowkay would have taken other teams years. But now the Afghans have seen what the Americans did to us. Trust is gone, and security will go bad."

Dan poured Imran some more tea.

"It will be a little while before I drink chai with you again," Dan said.

Soon it was time to go. Azmat and Niq and several of the arbakai came to say goodbye.

The helicopter was inbound. Imran and Dan walked together to the landing zone.

Dan watched the chopper flying up the valley and shook his head.

"Look at that. How many attacks have we been in and asked for air and never got it?" he said. "They didn't support us, but now they are bringing in separate birds for me and Jim."

Tears again started streaming down Imran's face. "You know, if I could have done anything for you, I would have done it," he said.

"I know," Dan said over the noise of the helicopter landing. "Be strong. We will get you out of here." Then he hugged Imran and walked to the aircraft.

Imran watched the helicopter until it flew out of sight, and then returned to his room for another long, lonely night. Roberts would not give him back his phone, so he could not call his family. Imran was doubly isolated, because Roberts had ordered all the U.S. soldiers not to speak with him.

The next morning, military investigators scoured the camp. Imran packed his bags, quietly wrote his resignation letter, and walked to the operations center to give it to Roberts.

"There is only one thing I am going to ask you," he told Roberts. "Take care of those guys for us, the ones who stayed in Chowkay," he said.

As he prepared to leave, Imran heard a knock at his door. It was one of Roberts's soldiers—Spec. Trevor Iler, who worked in the operations center. A lanky and soft-spoken twenty-three-year-old from McBain, Michigan, Iler never went on combat missions, but he did the best he could at his administrative job.

"I don't care if Roberts doesn't want me to talk with you," Iler told Imran. "They can't charge me with anything for talking to an Afghan. That is our goal here," he said.

Imran invited Iler in, and then Iler began to cry.

"Whatever they did, they did," Iler said. "But I always wanted to earn a lambda patch. I never had the chance."

Imran thought for a minute. Then he reached onto the shoulder of his jacket and pulled something off.

"Bro, here is my ghairat patch," he said, handing Iler a red, green, and black badge with the gold Pashto letters that meant "honor." It was a patch prized by arbakai as well as Jim's men.

"This was the most important patch to Jim," Imran said, placing it on Iler's right shoulder. "If he knew you were this kind of guy, Jim would have given it to you. I will tell him what you said."

"Thank you," Iler said. "I will never forget you."

Iler and another soldier helped Imran carry his bags out to the gate. On the way, Imran saw Roberts. He had a parting shot, and he took it.

"I know you don't like me," he told Roberts, looking him straight in the eye. "But there is something I want you to know: I don't like you, either."

Then Imran went on his way, headed for Jalalabad.


ABE DARTED IN AND out of the Jalalabad traffic, his torso almost immobile as his hands flicked left and right on the steering wheel. He passed bicycles and flatbed trucks and brightly painted rickshaws in streets lined with sundry shops. Then we drove down a narrow side road and pulled up to the blue metal gate of the Khans' large Jalalabad home. It swung open, and we drove into the inner courtyard.

Jalalabad was not a safe city—there were frequent explosions and other insurgent attacks, such as a bicycle bomb detonated just down the street during one earlier visit. But behind those high walls I felt protected from such dangers. What I feared most was that the U.S. military would somehow track me down. I listened intently to the vehicles passing on the street below, my ears alert to anything that sounded like a military patrol.

Abe led me up a long staircase from the entrance to a hallway and Ish's bedroom, where Ish's wife greeted me warmly. After Abe left, she and I sat on pillows placed on the carpet and drank tea as her three-year-old son Haroon played nearby. I was acquainted with Ish's wife, having spent a day with her there soon after the birth of her second son, who tragically died of illness as an infant. With an almond-shaped face and pale skin, she was gentle and slightly frail but had a happy disposition. As I spoke to her using my basic Pashto, she sensed I was upset and urged me to rest. Soon Abe's mother and sisters arrived and spread out a plastic sheet on the floor, then served a meal of flatbread, chicken stew, yogurt, and spinach. Ish's wife brought me a plastic bucket of warm water to wash in, and a matching green and magenta dress and pants in the Jalalabad style—more fitted than those worn by rural Konar women. I showed them photographs and videos I had taken of Ish and Abe in Chowkay, and saw the pride in their faces as they watched them.

Amid this outpouring of hospitality and kindness, though, I fell suddenly and violently ill, vomiting into the Afghan-style toilet in Ish's room several times before I finally fell asleep, exhausted.

The next morning, I woke up nauseous but knowing clearly what I had to do. First, I had to back up all my notebooks, tapes, and photographs so that no one could stop me from telling the story of what Jim and his teams had done. After that, I had to comb through my materials and get rid of anything that could possibly be detrimental if it fell into the hands of the military.

Imran arrived from COP Penich, and he and his younger brothers started scanning all of my notebooks while I backed up photographs and tapes on external hard drives—a process that would take us several days.

I also started sorting through electronics and papers, burning in a blazing furnace on the roof any material that might be used against them. We took precautions to keep me hidden from the U.S. military—switching out SIM cards on phones whenever I made a call, and avoiding using the Internet. I moved between Abe's house and that of another relative in Jalalabad several times, riding in the backseats of various cars covered from head to toe in my blue burkha.

Sometime that afternoon, we got word that Jim was disarmed and under escort at Bagram, and under a direct order to have no contact with me.

A COLD WIND WHIPPED at Jim as he rode in the Black Hawk helicopter on the long flight to Bagram with Lt. Col. Corley, who was assigned to escort him everywhere.

When they landed, Jim was driven directly to a room at the back of Camp Montrond, the SOTF-E compound at Bagram. His guns, knife, and other gear were taken away—he could no longer protect anyone, not even himself.

"You need to get cleaned up right now," Corley said. "Here are some scissors," said the balding and heavyset lieutenant colonel, handing Jim a pair of shears and following him into the bathroom.

As Corley watched, Jim stood in front of the mirror and slowly began cutting off his long, graying beard.

With each snip, Jim felt his honor falling away, and struggled to keep his hands from shaking. It was the ultimate act of submission. He made a ragged mess of his beard and then picked up a razor and scraped the rest of it off. Looking at himself in the mirror, he barely recognized the person in the reflection. It was not only demeaning; it was the beginning of the obliteration of his identity.

"Now get in the shower," Corley demanded, handing Jim a uniform, T-shirt, and size thirteen boots—clown-like because they were too large.

Jim took off his clothes and felt the water running over his body. Then he dressed and returned to his room, which adjoined Corley's, and sat down numbly on the bed.

Soon afterward, the command doctor arrived.

"I have to ask you some questions," he said. "Are you going to hurt yourself, or others?"

"No," Jim replied, although his mind was screaming otherwise.

After that cursory encounter, Corley took Jim to see Linn, who was waiting in his Camp Montrond office with the command judge advocate, or JAG.

With perfect hair and a pressed uniform, Linn struck Jim as a Captain America type of officer. The meeting was extremely formal and by the book. The JAG read Jim his rights again. He advised Jim that he was confined to Camp Montrond and Corley would escort him at all times—he could not go to the bathroom without informing Corley. Moreover, Jim was forbidden to speak about the case or to have any communication at all with me. He was ordered to stay at least fifty feet away from me at all times.

Linn advised Jim again of the allegations against him: alcohol and drug use, misappropriation of fuel, misuse of government funds, and an inappropriate relationship with me.

Afterward, Jim returned to his room and tried to sleep.

The next day he was isolated and alone. The silence in the small, seven-by-ten-foot room was deafening, and the loneliness painful. His head was throbbing from a headache. He took out a dark blue bound journal and started writing to me, citing a Pashto proverb that we had often shared.

"Behind the door, behind the mountain," he wrote. It meant that even if I were just on the other side of the door, my absence felt as though I was far away behind a distant mountain.

BACK AT THE QALAT in Chowkay, the new Special Forces team was facing a virtual mutiny, as it further alienated the Safi tribesmen who were loyal to Jim and outraged after watching how the U.S. military had removed him from command.

As soon as the armored vehicles pulled out with Jim and Dan, the arbakai swarmed around Ish.

"We're done," they said. "We're not doing this anymore." Sadiq was particularly angry, and wanted to lash out.

Ish appealed to them all to stay on. It took great fortitude for Ish to do so, because he hated being there himself, but he did it for Jim.

"It's the right thing to do," he urged. "Stay for now."

Reluctantly, they agreed.

But Capt. Fleming and his team so mistrusted and neglected the Afghans living with them that the situation quickly deteriorated. They immediately broke all of the basic rules that Jim had drilled into his men for securing themselves in the qalat. Fleming's team was keeping a distance from the Afghans, refusing to place trust in them, and failing to provide for them.

The first night, Ish bought a lamb and had it cooked for the arbakai, and stayed up with them talking until two in the morning to try to calm things down. But Fleming and his men, fully geared up in body armor and wired up with tactical intrateam radios, stayed up on watch all night.

"They didn't do anything but pull guard all night long because they were so scared and uptight," observed Ghulam Hazrat, one of the mercenaries, who remained at the qalat with the team for a few days. "No human being could possibly love those new team guys, they were like morons," he added. "We were so used to Jim. Every morning we would wake up for prayers and Jim would be in the qalat hugging us and greeting us. Jim and his men were like Pashtuns and acting normal."

Sadiq was particularly upset when the team members showed up for dinner with their armor and radio headsets on.

"When they came to dinner they were wired," he said. "I couldn't stand these new guys."

Fleming imposed new restrictions on where the arbakai could go within the qalat, and soon expelled them from the qalat altogether. He and U.S. Army investigators questioned commanders Zia Ul Haq, Sadiq, and others, fingerprinting them and making them take lie detector tests, estranging the tribesmen even more.

One of the first nights, Fernando was in the operations center when Mahmud Dwaher, the arbakai who had carried Abdul Wali to safety after he was shot, came to get the extra pair of night vision goggles and its battery that Jim had always let the ALP use to assist them in guarding the qalat. But that night, when Dwaher came into the operations center to pick them up, Fleming bristled.

"No more," he told Fernando. "That's a no-go. They don't come in here."

So the following night when Dwaher showed up, Fernando had to hold his arm out to stop him before he stepped inside.

Dwaher gave him a surprised look, and Fernando felt so ashamed he turned away.

The new team further unraveled Jim's empowerment of the tribe and local people by insisting upon handing the ALP weapons and ammunition back to the District Center, and retraining and re-vetting the local police, according to several Afghans involved.

In perhaps the most serious misstep of all, Fleming fired Abdul Wali, who had fought so bravely and suffered a gunshot wound as he risked his life to protect the qalat and everyone in it. Abdul Wali was still limping from his leg wound at the time he was released.

Fleming and his men refused to buy food and water for the arbakai. Unaccustomed to the austere living conditions on the qalat, they would themselves drive to FOB Fortress to eat in the chow hall and take showers there, or to pick up meals, according to U.S. soldiers and Afghans who remained there. Fleming was the same captain who had asked Jim details about a supposed chow hall and gym on the qalat.

Seeing the growing divide, Jim's U.S. soldiers who remained there grew increasingly worried about some sort of retaliation by the Afghans.

Frustrated, Fernando went to speak with Fleming.

"I don't recommend you guys throw that divide in there," he said. "What we did with Major Gant worked. We slept with them, ate with them, trained with them, fought with them—what you are doing is offensive for us. You are Special Forces—isn't this what you signed up to do, to train indigenous forces? Why aren't you doing it?" Fernando asked.

Fleming was particularly suspicious of Ish, a miscalculation that stunned Fernando, as Ish was doing everything he could to avert a disaster at Chowkay.

When Ish called Fernando, Fleming grew irate.

"Why is he calling you?" Fleming asked Fernando. "Why does he have so much control?"

"Dude, don't you realize these guys are about to walk?" Fernando asked him.

"Ish is the only guy who can glue these guys back together," Fernando said. "He is willing to help keep them on board."

Worries about the growing schism on the qalat were so pronounced that some of the U.S. team members voiced fears for their safety. Soon after Jim and Dan were removed from their positions, the two Special Forces soldiers on the team, Staff Sgt. Ed Martin and Sgt. 1st Class Tony Carter, whom Jim was recommending for an award for valor, were recalled from the Chowkay post.

As they prepared to leave one morning, Ed and Tony pulled Ish aside.

"You know the lieutenant is coming to pick us up and take us to Penich," Ed said, referring to Roberts. "I know you guys are really pissed off, but please don't attack, because we will be in that convoy," he said.

"Are you crazy? Of course we won't attack!" Ish replied.

Meanwhile, there were no Taliban attacks on the qalat, but the observation posts up on the ridgeline were hit often by insurgent fire. Fleming's team did not fire in support of the posts, and did not provide adequate supplies—including ammunition—for the arbakai who were manning them, according to members of the arbakai, Ish, Sadiq, and Zia Ul Haq.

Sadiq and the arbakai were at the end of their ropes. Soon afterward, the arbakai abandoned the observation posts, and insurgents started again shooting directly at the qalat. One of Fleming's men was shot, as was a young Afghan girl who lived in the village.

"They will not stay in Chinaray," Sadiq said.


ON THE MORNING OF March 16, Jim was escorted to a sterile, Bagram clinic. Doctors drew his blood and took a urine sample for mandatory alcohol and drug tests—both returned negative. Then he went to a command-directed psychological evaluation.

The psychologist, an Air Force lieutenant colonel, peered across a desk at Jim and quizzed him on his drinking habits and whether he was suicidal.

"I was in combat. I drank and took pain medications," Jim said matter-of-factly.

For much of the rest of the day, he was left alone.

Then suddenly, at about 7:00 p.m., Corley burst into Jim's room.

"Something is wrong. You need to see Linn right away," he said.

Corley rushed Jim to Linn's office. There, Linn stood behind his large desk. Normally cool and collected, Linn was visibly shaken.

"We have information that you are planning attacks on VSP Chowkay!" Linn said, referring to the Tribe 34 qalat. Linn apparently had received intercepted conversations from arbakai in the observation posts above the Chowkay qalat, in which the arbakai voiced anger at the U.S. military for taking Jim away. That, coupled with the paranoia of Fleming's team, led Linn to make the unfounded accusation. Linn handed Jim a piece of paper—a counseling statement—that said Linn had reason to believe Jim was planning and coordinating attacks on the qalat in Chowkay. Linn then told Jim to sign it.

Jim felt as though he had been punched in the chest, recoiling even as his heart filled with rage.

"You must be out of your minds!" he said. "I'm not signing that shit! I would never plan attacks on U.S. soldiers. I would die for those men out there," he said as tears of anger began streaming down his face. "I would die for the men out there now."

The room was silent.

Jim struggled to regain his composure. The thought struck him that his commanders had an utterly distorted picture of him—they had no comprehension of what he had done in Chowkay.

"I asked the Afghans in Chowkay to protect those soldiers, and if you let me, I will call right now and reiterate that," Jim said.

Linn agreed, and an aide brought Jim a phone.

Jim called Ish, while Linn and the others listened in.

Ish answered the call in his room in Jalalabad. I was sitting beside him. My heart leapt when I heard Jim's voice. I put my ear close to the phone and listened. I had no idea who else was with Jim, but given the protective order, I dared not speak.

"Haji Jan Dahd and Haji Jan Shah know that they have to protect those soldiers, don't they?" Jim asked Ish.

"Yes, sir, I told them," Ish said.

"I need you to tell them again," Jim said. "You must make sure they know they have to keep them safe."

His voice was urgent, almost desperate. I realized if anything happened to one of his men, Jim would never recover.

"How are you, sir?" Ish queried.

"I am all right, brother. Please, just do what I asked," he said.

"I will go to them tomorrow, sir."

Ish hung up the phone, and immediately made plans to return to Chowkay to see the tribal leaders on Jim's behalf.

In the office, Linn looked at Jim, his eyes narrowed.

"Everything in my body is screaming at me not to believe this is true," Linn said. "But I am going to resist that and accept what you say."

Linn reasserted that Jim was to be under escort at all times, and sent him away.

THE NEXT MORNING, ISH made the dangerous trip alone back to Chowkay, where he took Capt. Fleming to meet with Jan Dahd and Jan Shah.

"The captain is so worried that the ALP will put down their weapons," Ish explained. "They need to understand that you will protect them with your lives."

The Safi tribal leader stroked his long gray beard.

It was then a matter of Pashtunwali, and specifically of nanawati, by which a person must grant shelter and refuge to anyone who enters his house—even an enemy. These new Americans, however distasteful and culturally ignorant, were living on his land, with men from his tribe. He was duty bound to be hospitable and protect them, and especially if Jim, through Ish, was asking him to.

"You are new here," Jan Dahd said. "At one time, we didn't know Jim. He was an American like you. He was so good with us, asking our opinion about everything. That is why we miss him. We want you to do the same things, so when you leave we will miss you, too."

Fleming appeared relieved.

Then Jan Shah told Ish that he and a large group of elders were going to see Konar governor Fazlullah Wahidi and after that Afghan president Karzai to ask for Jim's return to Chowkay.

"I don't think that is a good idea," Fleming told Jan Shah.

"You should be proud," Jan Shah countered. "Jim is an American and it's not every American that we support like this."

A day later, Jan Dahd, several ALP commanders, and about forty tribal elders from Chowkay and Khas Kunar descended on the office of Wahidi in Asadabad to appeal for Jim's return. Wearing blankets and turbans, they sat in a circle in chairs set out on Wahidi's lawn and stood in turn to speak on Jim's behalf.

"Jim was the only American who really wanted to help the Afghans," said Jan Dahd, who sat next to Wahidi. "If the Americans want to take their forces out of Afghanistan, they should take out the Americans who are killing innocent civilians, not the ones who are helping us.

"Jim knows the situation in Khas Kunar and Chowkay. He trained the ALP, and if you don't send him back, the ALP may quit and never work with the Americans again. We are really happy with Jim. We have security because of Jim. We want him to come back, or everything will fall apart."

Wahidi was impressed. "I have never seen so many tribal elders come and ask me about bringing back an American," he said.

The group was planning to arrange a meeting with Karzai, but Ish asked them to hold off because of concern that at the time such a meeting could spark a worse backlash by the U.S. military against Jim and Dan.

AS AFGHAN ELDERS CLAMORED for Jim's return, Lt. Col. Kirila, the man leading the probe of Jim and Dan, sat down to draft the report of his Article 15-6 investigation.

> MAJ Jim Gant had been in Afghanistan for 22 months straight.
> In that time, MAJ Gant has performed very well in support
> of CJSOTF-A Operations. He is also a Silver Star winner and
> has been awarded for valor on two other occasions. It is the
> opinion of the Investigating Officer that his previous successes
> in Iraq and his mission success in Mangwel discouraged others
> from questioning his increasingly erratic behavior. In fact,
> it is my impression that his incredibly emotional response to
> the Suspension from Duty letter from COL Schwartz was
> indicative of a sense of relief that his Afghan odyssey was over.
> The resultant situation is one of considerable disappointment
> for the Special Forces Regiment as well as the officer and the
> men he led.

Among other recommendations, Kirila called for a "detailed Psychological Assessment for MAJ Jim Gant and possible re-evaluation for

TBI. The 16 JAN IED strike that MAJ Gant experienced may be a contributing cause to some of the observed symptoms of inability to sleep and extreme mood swings."

ON MARCH 20, JIM was taken for his second command-directed psychological evaluation.

The following day, his escort came to his room to take him to the office of the Criminal Investigation Division (CID) at Bagram.

"Get your uniform on and come with us," the escort said. They walked Jim to the CID office, where a CID agent read the charges against him. Jim said he would not answer any questions, and reserved his right to consult a lawyer.

Then the agents took him into a coffee area where young enlisted soldiers were socializing, had him take his shirt off, and photographed him and his tattoos—for identification, they said.

His headache got worse. The Internet was down. He was cut off from the world. He felt incredibly spent and off-kilter. His mind was trying to wrap itself around what was happening. Like a Global Positioning System, it kept loading and loading, but was unable to get a read on where he was—or where I was.

"I feel like I have lost you," he wrote to me in his blue journal, "and that all we had was a dream in a distant world, in the deepest ocean, in the farthest space."

IN JALALABAD AT NIGHT, I felt intense loneliness as I listened to the wind that marked the coming of spring rattle the dusty storefronts, billow through the curtains, and howl into the big, high-ceilinged room I stayed in at Ish's house.

The house was dark, the wind alive, like a spirit swirling around me. I could not sleep. I missed Jim terribly. As the dawn drew closer, I listened to the clip-clop of donkey hooves and rickshaws passing on the road, alert for the sound of heavier vehicles. The atonal voice of a mullah rose from the nearby mosque with a call to prayer.

By day, I had a front seat to the tragedy that was unfolding all around.

I spoke with Ish, Abe, and Imran and through them interviewed as many Afghans as I could about the events in Konar. Jan Shah came to visit the house, and I spoke with him for hours one night about his life and that of his father, as well as the Safi tribe's reaction to the removal of Jim. From brief phone calls from Dan and other U.S. soldiers, I pieced together what was happening with the investigation. I had only fleeting news about Jim.

But while he was isolated in a virtual prison behind the walls of the U.S. military base, I was in a position to fight, for all of us.

As I worked every day to stay free of U.S. authorities, to safeguard my material, and to report the rest of the story, I felt incredibly fortunate to be surrounded by my Afghan family—under the protection of Abe, Ish, and Imran, who treated me as their sister.

Each morning as we drank milk tea and ate flatbread together, I gained the strength to face whatever lay ahead. I was welcomed as a guest but also made to feel like a member of the household. The female relatives oiled and braided my hair and decorated my feet and hands with henna. We spoke Pashto together and I taught them some English words. I watched them cook, and learned about the division of labor in their huge household. I played catch with their children and showed them how to ride a bicycle in the courtyard. I experienced for a short time the sequestered life of an Afghan woman, remaining indoors virtually at all times. Only at dusk was I allowed to climb the stairs to the roof and peek out, my head and face covered by scarves, and drink in the view of distant green fields and mountains.


ON THE NIGHT OF March 24, Schwartz summoned Jim to his office.

Under escort, Jim arrived in the headquarters building, in the hallway that was lined with plaques to the fallen Special Operations personnel. He immediately noticed that the faulty award for his late teammate, Chris Falkel, had never been corrected in spite of him asking for it twice before. It infuriated him.

Jim walked into the office, looked straight at Schwartz, who was sitting behind a big desk, and saluted.

"Major Gant reports as ordered," he said.

Schwartz did not even look Jim in the eye. Instead, he began reading from a piece of paper.

"You are a disgrace to Special Forces," Schwartz began, and proceeded to dress Jim down in a loud and aggressive voice. "I am recommending a general court-martial, and the revocation of your Special Forces tab."

Jim was to return under escort to Fort Bragg, where he was to report to the United States Army Special Operations Command. The no-contact order with me remained in effect.

INDIRECTLY, I GOT WORD that the Article 15-6 investigation against Jim and Dan had been completed. About ten days after I arrived in Jalalabad, I started making preparations to get out of the country. One obstacle was that I had no entry stamp on my Afghanistan visa. Abe began calling his contacts in Kabul airport security forces to help me slip through.

One morning I said goodbye to Imran, Ish, and the rest of their family. Then Abe, Shafiq, and Abe's cousin, the mercenary Basir, loaded my bags into their small navy blue hatchback. I pulled on my burkha and got in the backseat. We drove through Jalalabad toward Kabul, along one of the most dangerous roads in the country, in part because of eighteen-hundred-foot cliffs. We followed the Kabul River, and after some time the road began to climb steeply as we went through the Kabul Gorge, known for insurgent ambushes. We passed by the carcasses of destroyed Russian tanks and also some U.S. military vehicles.

As we got closer to Kabul, police checkpoints became more frequent. Abe had allowed me to lift the head covering of my burkha for some stretches but warned me to replace it before each checkpoint. We managed to pass through all of them without a problem and then drove into the streets of Kabul, which were relatively deserted because of a holiday.

After buying my plane ticket and having lunch, we pulled into the airport parking lot with a short time to wait. We had decided that I would try to talk my way through the passport inspection, as I had done that once before. My worry about getting through security was eclipsed by the gut-wrenching sorrow of having to say goodbye. I realized what I needed to do, to make it all right. As Abe walked me alone to the airport terminal, I took from around my neck the Spartan shield that Jim had given me in Washington more than two years before.

"This is from Jim and from me," I told him as I placed it in his hand. "You have proven your bravery and your friendship to us every day. You are our brother and you have protected me with your life. We love you and we will never forget you. I have worn this shield for a long time, and before me Jim wore it, too. I am not giving it to you, I am lending it until we see you again."

Tears began running down my face, and I could not stop crying.

"Thank you. I know what this means," Abe said. "It will be okay. I will see you soon."

Abe took me as far as he could, and watched me as I walked to the glass-windowed booth of the passport officer. After a few minutes of tense discussion, the officer stamped my visa with a loud thud and let me through. I walked through and did not look back. I had so many things on my mind, but one stood out. Jim had asked me dozens of times if I would be there for him when he returned to the United States and got off the plane. He'd arrived before and found no one, and was terrified it would happen again. "Don't be late," he would tell me.

I would be there.

# CHAPTER 34

THE BLACK TRACFONE ON the bedside table rang, waking Jim from his dream—a recurrent one about walking in the mountains. It was 5:00 a.m. on March 29. For a moment, in the darkness, he had no idea where he was.

"Hey, I'm about twenty minutes away," I said over the phone. I was driving Jim's black Toyota pickup truck down I-95 in the woodlands of North Carolina, trying not to speed. A state trooper in Virginia had pulled me over for going 89 mph in a 70 mph zone. When I told him I was going to meet my fiancé, who had just returned from Afghanistan, he looked at me knowingly, alerted me to other police down the road, and waved me on.

"Okay, baby," Jim said. "Be careful."

Jim put down the phone, sat up in bed, and rubbed his hands over his clean-shaven face. He was in the upstairs spare room of another Special Forces officer's house on a quiet cul-de-sac in Fayetteville, North Carolina. He had arrived in the United States from Afghanistan under a military escort three days earlier, and was confined to the area around Fort Bragg. Jim faced a possible general court-martial and, in the worst-case scenario, prison time. The Army protective order prohibiting him from contacting me was still in place, and we had no idea whether it would be lifted. But risk had always been the norm in our relationship, and we decided to break the order.

"I can't handle this anymore," I told him late the day before. I was standing in the front yard of my Bethesda house. I had been back from Afghanistan for about a week, but I was reminded of what seemed like an endless separation from Jim by the yellow ribbons, faded from almost two years of sun and rain, that hung from the posts of the wooden porch. All that time apart, I had lived with an ache in my chest—it fluctuated between dull and sharp but never went away. At times, I was so sure Jim had been killed that I stayed away from the front of the house, afraid of looking out the window and seeing uniformed men coming down the walk. It had grown harder, not easier, to cope with the pain of separation. After everything that had transpired, it was crushing me.

"I am not okay," I said, my voice trailing off.

Jim didn't hesitate.

"Come on," he told me. "Come to me, baby."

Several hours later, it was still dark when I drove up the cul-de-sac and walked into the house, as the door was left open for me. As a precaution, he did not come outside to meet me. I knew the house and walked to the foot of the stairs that led to the study and upper bedroom. There in the shadows, I saw Jim.

He looked like a different man with his face soft and beardless—it was the first time I had ever seen him that way. His head was tilted downward, and he stood still, waiting for me to come to him. I rushed up the stairs and we embraced. He kissed me, his smooth skin unfamiliar to the touch, but his scent and arms around me comforting as always. Tears filled our eyes. He told me again and again that he loved me, and asked me over and over if I would still marry him.

"Yes. Yes, I will," I answered.

The world around us disappeared as we lost ourselves in each other.

As we lay together, he asked me to recite a verse for him. Ever since our first week together, he had loved to listen to me tell him poems by memory.

I took a breath, and painted for him a favorite scene from Homer's epic *The Odyssey*: Odysseus's reunion with his wife, Penelope, after the Greek king's ten-year struggle to return to Ithaca after the Trojan War.

> *The more she spoke, the more a deep desire for tears*
> *welled up inside his breast—he wept as he held the wife*
> *he loved, the soul of loyalty, in his arms at last.*
> *Joy, warm as the joy that shipwrecked sailors feel*
> *when they catch sight of land—Poseidon has struck*

*their well-rigged ship on the open sea with gale winds,*
*and crushing walls of waves, and only a few escape,*
*swimming, struggling out of the frothing surf to reach the*
*shore, their bodies crusted with salt, but buoyed up with joy*
*as they plant their feet on solid ground again,*
*spared a deadly fate. So joyous now to her*
*the sight of her husband vivid in her gaze,*
*that her white arms, embracing his neck,*
*would never for a moment let him go.*

Jim smiled weakly. He was depleted, having returned from Afghanistan only a few days before. After a silence, he began to speak about what had happened.

"Ann, this has changed everything," he said. "They have taken my honor. They took everything from me but you," he said. He felt he was no longer a warrior; he had been stripped of his rifle and his ability to fight for his soldiers and his Afghan friends and family.

"But through all of this, you have been there. You know that other person," he said.

"Yes," I said. "I know him."

I looked at the walls of the room we were in, and in a flash I was back in Mangwel. It was a warm summer night, and Jim and I were resting in our room in the shipping container.

"Do you see these walls?" Jim had asked me, looking around that room in Mangwel. "They are keeping many things away from us. One day, maybe it will be five years from now, or maybe ten, we will be somewhere and we will miss these walls, this room, this time. You will miss the man I am here," he told me. "And I will miss the person you are here, too."

In coming days and weeks in Fayetteville, those words resonated in my mind as Jim's crisis of identity deepened. We stayed together out of sight much of the time. To our relief, the protective order separating us expired and was not reinstated.

One afternoon, sitting on the floor in the one-bedroom apartment he was renting, Jim said he no longer felt that his names—Jim to his friends, Kirk to his family—suited him. He decided to adopt the Spartan lambda symbol as an alternative, and started signing his emails with "Λ."

He did not tell me at first, but I saw the symbol and understood. From that day on, at his request and out of respect for his feelings, I was unable to call him by any name, and only wrote to him as "^." I found myself strangely speechless in the course of our daily life, unable to call him to dinner or get his attention from another room. It was a painful void, one we both felt.

Not only did he no longer know himself, he was utterly detached from his surroundings. He grasped for ways to explain it.

"It's like watching a movie that is in color, but my eyes only see black and white," he said. "It is as if I have a sharp knife, but it cannot cut my hand. I have a dull knife that will slice right through me."

AT FORT BRAGG AND within the Special Forces community, wild rumors began circulating about Jim. Even before he was pulled out of Konar, gossip had spread about Jim inside the USASOC headquarters, a multi-story high-security building surrounded by a grassy parade ground and pine trees.

"There is a belief in that building that you went COL Kurtz and went totally native," a Special Forces officer wrote to Jim in early March, referring to the rogue Green Beret colonel in the 1979 Vietnam War epic film, *Apocalypse Now*. "There is even a rumor around USASOC headquarters that you married an Afghan girl."

As word of his downfall spread, Jim's already exaggerated mystique began taking on larger-than-life proportions. Detractors and desk jock-eys jealous of Jim smelled blood. One story making the rounds was that he was the ringleader of a drugs, prostitution, and gunrunning opera-tion in Afghanistan. When Jim was flown back to Bragg under escort, the Special Forces officer sent to pick him up at the airport was advised to take at least three men, or better yet military police. "They said he was a maniac," said Maj. Travis Worlock, the headquarters company commander for USASOC. "It was horribly blown out of proportion."

In April, Jim obtained a copy of Lt. Col. Kirila's complete Article 15-6 investigation into the alleged misconduct by him, Dan, and the rest of his team.

Reading the two-hundred-page document, which contained dozens of photographs taken as evidence from our room in the shipping container, was surreal. The investigation contained facts but also many false or inaccurate statements. It recognized the achievements of Jim and his team, but also created a sensationalized, tabloid picture of Jim's misdeeds.

"The activities of ODA 3430G in the Konar since January 2012 must be characterized as being operationally and tactically significant successes," Kirila wrote. "However, there have also been a series of unethical activities."

He concluded that Jim had violated General Order 1B by possessing alcohol and controlled medications and consuming alcohol—which was true—as well as by having "sexually explicit signage and devices as well as unexploded ordnance" in his room. He did have some mortar rounds and grenades in our room. The "sexually explicit device" (not devices) mentioned was a gag Christmas gift sitting packaged in a box under his bed—a ridiculous, sixteen-inch-long dildo. A photograph of it, in its box, was contained in the investigation. Similarly, the "signage" consisted of a couple of raunchy jokes.

Although Jim had clearly broken the rules by drinking, by those standards the vast majority of Special Forces teams would have to be removed immediately from the battlefield.

"Eighty to ninety percent of teams downrange have alcohol," said a senior Special Forces noncommissioned officer. "It is rare not to." The fact that alcohol is epidemic among deployed Green Berets is common knowledge inside the Special Forces community. "I would say it is more like a hundred percent of teams," said a field-grade Special Forces officer who, like Jim, had just returned from Afghanistan. "Most teams have a knack for acquiring alcohol," he said. It was also well-known that a significant but smaller percentage of Special Forces soldiers used steroids while deployed. Many teams also hoarded prescription narcotics in case of injuries, said another Special Forces officer.

My own firsthand experience as a reporter had taught me that drinking was practically the norm on Special Forces teams. Every Special Forces unit that I had ever spent time with in Iraq and Afghanistan possessed alcohol.

Similarly, pornography and explicit signs were prevalent in Special Forces camps. Kirila had only to walk a short distance from Jim's room to another barracks that was covered with photographs of naked women. No one was charged with misconduct for that, however.

"There is a huge discipline problem within the force" after a decade of war, said a senior noncommissioned Special Forces officer. "It's worn-out," he said. As a result, the number of Special Forces tab revocations has increased significantly in recent years, he said.

Kirila found that, as Roberts alleged, the team did routinely provide the Afghan Local Police with fuel for their Ford Ranger trucks, despite a directive that the police are supposed to obtain fuel from the Afghan government. However, Kirila noted "it is a common practice throughout Afghanistan for BSO [battlespace owner] and SOF [Special Operations Forces] assets to provide fuel for ALP in support of mission objectives." He said that "the allegations of improper disbursement of fuel are technically correct; however, it has been CJSOTF-A practice to never let a mission fail because the Afghan logistics process or institution was unequal to the task. A preliminary estimate of this situation yields a prevalence of this activity."

Similarly, Kirila discovered that Roberts's accusations that Jim had falsified the number of workers on the camp were incorrect, and resulted because Roberts did not know about the workers at Chowkay.

Roberts was also wrong in accusing Jim of failing to report his injuries, Kirila said. "The investigating officer was able to find the concussion reports and found them to have been submitted to higher appropriately," he wrote.

Overall, Kirila found that "1LT Roberts's initial sworn statement although mainly accurate, reflects a measure of speculation and failure to adequately check all his facts prior to writing his initial statement. Subsequent investigation revealed that some of his allegations were inaccurate or at a minimum not based upon evidence."

As for me, Kirila found indications that Jim had an "inappropriate relationship" with me. As evidence, the investigation included a photo of a handwritten sign above our bed in which Jim told me he loved me, and photographs of me including two in which I was partly nude. The investigation incorrectly asserted that Jim was married, despite the fact that he divorced in 2010. Kirila reported that I had lived with Jim in Konar and traveled with the team, but much about my status remained unclear, he acknowledged. "According to Honeylee Tembaldor, US Embassy—Afghanistan, Ms. Tyson has no authority to be here, however her status is questionable."

In fact, I did have U.S. government approval to work in Afghanistan.

Unknown to Kirila, I had official letters of authorization from the Department of Defense as part of my contracting work conducting educational assistance and outreach for rural women in Konar. I was employed in that work by Dr. Dave Warner, who operated the Taj in Jalalabad and worked for the company MindTel. I was also supported in those activities as a volunteer for the Rotary Club of San Diego.

Despite its weaknesses, the Article 15-6 investigation served as the basis for Schwartz and other commanders to crack down on Jim. With each successive action, it seemed the accusations against Jim grew more spectacular.

Schwartz and Linn wrote a new "relief for cause" Officer Evaluation Report (OER) for Jim, replacing an OER a month earlier that had praised his work. In February, after Schwartz ordered Jim to carry out the challenging and deadly assignment in Chowkay, he described Jim as "an outstanding officer who has performed brilliantly in Afghanistan." "Jim is a true combat leader, and his passion for the Village Stability Operations mission is unmatched by any officer in the CJSOTF-A. His unprecedented success conducting VSO, as well as his ability to articulate his mission to senior leadership, has had a strategic impact on our operations in Afghanistan," he wrote. Schwartz may have disliked Jim, but given Jim's performance he had little choice but to recommend him for promotion to lieutenant colonel. He said Jim should be placed in "the most demanding assignments in the Special Forces Regiment, preferably in combat."

But after the March investigation, Schwartz completely changed his tone to one of shock and disgust. "MAJ Gant violated every value that the Army stands for," he said. Breaking General Order 1B was "despicable" and "inexcusable," he said. He referred to me as Jim's "paramour" and cast my presence as a strategic risk. "By providing his paramour unimpeded access to classified documents in a combat zone, MAJ Gant compromised the US mission in Afghanistan," he wrote.

However, during my years as a Pentagon reporter, I had been given access to classified information by military personnel on countless occasions. When Petraeus was the top commander in Iraq, for example, he allowed me and other reporters to sit in on a highly classified briefing at his Baghdad headquarters that lasted about an hour—all I had to do was sign a nondisclosure agreement. Then as well as with Jim, I had not compromised that information. Moreover, the harsh conditions in which we were living in Mangwel and Chowkay made it impossible to secure classified materials in accordance with Army regulations. Afghans and Americans without security clearances had access to those materials, as they did at other remote Special Forces outposts.

The officer evaluation also accused Jim of putting the lives of his U.S. soldiers and Afghan partner forces at extreme risk by using members of his team to "facilitate" my movement in Afghanistan.

Yet while I traveled with Jim and his team on many occasions, he never conducted a mission with the sole purpose of transporting me somewhere. Any risk involved with those missions would have existed whether I was there or not. One could argue that my presence with Jim's team on patrols in villages and other missions increased the safety of the team because it demonstrated Jim's great trust in the Afghans who were protecting us.

Nevertheless, based on the Kirila findings, Schwartz recommended in late March that Jim be court-martialed, stripped of his Special Forces tab, and forced out of the Army.

A few days later, Brig. Gen. Christopher Haas moved to do just that.

On April 4 at his command in Kabul, Haas signed a memo initiating the revocation of Jim's Special Forces tab. In addition to the charges related to drugs and alcohol and having an inappropriate relationship with me, Haas added something new: Jim, he said, had used "unconventional" tactics against the enemy.

"During your time in command, you purposefully and repeatedly endangered the lives of your Soldiers. You taught, and ordered executed, unconventional and unsafe 'figure-8' immediate actions in response to enemy contact. You painted inappropriate and unauthorized symbols on Government vehicles, painted the symbol on your vehicle a different color, then challenged the enemy to try and kill you without consideration to your Service Members' lives or well being. You sent 'night letters' to the enemy, further drawing dangerous attention to yourself and subordinates. These are the same Soldiers that you have the duty to properly train, mentor, lead and most importantly, defend."

The irony was that Jim, while conducting highly aggressive operations, had never lost a man. In his view, using unorthodox, provocative tactics was the best way to outwit insurgents and keep them on their heels. The "unauthorized symbols" Haas referred to were Spartan lambdas.

As charges mounted against Jim, Dan and others who had served under him were being drawn into a widening witch hunt by the command in Afghanistan. Dan had been sent back to Fort Bragg in April, but the rest of the team remained in Afghanistan.

Dan was accused in the Kirila investigation of having a beer on Christmas, and hiring and paying Afghan contractors to fortify the qalat at Chowkay without awaiting the proper financial authorizations.

What the investigation failed to recognize was that Dan's employment of contractors was critical to helping keep us alive at Chowkay. Jim's team was ordered to rapidly embed in Chowkay but not advanced any funds to do so. Had Dan not hired contractors to work and promised to pay them later, for example, we would have gone weeks without even a gate or finished walls on the qalat. Funding delays were a chronic and longstanding problem faced by Special Forces teams. Indeed, soon after his arrival in Afghanistan in 2010, Jim had criticized the cumbersome accounting process that tied up team members for weeks drawing funds. In a memo to then top commander Petraeus, Jim had proposed creating "money teams" to streamline the process and deliver the funds to the ODAs. Petraeus endorsed the idea, but the teams never materialized.

On the charge of possible misappropriation of funds, the investigation turned up only some blank receipts in Dan's room and statements that he and Imran fabricated one vendor's signature on a receipt. However, according to Dan the blank receipts had been left in the operations center by a previous Special Forces team, and their use was common. Kirila also found that a "thorough 16-month review of all the previous fiscal clearing documents associated with 3430G and CPT McKone found no obvious discrepancies." According to some officers, it is common for Special Forces teams to engage in the misappropriation of funds—stating for instance that they are buying gravel or lumber but actually spending the funds on other things that were needed, or simply pocketing the money. "Everybody who has been a team leader in Afghanistan has done that fraudulently," according to a field grade Special Forces officer.

One of the main complaints lodged against Dan, Fernando, and other noncommissioned officers and soldiers on the team was that they had not reported on Jim.

"CPT McKone was fully cogent [*sic*] of the misconduct by the ranking officer," but "failed to report the conduct," Linn stated in March in a highly critical officer evaluation of Dan. "CPT McKone lacks the morale [*sic*] compass and integrity expected of an officer in the United States Army and therefore, it is my assessment that CPT McKone should not remain in the United States Army."

Unlike Jim, Dan had not completed twenty years of service, and so being forced out of the Army would mean losing his retirement benefits. He decided to fight the charges at every turn. Meanwhile, Dan was diagnosed with PTSD and a serious back injury and advised he could not return to combat for medical reasons. The Army could have simply allowed Dan to go through a medical evaluation board to determine if he should be separated or retired for health reasons. Instead, it revoked his Special Forces tab and moved to force him out by using a formal Board of Inquiry, composed of three senior officers who would rule on whether he should be allowed to stay in the Army.

In December, the Board of Inquiry convened. An Army Judge Advocate presented the government's case against Dan, claiming he was a renegade officer unsuited to fight the nation's future wars. "How can we retain such officers when we are trying to teach them the rule of law," the Judge Advocate argued—"them" referring to foreign peoples U.S. troops would encounter. At one point, one of the three board members asked Dan why he was contesting the charges against him. By then Dan was angry. He had nothing to lose. "I am here to salvage my dignity and my honor," Dan shot back. When the team moved into Chowkay with no advance funds, Dan said, "we had to be ready to fight. We didn't want a situation where we lost nine people. I kept asking the pay agents when the money would come in," he said. "The bureaucratic systems were designed to make sure people are not pocketing money, buying Harleys, or anything else that anyone in the room knows has happened. The problem is that it gets in the way of the combat mission." After five hours of proceedings, the board reached a verdict: Dan was guilty of one act of misconduct, drinking a beer on Christmas 2011 in Mangwel. In a minor vindication, the board voted to retain Dan in the Army.

Fernando, the skilled intelligence sergeant on the team in Chowkay, was reprimanded by Gen. Haas for being "oblivious to what was occurring" at Chowkay, Penich, and Mangwel. "You are hereby reprimanded for dereliction of duty due to your complete lack of awareness," Haas wrote in the memo, signed April 3. "It is disturbing that you claim so much ignorance about the blatant and deliberate violations . . . which were occurring on the VSP on a nearly daily basis." Haas offered no specifics as to what Fernando was ignorant of. In fact, Haas showed a lack of awareness in his own letter—Fernando had never served in Mangwel. Nevertheless, the reprimand had a serious negative impact on Fernando's military career.

Fernando had been moved to a job at the U.S. military base at Bagram when he learned of the action against him from Sgt. Maj. Edward Mongold, the senior noncommissioned officer in Linn's command.

"You were in Chowkay. You are being relieved of your duty and will be leaving the country," Mongold told Fernando in a meeting in his office. "This is being command-directed by Haas. You are not the only one. It is everybody who was there."

The Army also took action against Tony, the weapons sergeant from the 19th Special Forces Group whom Jim had recommended for an Army Commendation Medal for valor for his bravery manning the .50-caliber machine gun on the roof during the Chowkay fighting.

Even three lower-ranking soldiers who had served with Jim at Chowkay were reprimanded, according to Fernando, Bird, and others. Most of the infantry soldiers were moved from Chowkay to the Special Forces headquarters at Jalalabad, where they were treated as though they had been indoctrinated. "They said we were brainwashed, but we were sad because they took our commander from us," said Pfc. Chad Armstrong, a twenty-four-year-old from Grand Rapids, Michigan, who worked at the operations center in Chowkay and had the courage to volunteer to be Jim's driver. The Chowkay soldiers voiced resentment against Lt. Roberts for turning in Jim and Dan.

For Jim, watching the Army go after Dan, his closest comrade, and others who had served under him was the worst punishment he could face. He felt responsible for whatever harm came to them and their careers. Once again he was unable to protect those who had followed him and sacrificed for him and the mission.

Roberts, meanwhile, was hailed by the chain of command as a whistle-blower and paragon of moral courage.

ON JUNE 11, JIM got up early, shaved, and pulled on the only uniform and boots he had—the oversized ones he was ordered to wear at Bagram Air Field in Afghanistan.

Then he climbed into his truck, drove through the gate at Fort

Bragg, and turned up a side road to the headquarters of the United States Army Special Operations Command (USASOC), where he had been summoned to a meeting with its three-star commander, Lt. Gen. John Mulholland.

After Schwartz recommended that Jim face a general court-martial and sent him from Kabul back to Fort Bragg, responsibility for Jim fell to Mulholland. A big Irishman known among other officers for his stubborn streak, Mulholland knew Jim and had visited Mangwel and met Noor Afzhal in 2011. He also facilitated Dan's assignment to Jim's team. I was acquainted with Mulholland. I had first met him in 2002, interviewing him about a major battle in Afghanistan when I was a novice Pentagon reporter and he was a Special Forces colonel. Later, I reported on Mulholland's soldiers as they trained Afghan commandos and earned awards for valor.

Now Jim's fate lay in Mulholland's hands. Still, we had no idea how Mulholland would rule on the case, or what kind of internal pressure he faced.

Jim pulled into the parking lot outside the towering USASOC headquarters, and stopped to gather his thoughts. A light rain fell on the windshield and pavement.

As he stared out the window, Jim recalled the anticipation he had felt three years earlier, when he arrived at the same building just after the publication of "One Tribe at a Time" for his teleconference with Special Operations Command (SOCOM) chief Adm. Olson.

He had left the meeting with Olson elated, and set off to help win a war.

He shook his head.

*You were a fool to believe them*, he thought.

Jim wanted to lash out at his accusers, but he did not. He had long planned to retire from the Army after he returned from Afghanistan, knowing he would never have a mission that could compare with his last. So he drew a line in the sand. He let Mulholland know through subordinates that he would be willing to retire as a major with an honorable discharge and full benefits. Otherwise, he planned to demand a trial by court-martial, something he believed the command wanted to avoid because they sought to keep the matter quiet.

Jim decided not to ask any of his high-ranking supporters, such as Petraeus, to intervene on his behalf. He felt he had let them down and did not want to involve them in his troubles. As a matter of honor, he wanted to bear the responsibility alone.

So he wrote a three-page statement, which he placed in a manila folder and tucked under his arm as he walked toward the USASOC building. At the entrance, he was barred because his security clearance had been suspended.

"Sir, you are blocked," a soldier said, requiring him to wear a visitor badge marked with "Escort Required." Maj. Worlock, who handled personnel for USASOC, escorted Jim to Mulholland's office on the top floor of the USASOC building. Outside, they waited together with Judge Advocate Col. Steven Weir until they were called in.

Mulholland was sitting behind a large desk in a spacious office furnished with leather chairs.

"Sir, Major Gant reporting as requested," Jim said, standing at attention and saluting.

Mulholland looked at him sternly and returned the salute.

"Stand at ease," he said.

Jim relaxed his stance and stood at parade rest.

"Major Gant, do you know why you are here today?"

"Yes, I do, sir," Jim said.

"I am considering whether you should be punished under Article 15 UCMJ for the following misconduct . . ."

Mulholland read the by then familiar charges against Jim: Alcohol use, possessing pain medicine and other controlled substances. In addition, he said, Jim had given me access to classified information and used his team to transport me in Afghanistan—thereby putting his men at "extreme and unnecessary risk . . . likely to cause death or grievous bodily harm."

"Do you want a trial by court-martial?" Mulholland asked.

"No, sir," Jim said.

"Do you have anything to say in your defense?"

"Sir, here is my statement," Jim said.

Mulholland took the document and began silently reading Jim's words.

> I went to Afghanistan in 2010 to try to win the war. At the
> time, I believed that defeating the insurgency in Konar

Province would prove strategically significant for the broader
war effort in Afghanistan, and that our success there could tip
the outcome in our favor. I am deeply pained that my efforts
to serve my country, my Regiment, my men, and the people of
Afghanistan are concluding in this manner.

Mulholland read on. A summer rainstorm blew in and drenched the
lush landscape outside the window behind him. He finished, quickly set
the papers down, and looked at Jim, his face tense and angry.

"Jim, there is nothing in this statement that in any shape, form, or
fashion mitigates or excuses any of your conduct on the ground," he said
curtly. "There is nothing special at all about what you did or what you
were asked to do. Absolutely nothing you did in your military career, in
particular what you did over there, matters any more. Reading your
statement, it's very apparent to me that you have a skewed perception of
yourself and your importance to the war and to the Regiment. If anything, your statement leads me to believe that you are delusional. I have
been in Special Forces thirty-three years and served with a lot of great
men, and you are not one of them. You may perceive you are, but you are
not. As a matter of fact, I am ashamed of you, and you are a disgrace to
the Special Forces Regiment."

Mulholland was sitting straight up in his chair, looking Jim in the
eye, growing more agitated by the minute as he spoke.

"Do you understand the situation that you have put me and your
entire regiment in? The politics of this are an absolute nightmare!" he
fumed, his face flushed. "Do you know what would happen if this three-
and-a-half-inch investigation became public in the *New York Times* or
the *Washington Post*? If it did, you would be personally responsible for the
destruction of eleven years of the hard work of the Special Forces.

"Do you know what the Army Chief of Staff said about Special
Forces after Vietnam? They had received nineteen Medals of Honor,
and he said they were a bunch of lawless renegades. That is what we have
been battling against ever since. It has taken us eleven years to get where
we are, and this one incident could destroy all that. How many missions
do you think Special Forces will be given if they know this is how Special Forces officers behave in combat?

"You used your top cover and access and isolation to live out a fantasy. You lived a movie and not a real movie. You made your own. It was extremely selfish. When I came and visited you I was very impressed, but there were things you said that gave me pause, and in retrospect they should have made me look at your situation differently. You directly quoted Colonel Kurtz. Obviously, there was something wrong with you.

"I want to make something very clear. You are a shame and disgrace to the Special Forces Regiment. You drank alcohol downrange. You took drugs and had unauthorized drugs in your room. You put your men at risk—you say in the statement you did not and your men would not say that, but they did—and you gave Ann Scott Tyson access to classified information and moved her around the battlefield. Is that true?"

A long pause followed.

"Yes, sir," Jim said.

Mulholland took a deep breath. "There is a famous British saying," he said. " 'The further you move into the jungle, the nicer the clothes you should wear to dinner.' You failed to do that. You let us all down," he said. "Now get out."

Jim turned and walked out.

Worlock followed him, closing the door.

"Damn, dude, are you okay? That was brutal," Worlock said.

"Yep," Jim said.

They sat down. But Mulholland was not finished yet. Behind the closed door, he was consulting with Col. Weir, the JAG, on what punishment to mete out. About ten minutes later, the general called them back in.

"Did anything I said resonate with you?" he asked Jim. "Do you feel any remorse? One of the first things I noticed when you walked in here is that you are out of uniform and you need a haircut, which makes me think you just don't care. You don't have any discipline whatsoever."

"Yes, sir, I do feel very bad about the situation and I have from the beginning. It is hard for me to get up in the morning, sir."

"Someone who loves his regiment does not do this," Mulholland said. "Your statement does not show any remorse."

Jim was silent.

"All right, Major Gant," Mulholland said, almost mockingly. "Here is your punishment. I will give you this Article 15 in your official military file. I will give you a general officer letter of reprimand. I will support your tab revocation. Do you have anything else to say?"

"Yes, sir. I would ask that you look very carefully at Captain McKone's situation. I was the commander."

"Captain McKone is a seasoned Special Forces soldier, commissioned officer, and grown man," Mulholland said. "He could have approached you or this command at any time. He could have stopped all of this with a phone call or an email. That was his duty. I am embarrassed that the officer I sent to you didn't do that, and we had to have a first lieutenant in the Infantry show that moral courage."

Mulholland looked at Jim, his face a mixture of anger and disappointment.

"Major Gant," he said, "this is a sad day."

Mulholland picked up a pen and signed the Article 15.

"You are dismissed."


SOON AFTERWARD, MULHOLLAND ISSUED Jim this reprimand:

> While fully acknowledging your record of honorable and valorous service to the Regiment, our Army and our country, the simple truth is that your subsequent conduct was inexcusable and brought disrepute and shame to the Special Forces Regiment and Army Special Operations. You were entrusted to maintain the highest standards of discipline, operational deportment, and leadership in an environment of austere conditions and high risk; the very conditions in which Special Forces is intended to thrive. Instead, you indulged yourself in a self-created fantasy world, consciously stepping away from even the most basic standards of leadership and behavior accepted as a norm for an officer in the U.S. Army. In the course of such self-indulgence, you exposed your command and the reputation of the Regiment to unnecessary and unacceptable risk. In short, your actions disgraced you as an officer and seriously compromised your character as a gentleman.

BEHIND THE SCENES, JIM received encouragement from several more senior Special Forces officers who valued his talents and considered his punishment over-the-top. "What went down with you was ridiculously shameful," said Lieut. Col. Scott Mann, a retired Special Forces officer who served three tours in Afghanistan and was an architect of the VSO program. "I consider you a modern day Lawrence of Arabia," he said. "History will reflect your impact. The [Special Forces] Regiment will recognize that one day."

Bolduc, who returned to Afghanistan in mid-2012 and oversaw the expansion and strengthening of the Afghan Local Police and VSO programs, told Jim that his work in Konar was pathbreaking. "If you had not done what you did early on, there would not be two thousand Afghan Local Police in Konar now," said Bolduc, who was promoted to brigadier general in 2013. Despite major unresolved problems with pay, logistics, communications, and financing, the Afghan Local Police are the most successful and cost-effective grassroots security force in Afghanistan, Bolduc said. As of 2013, he said, the approximately twenty-five thousand ALP patrol rural areas inhabited by about 20 percent of Afghanistan's population. The ALP are the number one target of insurgents and suffer a higher rate of members killed in action than the Afghan national army or police, he said. Bolduc helped secure Afghan government approval to increase the ultimate size of the force from thirty thousand to forty-five thousand and extend the duration of the program from 2015 to 2025.

Other higher-up officers in Jim's chain of command told him they felt collectively responsible for his undoing. "You altered the course of the war in the northeastern region of the country reaching into Pakistan," said one officer who had served with Jim. "Commanders take advantage of you over the years. We keep you out there when you obviously need a break and are exhibiting symptoms of stress that we as a system placed you under. Even after you were hit with an IED, we just left you out there. Then we punish you for the results of our own decision making."

In late June 2012, Bolduc signed an officer evaluation saying Jim had made "invaluable" contributions to the Afghanistan mission under his command in 2010 and 2011, and was a "must select" for promotion to lieutenant colonel.

In late July, the Department of the Army, unaware of the adverse actions, announced Jim's promotion to the rank of lieutenant colonel. Congratulations began pouring into his official email.

Jim, knowing the promotion would eventually be reversed, declined it—a rare step that required the approval of Army Secretary John McHugh.

A few months later, the Army acted again, this time demoting Jim for the purpose of retirement to the rank of captain.


LIVING WITH JIM IN Fayetteville, I stayed close by his side as the arrows kept flying at him, one after another.

His mind and body were battered and broken after nearly two straight years in combat. A series of medical and psychological exams unsurprisingly diagnosed him with PTSD and traumatic brain injury (TBI). Ever since I had known Jim he had shown anxiety in confined or crowded spaces, an extreme startle response to sudden or loud noises, and other textbook symptoms of PTSD. Now, I had to be more careful than ever not to walk up behind him without him seeing me; we had three close calls. New concerns emerged. His brain was not functioning as well as it had in the past; he felt confused and forgetful and sometimes had trouble connecting thoughts and emotions.

One warm, humid morning in mid-June, Jim was scheduled to have four MRI scans of his brain, hip, shoulder, and back. He doubted he could tolerate two hours strapped down in the imaging machine bombarded by loud noises, but the tests were critical, so I went with him to see if he could stand it. An Army doctor, a sympathetic Vietnam veteran "Doc" Earl Benson, sent him to a civilian facility that had open-ended MRI machines. Jim took a sleeping pill and changed into a surgical gown. Then we went into the chilly, sterile MRI room and he lay down as instructed on a table. A technician gave us both earplugs. Then she fastened a mask—similar to a baseball catcher's mask—over his face and used a cushion to keep his head straight. I held his hand, and a technician pressed a button that moved the table inside the MRI machine.

Then came the loud, jarring noise of the machine, which seemed to be mimicking gunfire.

Jim's eyes popped open. He squeezed my hand so tightly that it hurt.

I held on. He squeezed harder.

So it went, off and on, for about two hours. Finally the noise stopped.

Doctors would have to remove some small metal shards from his head, and also scrape away scar tissue that was impairing his hearing from multiple perforations of the eardrum. In addition, the exams helped diagnose severe shoulder pain—likely caused by the January 16 IED blast—that was preventing him from using his shoulder.

Jim had sometimes described himself to me as a kind of Don Quixote, riding an old horse and carrying a scratched and dented shield.

"You are my knight in armor," I would tease him gently.

He smiled. "Yes, that is exactly what I am."

We had many happy times, when the sun would break through the clouds. We were living for the summer in his old house in Fayetteville caring for his two younger children, Tristen and Scout, until school started. Jim enjoyed doting on his son and daughter, taking them to the movies and buying them their favorite foods. He signed up online for a minister's certificate and performed a wedding ceremony for Tony Siriwardene, his comrade from ODA 316, who married the sister of another 316 member, Mark Read. We laughed and cried together with Dan and reminisced with him at the dinner table. We started watching the Academy Award Best Picture-winning movies of the past fifty years. Most days we worked out at the gym or went running together on the rust-colored dirt trails around Fort Bragg. In the warm summer evenings, Jim would ask me to put on a dress and go out to shoot a few games of pool with him in local bars. Or he would play music and slow dance with me in our room. He liked to listen to me play the piano, never mind that it was out of tune and had some broken keys.

But all around us, darkness kept pressing in. Professionally, physically, and emotionally, Jim felt beaten down. He was so angry, depressed, and exhausted.

One day, he was watching one of his favorite video scenes, the boxing match between Muhammad Ali and Joe Frazier for the world heavyweight championship in Manila in 1975. Ali won the match by answering the bell for the fifteenth round, when Frazier could not.

"All I have to do is stand up, and I will win," he told me. He would stand, stagger, fall, and get up again.

He needed to rest, but he could not—apart from his own battles with the Army, he felt obligated to help Dan, Fernando, and other comrades who were struggling against the Army bureaucracy.

Jim was drinking more heavily, including just prior to his mandatory appointments in the Army Substance Abuse Program known as ASAP. He knew he needed to cut back or stop. "Just not yet," he would say. "I know this may sound bad," he told me. "But right now, alcohol is like my friend."

Outwardly, he voiced confidence that he would recover, but he admitted to those closest to him that he might not make it. He tried to find a psychologist in the Army and met with two or three but encountered no one he felt he could completely trust.

Often he told me he wished he had died fighting in Afghanistan.

"Not a cheap death, something hard," he said. "Then I could have proven to everyone, in that one action, that I am who I say I am."

He spoke of suicide, and told me many times that were it not for me, he would already have killed himself.

For my part, the urgent responsibility I felt to recount Jim's story grew as his grip on life became increasingly tenuous. I wanted to relate what happened to Jim in a way that would help heal him and other similarly haunted men, by letting the broader society know what they had gone through, so their sacrifices would never be forgotten. To be able to do that I had to listen without judging him and be willing to experience some of the terror, grief, and rage that he had.

We slept on a foam mattress covered with thick blankets on the floor, with a loaded pistol inches away and a knife in the closet. He didn't feel safe in his neighborhood in Fayetteville, and had the gun for our protection. But at one point he fell into such a depth of depression that I started to worry about the weapons.

On Friday, June 22, Jim drove to Worlock's personnel office on Fort Bragg at the appointed time to sign the paperwork for the revocation of his Special Forces tab.

Revoking the tab meant that Jim's service in the Special Forces Regiment would be expunged from his official military record. He could no longer wear the tab or Green Beret.

But when Jim arrived at Worlock's office, he was told the printer was broken. He would have to return on the following Monday.

Dan came over for dinner, and gave Jim his own Special Forces tab, which he had worn since he was a sergeant on ODA 316. Dan pointed out that because Jim had also earned a Green Beret as an enlisted soldier, he technically still could wear the tab—only his officer tab had been revoked. Jim placed the tab in a small picture frame over a bloodred image of Marlon Brando as the bald Colonel Kurtz. A short time later, Jim shaved his head.

The next day, Jim woke up in an especially dark mood. Starting early in the morning, he drank, more than usual. That night as we ate dinner with Scout, his twelve-year-old daughter, he got up and left the table. When he did not return after about fifteen minutes, I felt uneasy and went to check on him. I found him lying on the floor in the closet of our room, his body curled up in a ball. An arm's length away lay his pistol. I knelt down beside him and put my arms around him.

He started sobbing—hard sobs so full of anguish that it was almost unbearable to see and hear.

"Talk to me!" I said.

He shook his head.

"Please!"

"I am scared," he said. "I am going to die. I have a gun right there. It's right there!" he said.

"You are not dying," I said. "I am not going to let you."

He looked into my eyes, and I saw his were tear-filled but very cold—on the edge of fury.

"When I can't see anything, I grab on to you," he said. "I know you, I don't know anyone else. You are the only one left.

"I am so angry," he said, shaking. "But I have nothing to fight. Where is my enemy?"

I held him for several minutes. Then it was time to pick up his fourteen-year-old son, Tristen, from a friend's house across town.

For the first time, he asked me to drive. On the way, he started vomiting into a plastic bag that I pulled out of the backseat.

"It isn't worth it, is it?" he asked me bitterly. "I am not worth it."

Just then a summer thunderstorm struck, pelting the truck with rain. I turned on the wipers and kept driving. Lightning flashed, illuminating the rain-washed streets with a harsh bluish light as thunder crashed nearby. Jim ducked and held his arm up in front of his face. Then came more lightning and thunder, making Jim startle each time.

We made it back.

On Monday, he signed the tab revocation papers. That night in a light drizzle, we walked around the statues of Special Forces legends Dick Meadows and Bull Simons on the parade grounds of the USASOC compound.

The next day, without warning, he collapsed on his knees in the kitchen.

"I feel like I am losing control of my mind," he said. "Am I going crazy?"

"You act crazy sometimes. You have combat stress. But are you out of your mind? No," I said.

"Are you scared of me?"

"No, I'm not," I said.

The very next afternoon, we were in our room resting, when he turned to me.

"What would you think about that?"

"About what?"

"What would you think about that. Not how you would feel in your heart, but what would you think?"

"About what?"

"If I did that?"

"Did what?"

"If I killed myself, what would you think?"

I paused for a second.

"If you killed yourself, I could never separate how I felt in my heart and what I thought."

"Yes, you probably could not," he said.

As Jim seemed to go into free-fall, I racked my brain for what to do. What was I missing? Should I put away his gun? But I could not—especially in light of Pashtunwali and his honor. He needed something to fight for.

That evening, I decided to speak with him. We went to our room and sat on our bed on the floor. I took his hand.

"I am very concerned how you have been talking about killing yourself. I know you are in a huge amount of pain," I said.

His eyes welled up.

"I have been in this place for a very long time," he said.

"Do you think about it a lot?" I asked. "Should we put away the gun and knives and pills?"

"Are you afraid?" he asked.

"No."

"Don't you think I am capable of figuring out a way to kill myself?"

"Yes, but what if you suddenly went into an abyss of depression, it would be better not to have those things around."

"If I am going to kill myself, I am going to have a very detailed and complete plan," he said.

The sadness of the thought overwhelmed me. Then tears began to roll down his cheeks. I reached out to hold him but he pushed my arm away, and motioned with his hand for me to stay back.

I looked into his eyes. They were the same hard, cold, angry eyes from the night when he was curled up in the closet.

"The voices are not there anymore," he told me slowly.

His words sank in, their pain striking me deeply. I knew how much they meant to him.

"The ones that made me great, that would speak with me and tell me, 'Get up! Work hard! Prepare for battle!' The ones that would tell me I was a coward because I was not fighting, that I had to go to war. Those voices are not there. They have abandoned me, too," he said. "The silence is devastating."

IN MID-AUGUST, AT THE end of the Muslim holy month of Ramadan, we received a phone call from Imran saying his family was feasting and he, Ish, and Abe wanted to speak with us from Jalalabad over Skype.

Jim was excited to see the brothers for the first time since he left Afghanistan. He asked me to iron his best Afghan clothes—a brown tunic and pants—that Abe had given him. I put on a scarf and dress. Dan came to the house.

Suddenly, Ish, Abe, and Imran were sitting before us on the carpet in one of the rooms of their large Jalalabad home. We began talking and laughing, but the sight of our Afghan brothers was an emotional overload for Jim. He said a few words, then choked up. All he could do was reach out toward the computer screen, as if trying to touch them. Then he left the house. Later, I found him wandering through the neighborhood in his Afghan clothes with a large wooden staff.

A big part of Jim and me still resided in Afghanistan. We spoke about it every day. We had each spent years overseas, but our experiences during the past two years in Afghanistan fundamentally changed how we felt about being American. News of our comrades there came to us in bits and pieces, mostly in calls from Abe, Ish, and Imran.

WE LEARNED THAT THE qalat in Chowkay had been abandoned by Capt. Fleming and his team about a month after Jim and Dan were pulled out. After the team alienated the arbakai, who in turn stopped manning the observation points in the high ground, Taliban attacks intensified again on the qalat. The team lost critical intelligence on the Taliban that Jim had gained through his relationships with arbakai commander Sadiq and others. Fleming decided occupying the qalat was untenable, and blamed it on Jim by claiming it was in a poor location.

Meanwhile, the growing spate of killings of U.S. soldiers by Afghan forces was undermining the American mission in Afghanistan. Top military commander Gen. Allen responded by suspending joint operations and ordering a re-vetting of the Afghan forces—thereby increasing the alienation between the two groups. The trends underscored how critical and unique was the close rapport of Jim's team with the Afghans who protected them.

After Jim was pulled out by his command, the Taliban insurgents celebrated, claiming they had killed him.

"Your son is gone!" one of the Taliban commanders taunted Haji Jan Dahd.

"Pray that he has left! If he were here, you would be dead," Jan Dahd replied.

In May, Imran was the first of his brothers to travel back to Mangwel since Jim had been forced out of the country. He described the trip to me in an email.

I took the Spin Jumat Bridge and from there I took the road in front of Penich.

The area where we hit the IED in the tree used to have a lot of ALP in it—remember all Niq's guys were out there all

the time? Now there is no one around. I asked Niq what was
going on. He said the government told them if they went out
of their village and took casualties or faced any trouble, they
would be responsible, not the government. He said they don't
get any help from the US Army guys, such as fuel, water, or
ammo.

On the way to Mangwel I didn't see any ALP in the area
at all, and when I got into Mangwel it was like walking in a
graveyard. No one was out in the village.

Most of the schoolteachers and other random people
think that Jim was killed in Dewagal Valley. I told them that
I talk to Jim all the time, but they wouldn't believe me. I
don't know where they heard that Jim was shot in Dewagal.
Asif and Sitting Bull understand that he is alive. They are all
heartbroken, not that Jim did anything wrong, but because of
what Jim's own did to him. They know he will be back one day
to see them again.

I told them Jim's words: "Achilles absent, is Achilles still."

The phrase was one of Jim's favorites and came from a passage in
Homer's *Iliad*. The warrior Achilles slays his foe Hector in revenge for
the death of Patroclus, who was pretending to be Achilles. Standing over
Hector's dead body, Achilles says:

> *At least is Hector stretch'd upon the plain,*
> *Who fear'd no vengeance for Patroclus slain?*
> *Then, prince, you should have fear'd what you now feel;*
> *Achilles absent, was Achilles still.*

MONTHS LATER, WINTER BROUGHT the first dusting of snow on the mountains
around Mangwel. Noor Afzhal, his wooden cane in hand, watched a group
of workers as they lay one stone upon another, building a new qalat.

To everyone who asked, he said the same thing:

"You'd better not give me any trouble," he said. "Jim is coming back
to live here."

Yet by the beginning of January the air was so cold it took a man's breath away. Soon afterward, the cruel winter winds stole away the body of one man, and the spirit of another.

On the chilly evening of January 11, 2013, at about nine o'clock, I received a call from Imran. It was the middle of the night in Jalalabad.

"Just a few minutes ago, I got a call from Asif in Mangwel," Imran said. "Sitting Bull has passed away . . . a stroke."

I closed my eyes, no longer hearing what Imran was saying. I saw Noor Afzhal's jaunty wave from the back of the motorcycle, his boyish smile when he arrived at the qalat holding a bouquet of flowers for me, and his wise determination as he stood tall and spoke to the American soldiers and Afghan tribesmen, hoisting his cane to underscore his words.

When I told Jim, he bent over, put his head in his hands, and collapsed into sorrow. His body convulsed, his sobs so violent that I worried he was casting off whatever it was that held him together.

We wept, and then I helped him slowly to his feet. I guided him to the door, threw a jacket over his shoulders, and stepped with my arm around him into the bitter, moonless night. I had to keep him moving forward, so we just walked, saying nothing. One foot in front of the other.

From that day on, each night I lit a votive candle in front of a framed photograph of Noor Afzhal and our fractured family. In the picture, he smiled as he sat next to Jim, me, Dan, Abe, and Ish outside the kitchen of the qalat in Mangwel. Jim's hand rested comfortably on the older man's knee.

One by one, the candles burned and melted. Spring came to Bethesda, with its blossoms and rustling shade. The azaleas began to open. Then one sunny morning in late May, Jim and I climbed into the pickup truck and drove south, toward the outer banks of North Carolina and Cape Hatteras. Jim wore around his neck the golden lighthouse I had given him, and also a silver lambda that was a gift from his Iraqi comrade Mack. We drove through Norfolk, Virginia, and turned onto the road that traces the outer banks—a narrow, sandy spit less than a mile wide that juts into the ocean. Waves lapped the shore on either side, and waterfowl flitted among the gnarled trees stunted by the salty breeze. As we grew closer to Hatteras lighthouse, though we still could not see it, the anticipation of reaching the place that had meant so much to us overwhelmed me, and tears began rolling down my cheeks. The lighthouse, stately in its diagonal black and white stripes, had for more than three years symbolized the beacon that Jim had to keep in his sight, the one that could guide him back and bring us together at the shore— this shore.

We embraced later in the darkness as we stood for a long time watching its steady and hauntingly beautiful beam sweep across the vast blackness of the ocean.

At sunrise the next morning, we walked on the beach. The tide was going out, and a cloud bank lay on a pale pink horizon above the silver sea. I wore a white satin nightgown and robe, and he was in a khaki shirt and jeans. We sat on the sand near the shore and Jim took my hand in his. With a strong, knowing look in his eyes that I will never forget, he recited for me the promises that were his wedding vows, and I did the same. He put his hands on my cheeks and kissed me for the first time as my husband. Then Jim picked up a blue glass scotch bottle that held our folded, handwritten vows, a lambda patch, and verses from Homer's *Odyssey*, and hurled it into the receding tide. We watched the bottle go, but suddenly a big wave sent it bobbing back. Laughing, I ran into the surf in my nightgown, swam until I caught the bottle, and threw it out farther still. Soon it disappeared from view. We relaxed on the beach, and as the sun climbed higher in the sky, the crashing waves lulled us, soothing Jim's mind as never before.

All day, he played on the beach. He built an elaborate sand castle. He stood in the surf and tossed a plastic football into the waves and let the waves toss it back. The next day we climbed the 257 steps to the top of the lighthouse and looked out upon our peaceful domain. As we prepared to go, Jim sat by my side, bowed his head, and for the first time in a very long time, he prayed.

Our marriage was all we dreamed of. One thing, though, was missing in our relationship. For more than a year, at Jim's request, I had no name to call him by. It was an awkward absence that we both felt, and yet were helpless to fill. Everyone else could call my husband "Jim" or "Kirk," but I was strangely deprived of that element of intimacy. Nonetheless, I understood. The man I loved and was starting a new life with was still struggling to recover his identity.

As part of our plan to start afresh, we decided to move across the country to the Pacific Northwest. The move itself was unexpectedly hard for me. It was July and sweltering. Sorting through decades of belongings was bittersweet. Then a flash flood filled the basement with water and soaked boxfuls of my photos, letters, notebooks, children's artwork, and baby books. I strew them across the backyard to dry in the sun. Jim proved an anchor, shifting into "accomplish the mission" mode to keep me from losing it at the last minute. Finally one evening in mid-July, we again climbed into the truck and headed west. Jim had let me know that he had many things to share, but waited until we pulled onto the highway to begin to speak.

"I have something to tell you that I have realized in the last few days," he said. "When it first came to me, I was unsure. But something I read in your old letters, and something you said, let me know I was right."

He turned to look at me and I saw his eyes welling with tears.

"Ann, you have given me my name back," he said. "My name is Jim."

Several weeks later, on September 11, 2013, Jim and I boarded a flight and began our journey back to Afghanistan.

# EPILOGUE

RECENTLY A TRUSTED FRIEND inside the government contacted Jim. There was something he needed to know. It was sensitive.

Jim and I decided we needed to investigate and discuss the matter face-to-face. What followed were two meetings with separate individuals with access to government information who independently relayed the same news:

During the May 2, 2011, U.S. military raid that killed Al Qaeda leader Osama bin Laden in a compound in Abbottabad, Pakistan, piles of documents were seized. Analysts have since combed through the recovered materials, a process known as "sensitive site exploitation." One of the documents, found in bin Laden's quarters, was an English copy of Jim's paper, "One Tribe at a Time," with notes in the margins.

The notes made reference to the problems Al Qaeda was having in Konar. At the time, Al Qaeda was known to be seeking to reestablish itself in Konar in order to avoid the intensifying U.S. unmanned aerial drone strikes on its operatives across the border in Pakistan's tribal areas.

The handwritten notes were consistent with others identified as written by Osama bin Laden.

Another document uncovered was a directive from Osama bin Laden to his intelligence chief. The directive mentioned Jim by name, and said he was an impediment to Al Qaeda's operational objectives for eastern Afghanistan and needed to be removed from the battlefield.

# LIST OF PARTICIPANTS

Tribe 33: The call sign and unit designation of the small U.S. military team commanded by Major Jim Gant that embedded in Mangwel, Khas Kunar District, with tribal chief Malik Noor Afzhal and the Mohmand tribe.

Tribe 34: The call sign and unit designation of the small U.S. military team commanded by Major Jim Gant that embedded in Chinaray, Chowkay District, with tribal chief Haji Jan Dahd and the Safi tribe.


## AMERICANS

Armstrong, Private 1st Class Chad: Member of Tribe 34. Infantry soldier who worked in Chowkay operations center.

Bartlett, Private 1st Class Jonathan: Member of Tribe 34. Infantry soldier who was one of Dan's gunners.

Bird, Sergeant Danny: Member of Tribe 34. Noncommissioned officer for the soldiers of the Fort Lewis–based 2nd Battalion, 3rd Infantry Regiment, who were on Jim's team in Chowkay.

Bolduc, Brigadier General Donald: Special Forces commander assigned to work with future Afghan president Hamid Karzai during the U.S. military invasion of Afghanistan in 2001. In 2010 and 2011, Bolduc advanced the local defense initiative as commander of the Combined Joint Special Operations Task Force–Afghanistan (CJSOTF-A). Bolduc supported Jim's work as an advisor and ground commander in eastern Afghanistan and visited Jim's team in Mangwel on several occasions. Jim's battalion commander at 1st Battalion, 3rd Special Forces Group, in 2005–6. Awarded Bronze Star for valor and Army Commendation Medal for valor.

Boles, Sergeant Sonny: Member of Tribe 33. One of Jim's gunners.

Brooks, Senior Master Sergeant Wesley: Member of Tribe 33 and Tribe 34. Senior noncommissioned officer (NCO) and Air Force Joint Tactical Air Controller ( JTAC) for Jim's Mangwel and Chowkay teams.

Campbell, Lieutenant General John: Regional Command–East (RC-East) commander during Jim's embed into Mangwel. Campbell was deputy commander of the military division in charge of Baghdad during Jim's Iraq deployment in 2006–7, and was Jim's battalion commander when Jim was a platoon leader in 2nd Battalion, 5th Infantry Regiment, in Hawaii in 1996–97.

Carter, Sergeant 1st Class Tony: Member of Tribe 34. Special Forces weapons sergeant.

Chase, Staff Sergeant Robert: Member of Tribe 33. Acted as Jim's team sergeant in Mangwel.

Clement, Specialist Chris: Member of Tribe 34. Jim's driver. Wounded in January 16, 2012, improvised explosive device strike on Jim's vehicle.

Deahn, Staff Sergeant Andy: Member of Tribe 33. Joint Tactical Air Controller ( JTAC).

Falkel, Staff Sergeant Chris: Member of Operational Detachment Alpha (ODA) 316, the Special Forces team Jim commanded in Afghanistan in 2003 and 2004. Awarded Silver Star for bravery in a battle in Mari Ghar, Afghanistan, in 2005, during which he was killed in action.

Fleming, Captain Randy: Team Leader of Special Forces Operational Detachment Alpha (ODA) 3412, who took over in Chowkay after Jim was relieved of command.

Franks, Sergeant 1st Class Tony: Member of Tribe 33. Senior noncommissioned officer for soldiers on Jim's Mangwel team from 1st Battalion, 16th Infantry Regiment, based in Fort Riley, Kansas.

Gonzales, Sergeant 1st Class Fernando: Member of Tribe 34. Intelligence sergeant.

Gray, Private 1st Class Andrew: Member of Tribe 33. One of Jim's gunners.

Greenwalt, Specialist Chris: Member of Tribe 33. One of Jim's drivers.

Haas, Brigadier General Christopher: Commander of Combined Forces Special Operations Component Command–Afghanistan (CFSOCC-A) in 2011 during latter part of Jim's last Afghanistan deployment. Visited Jim's team in Mangwel.

Harvey, Sergeant Jeremiah: Member of Tribe 33. Medical noncommissioned officer.

Hicks, Private 1st Class Jeremiah: Member of Tribe 33. Awarded Army Commendation Medal for valor by General Petraeus for actions in Shalay valley, Afghanistan, on May 7, 2010.

Jimenez, Major Eddie: Commander of Special Forces Company 3310 in Jalalabad under Lieutenant Colonel Robert Wilson.

Kirila, Lieutenant Colonel Robert: Investigating officer for Army 15-6 investigation into alleged misconduct by Jim and Dan.

LeBrocq, Sergeant 1st Class Jean-Paul: Jim's driver in Iraq. Awarded an Army Commendation Medal for valor for actions on December 11, 2006. Died of illness in December 2009.

Lerma, Private 1st Class Richard: Member of Tribe 34. Driver and gunner.

Linn, Lieutenant Colonel William: Special Operations Task Force-East (SOTF-E) Commander who in March 2012 ordered health and welfare inspection of Jim's living quarters.

Martin, Staff Sergeant Ed: Member of Tribe 34. Special Forces medical sergeant.

McKone, Captain Dan: Member of Operational Detachment Alpha (ODA) 316. Member of Tribe 33 and Tribe 34. Jim's second-in-command, executive officer, operations officer, logistics officer, and development coordinator. Awarded Bronze Star for valor and Army commendation medals for valor.

Miller, Major General Scott: Commander of Combined Forces Special Operations Component Command–Afghanistan (CFSOCC-A) in 2010–11 during Jim's last deployment to Afghanistan. Visited Jim's team in Mangwel. Awarded the Bronze Star for valor for actions in Mogadishu, Somalia, in October 1993.

Mitchell, Specialist Francis: Member of Tribe 33. Jim's primary driver.

Mulholland, Lieutenant General John: Led 5th Special Forces Group (Airborne) during the invasion of Afghanistan in 2001. Commander of United States Army Special Operations Command (USASOC) during Jim's deployment to Afghanistan. Visited Jim's team in Mangwel. In charge of Jim's official punishment and reprimand.

Olson, Admiral Eric: Commander of United States Special Operations Command (USSOCOM) during the initial portion of Jim's deployment to Afghanistan. Visited Jim's team in Mangwel. Awarded the Silver Star for actions in Mogadishu, Somalia, in 1993.

Pelleriti, Lieutenant Colonel John: Operations officer (S-3) for Special Operations Task Force–East (SOTF-E) during Jim's initial embed into Mangwel. Visited Jim's team in Mangwel.

Petraeus, General David: Commander of Multinational Forces–Iraq (MNF-I), United States Central Command (USCENTCOM), International Security Assistance Force (ISAF), and United States Forces–Afghanistan (USFOR-A), and Director of the Central Intelligence Agency (CIA). Jim was under his command in both Iraq and Afghanistan. Visited Jim's team in Mangwel.

Porter, Staff Sergeant Ryan: Member of Tribe 33.

Pressfield, Steven: Author of several works of historical fiction including the bestselling novel about the Spartan battle at Thermopylae, *Gates of Fire*. Jim's friend and mentor.

Redden, Private 1st Class Kyle: Member of Tribe 33. One of Jim's drivers.

Roberts, 1st Lieutenant Thomas: Platoon leader of soldiers from 2nd Battalion, 3rd Infantry Regiment, who were assigned to Jim's team. West Point graduate who wrote the sworn statement that began the Article 15-6 investigation into alleged misconduct by Jim and Dan.

Ruiz, Specialist Fernando: Member of Tribe 34.

Salyer, Private 1st Class Jonathan: Member of Tribe 34. One of Jim's gunners.

Schwartz, Colonel Mark: Commander of Combined Joint Special Operations Task Force–Afghanistan (CJSOTF-A) during Jim's removal from Afghanistan. Was operations officer at CFSOCC-A in 2009 and sent Jim an email rejecting his plan to conduct "tribal engagement." Visited Jim in Mangwel.

Smith, Command Sergeant Major Thomas: Command Sergeant Major (CSM) for United States Special Operations Command (USSOCOM) when Jim's "One Tribe at a Time" was published. CSM Smith was instrumental in the early support of Jim. Visited Jim's team in Mangwel.

Solheim, Major Kent: Special Forces Company 3430 commander under Lieutenant Colonel William Linn. Awarded Silver Star for actions in July 2007 in Iraq.

Taylor, Sergeant Michael: Member of Tribe 33.

Thomas, Staff Sergeant Justin: Member of Tribe 33.

Warner, Dave, MD, PhD: Head of the Synergy Strike Task Force based out of the Taj guesthouse in Jalalabad.

Wilson, Lieutenant Colonel Robert: Commander of Special Operations Task Force–East (SOTF-E) during Jim's embed in Mangwel. Visited Jim several times in Mangwel.


**AFGHANS**

Abdul Wali: Sub-commander in Chowkay Afghan Local Police (ALP).

Abdul Wali: Insurgent commander in Konar and Pakistan.

Abu Hamam: Taliban commander in Shalay valley.

Amir Mohammed: Sub-commander in Chowkay Afghan Local Police (ALP).

Asif: Malik Noor Afzhal's eldest son in Mangwel. Khas Kunar Afghan Local Police (ALP) leader. Jim's tribal brother.

Assad: One of Jim's mercenaries in Chowkay.

Azmat: Malik Noor Afzhal's second eldest son in Mangwel. Mangwel Afghan Local Police (ALP) commander. Jim's tribal brother.

Basir: One of Jim's mercenaries in Chowkay and Abe's cousin.

Chevy: Afghan orphan whose father was killed by the Taliban. Taken under the wing of Operational Detachment Alpha (ODA) 2312 and Jim.

Dost Mohammed Khan: Malik Noor Afzhal's younger brother in Mangwel.

Fazil Rahman: Worked as a security guard at Camp Dyer at the Jalalabad Air Field (JAF) and helped repel a Taliban attack on the camp. Was later shot by the Taliban and had his qalat burned down. Jim attempted to save his life but failed. Jim supported his family for several months until members of his family returned to Mangwel to help them.

Ghani Gul: Member of Mangwel Afghan Local Police (ALP).

Ghulam Hazarat: One of Jim's mercenaries in Chowkay.

Haji Ayub: Leader of the Mushwani tribe located in northern Konar and Pakistan. Son of Malik Zarin.

Haji Jan Dahd: Leader of the Safi tribe located in central Konar.

Haji Jan Shah: Eldest son of Haji Jan Dahd. Jim's close friend and ally.

Hakim Jan: Member of Mangwel Afghan Local Police (ALP).

Ibrahim Khan: Jim's advisor, bodyguard, interpreter, and brother. Also known as "Abe."

Imran Khan: Jim's interpreter and brother.

Ismail Khan: Jim's advisor, bodyguard, interpreter, and brother. Also known as "Ish."

Khetak: Insurgent fighter. Paid as an assassin by Abdul Wali, the insurgent commander.

Little Malik: Malik Noor Afzhal's grandson and Asif's son.

Mahmud Dwaher: Member of Chowkay Afghan Local Police (ALP).

Maulawi Basir: Leader of an insurgent group located in the Dewagal valley.

Malik Noor Afzhal: Malik of the Mohmand tribe in Mangwel. Also known by many as "Sitting Bull," after the great Native American tribal chief. Jim's tribal father.

Malik Zarin: Leader of the Mushwani tribe prior to his assassination in April 2011.

Mohammed Ghul: Leader of the Chamaray Afghan Local Police (ALP).

Mohammed Hanif: Leader of a Tajik enclave located within the Mohmand tribal area near Mangwel.

Mohammed Jalil: Member of the Mohmand tribe. Member of Khas Kunar District Shura.

Mohammed Sadiq: Commander of Chowkay Afghan Local Police (ALP).

Niq Mohammed: Commander of Kawer Afghan Local Police (ALP).

Noor Mohammed: Commander of Northern Khas Kunar Afghan Local Police (ALP).

Obeidullah: Former insurgent. Member of the Sirkanay Afghan Local Police (ALP).

Raza Gul: Malik Noor Afzhal's youngest son. Member of Mangwel Afghan Local Police (ALP). Jim's tribal brother.

Sahib Zada: One of Jim's mercenaries in Chowkay.

Salim: Tribe 33's Afghan cook, later hired as a member of the Khas Kunar Afghan Local Police (ALP).

Shafiq: Member of Tribe 33 and Tribe 34. One of Jim's drivers, mechanics, and interpreters.

Sher Ali: Member of Mangwel Afghan Local Police (ALP).

Umara Khan: Member of Mangwel Afghan Local Police (ALP). Close friend and confidant of Jim's.

Zia Ul Haq: Commander of the Chowkay Afghan Local Police (ALP). Haji Jan Dahd's son. Jim's close friend.


**IRAQI**
Alsheikh, Mohammed Lateef Moustafa: Jim's advisor, bodyguard, interpreter, and brother in Iraq. Also known as "Mack."

Taher, Colonel Alasadi: Commander of the Iraqi National Police Quick Reaction Force (QRF) battalion that Jim advised.

# GLOSSARY

*These definitions reflect the meanings of the words as used by the
Pashtun tribesmen with whom I lived in eastern Afghanistan.*

*Arbakai* – A traditional Afghan tribal police force, especially prevalent
in eastern Afghanistan, that protects tribal territory and upholds the
decisions of tribal leaders. The arbakai system of local defense was the
social and cultural basis for the Afghan Local Police (ALP) program
set up by the U.S. military and Afghan government in 2010. The terms
"arbakai," "Afghan Local Police" ("mahali police"), and "militia" were
all used by Afghans to describe the tribal security forces, and are used
interchangeably in the book.

*Badal* – Revenge. An often violent or deadly exchange to avenge a breach
of honor. The requirement to seek badal, even for seemingly minor insults,
can lead to longstanding tribal blood feuds. A pillar of Pashtunwali.

*Ghairat* – An individual's honor or pride, often demonstrated through
acts of bravery. A pillar of Pashtunwali.

*Jihad* – A religious duty of Muslims to struggle to defend or spread Islam,
often used to mean "holy war" directed at nonbelievers, or infidels.

*Jirga* – A tribal council or assembly that meets to resolve conflicts
through consensus and compromise.

*Lashkar* – A tribal army, which historically could comprise thousands of
men, mobilized temporarily against an external threat.

*Mahali police* – Local police.

*Meena* – Love, used in an all-encompassing manner that means
"everything," as in "you are everything to me."

*Melmastia* – A Pashtun obligation to offer food, lodging, protection,
and other hospitality to any visitors—including enemies and non-
Pashtuns—to the tribe's territory. A pillar of Pashtunwali.

*Mujahideen* – A group of people fighting a jihad. Mainly used to describe Afghans who waged a guerrilla war to free their country from Russian occupiers in the 1970s and 1980s.

*Namoos* – The land, guns, and women in the family of a Pashtun male—those things considered possessions that he must protect in order to defend his honor. Namoos most often refers to women.

*Nanawati* – The act of asking a stronger opponent for mercy, pardon, or asylum. It can also describe a payment or offering aimed at correcting a wrong. Nanawati is not a sign of guilt or admission of guilt.

*Nang* – Collective (tribal) honor. A pillar of Pashtunwali.

*Pashtunistan* – A geographic area stretching from eastern and southern Afghanistan across the border into western Pakistan that is dominated by ethnic Pashtuns and is considered the homeland of the Pashtun people.

*Pashtunwali* – An ancient social code, pre-dating Islam, followed by Pashtun tribes.

*Peghor* – Shame or disrespect, a violation of honor.

*Qalat* – A typical rural Afghan walled compound or home, literally, "fortress."

*Shura* – A meeting of tribal and other leaders, such as religious figures or government officials.


**ENGLISH TERMS**
*Afghan Local Police (ALP)* – A paid, armed local security force recruited from Afghan villages under a program initiated and funded by the U.S. military and authorized in 2010 by Afghan president Hamid Karzai. The program is part of the Village Stability Operations (VSO) strategy, aimed at promoting security, development, and governance in rural Afghanistan. Under it, small U.S. military teams work with local leaders to recruit and train groups of ALP—up to 400 police for each rural district—and carry out development projects. ALP fall under the Afghan Ministry of Interior. The program was inspired in part by Maj. Jim Gant's 2009 paper "One Tribe at a Time," which drew upon Afghanistan's history of

decentralized rule and tribal security forces, known as arbakai, in Pashtun areas of eastern Afghanistan.

*Battlespace owner (BSO)* – A term commonly used to describe the military unit that has overall control over the conduct of operations in a designated geographic area.

*Counterinsurgency (COIN)* – A counterinsurgency operation involves actions taken by the recognized government of a nation to contain or quell an insurgency, or an armed rebellion, in part by winning over the population to the side of the government.

*Combat outpost (COP)* – A reinforced observation post, usually manned by a platoon or company of U.S. service members who use the base for limited combat operations in a conflict zone.

*Combined Forces Special Operations Component Command–Afghanistan* (CFSOCC-A) – A headquarters created in Afghanistan in 2009, led by a U.S. brigadier general, in charge of all U.S. and international Special Operations Forces (SOF) in Afghanistan with the exception of classified counterterrorism units.

*Combined Joint Special Operations Task Force–Afghanistan (CJSOTF-A)* – A headquarters led by a U.S. colonel in charge of day-to-day special operations missions by U.S. and other NATO and international forces in Afghanistan.

*Counterterrorism* (CT) – The conduct of military and other government operations to attack and degrade or destroy terrorists and terrorist organizations.

*Durand Line* – The line drawn in the Hindu Kush in 1893 that divided Pashtun and other tribal lands in Afghanistan and British India, marking their sphere of influence. The line later delineated the border between Afghanistan and Pakistan, although Afghanistan does not recognize it.

*Federally Administered Tribal Areas (FATA)* – A semi-autonomous group of small administrative units in northwestern Pakistan populated by about three million people, mainly from Pashtun tribes.

*Field Manual 3-24* – The U.S. Army and Marine Corps manual published in 2006 intended to put the best counterinsurgency practices and lessons into current U.S. military doctrine, the first major update in counterinsurgency doctrine in twenty years.

*Forward operating base (FOB)* – A military base or position used to support tactical operations.

*Fragmentary Order (FRAGO)* – A military order given when changes to an Operational Order (OPORD) are needed.

*Humvee* – A High Mobility Multipurpose Wheeled Vehicle (HMMWV), a four-wheel drive vehicle commonly used by U.S. military forces in Iraq and Afghanistan.

*Hilux* – A model of Toyota pickup truck popular with the Taliban and other insurgent groups.

*Improvised explosive device (IED)* – A bomblike device used effectively by insurgent fighters. The device can be constructed using military explosives or homemade materials and set off by remote control, command-wire, or sensors, or by the victim stepping onto or driving over it. IEDs have inflicted a majority of the wounds and deaths among service members in Iraq and Afghanistan. Also commonly referred to as a "roadside bomb."

*Green on blue* – A phrase used to describe attacks on U.S. and other NATO forces by members of the Afghan National Security Forces (ANSF).

*International Security Assistance Force (ISAF)* – The NATO-led military coalition in Afghanistan.

*Jaish al Mahdi (JAM), Mahdi Army* – A primarily Shiite Iraqi paramilitary force created in 2003 by the Shiite Islamic cleric Muqtada al-Sadr.

*Judge Advocate General's Corps (JAG)* – A military branch that provides legal services for military members, prosecutes courts-martial, and provides legal advice to commands. A member of the corps is often referred to as a judge advocate, or JAG.

*Khyber Pakhtunkhwa (formerly known as North-West Frontier Province)* – The northernmost province of Pakistan with a population of about twenty million, mainly from Pashtun tribes.

*Mine-resistant ambush protected (MRAP)* – A heavily armored vehicle used by U.S. forces to counter improvised explosive devices (IED) by providing protection against the explosions.

*National Directorate of Security (NDS)* – A national intelligence agency of the Afghan government.

*Operational Detachment Alpha (ODA)* – A small tactical team organized to have twelve U.S. Army Special Forces soldiers (Green Berets) highly trained in weapons use and communications, and medical, intelligence, and other skills, which plans and conducts a wide variety of missions.

*Post-traumatic stress disorder (PTSD)* – A severe condition that develops after a terrifying, traumatic event. Many current and former U.S. military members struggle with the disorder as a result of their combat experiences. The main problems associated with PTSD are anger, depression, mood swings, loneliness, insomnia, graphic and violent dreams, difficulty in relationships with other people including loved ones, alcoholism, and drug abuse.

*Quick Reaction Force (QRF)* – A military force used to react to emergency situations in combat.

*RG-31* – A large armored personnel carrier used by U.S. forces to counter improvised explosive devices (IED). Also a type of mine-resistant ambush protected (MRAP) vehicle.

*Robin Sage* – The field exercise that culminates U.S. Army Special Forces training, considered one of the best unconventional warfare (UW) exercises in the world.

*Rocket-propelled grenade (RPG)* – A shoulder-fired weapon that shoots an explosive round.

*Rupees* – The basic unit of Pakistan's currency.

*SEAL Team Six* – Formally known as the United States Naval Special Warfare Development Group (NSWDG), or DEVGRU, SEAL Team Six is a counterterrorism unit composed of Navy SEALs that is part of the Joint Special Operations Command (JSOC).

*Spartan* – A native or inhabitant of the ancient Greek city-state of Sparta.

*Special Forces* – A branch of the U.S. Army composed of elite soldiers selected for high-risk missions such as unconventional warfare (UW). The Army's premier soldiers for training and advising indigenous forces. Special Forces soldiers are also known as Green Berets for their distinctive service headgear.

*Special Operations Command (SOCOM)* – The United States Special Operations Command trains and equips all Special Operations Forces (SOF) to prepare them for overseas missions. SOCOM is also the primary command in charge of fighting terrorism.

*Special Operations Task Force–East (SOTF-E)* – The command in charge of conducting all special operations missions in eastern Afghanistan.

*Taliban* – An Islamic fundamentalist movement composed largely of ethnic Pashtuns fighting under a variety of different commands and organizations in Afghanistan and Pakistan.

*Traumatic brain injury (TBI)* – A condition suffered when an external force damages the brain, causing cognitive, social, and behavioral problems. TBI in U.S. military members is often caused by IED blasts.

*Unconventional warfare* – Activities conducted to enable a resistance movement or insurgency to coerce, disrupt, or overthrow a government or occupying power by operating through or with a guerrilla force and its supporters in a denied area.

*Village Stability Operations (VSO)* – Military operations at the village level to raise local defense forces, bring in development opportunities, and create ties to district governments in Afghanistan. A grassroots initiative, it became the focus of U.S. Special Forces, Navy SEAL, and Marine Special Operations units throughout Afghanistan beginning in 2010.

*United States Army Special Operations Command (USASOC)* – The command in charge of training and equipping all Army Special Operations Forces, including Special Forces (Green Berets), Rangers, and Civil Affairs soldiers.

*United States Central Command (CENTCOM)* – The military command in charge of the Middle East and Central Asia.

*Zombieland* – The contested road between the village of Mangwel and Combat Outpost Penich in Afghanistan's Konar Province.

# ACKNOWLEDGMENTS

The heroes of this book are the brave and loyal Pashtun tribesmen who joined forces with Jim and his U.S. soldiers, welcomed us into their villages, and protected us with their lives. Their trust underpinned Jim's mission, and this book. First and foremost among them was Mohmand tribe leader Malik Noor Afzhal, a wise, honorable man who loved and served his people. Together with his wife, Hakima, and sons Asif, Azmat, and Raza Gul, Noor Afzhal treated Jim and me as family and gave me the incredible gift of his life story. Strong support also came from Safi tribe leader Haji Jan Dahd and his son Haji Jan Shah, Afghan Local Police members Umara Khan, Niq Mohammed, Zia Ul Haq, Sadiq, Abdul Wali, and Salim, and mechanic Shafiq. Brothers Ibrahim, Ismail, and Imran Khan worked tirelessly and fearlessly as advisors, interpreters, and comrades. This story could not have been told without them. Special thanks goes to Senator John McCain for working to get the brothers out of Afghanistan.

Many current and former members of the U.S. military and Special Forces contributed vital information, interviews, and assistance, including Jim's teammates and commanders, and others whom I had the privilege of covering in combat. I wish to thank all of Jim's U.S. soldiers on Tribe 33 and Tribe 34 and second-in-command Capt. Dan McKone. They embraced the mission, executed it with hard work, courage, and heart, accepted me as a teammate, and opened their lives to me. I am beholden also to Brig. Gen. Don Bolduc, Gen. John Campbell, Lt. Col.

John Pelleriti, Lt. Col. Scott Mann (Ret.), CSM Tom Smith (Ret.), Maj. Gen. Scott Miller, Col. Chipper Lewis (Ret.), veterans of ODA 316 and ODA 2312, interpreter Khalid Dost, and Jim's Iraqi comrade Mack. I am especially grateful for the insights and encouragement of Gen. David H. Petraeus (Ret.), a bold and inspiring leader who launched the local security strategy in Afghanistan, championed his ground commanders, and tried to win the war.

This book is based on hundreds of interviews and years of reporting, research, and writing. I alone am responsible for the content and any errors. But I leaned heavily on a few key people in the process: my brilliant agent, Shawn Coyne, who had an uncanny ability to help conceptualize the story; an anonymous friend and reader, who kept me going through some dark days; my ever enthusiastic editor David Highfill; author Steven Pressfield, the godfather of *American Spartan*. I benefited greatly from the suggestions of early readers: Dan McKone, who carefully pored over several drafts; Kalev Sepp, who offered extensive and meticulous comments, Callie Oettinger, who gave expert publicity advice; and my daughter Sarah Tyson, who helped check facts. Dave, Jeff, Barbara, Ray, Matt, and Debbie provided financial, logistical, and moral support.

I am grateful to my entire family, especially to my parents, Joy and Haney, for their steadfast love and devotion, and to my children James, Sarah, Scott, and Kathryn, for loving and understanding their stay-at-home mother who became a war correspondent. My deepest thanks goes to my husband, Jim, for sharing with me the mission of a lifetime, and for his love, trust, and companionship—now and in the many happy days ahead.